D0142731

EMERGING BIOLOGICAL THREATS

EMERGING BIOLOGICAL THREATS

A Reference Guide

JOAN R. CALLAHAN

GREENWOOD PRESS
An Imprint of ABC-CLIO, LLC

A B C CLIO

Santa Barbara, California • Denver, Colorado • Oxford, England

Library of Congress Cataloging-in-Publication Data

Callahan, Joan R.
Emerging biological threats : a reference guide / Joan R. Callahan.
p. ; cm.
Includes bibliographical references and index.
ISBN 978-0-313-37209-4 (alk. paper)
1. Emerging infectious diseases. 2. Health risk assessment. 3. Food security. I. Title.
[DNLM: 1. Disease Outbreaks—prevention & control. 2. Food Contamination—prevention & control. 3. Hazardous Substances—adverse effects. 4. Public Health—history. WA 105 C156e 2010]
RA643.C263 2010
362.196'9—dc22 2009046252

14 13 12 11 10 1 2 3 4 5

This book is also available on the World Wide Web as an eBook.
Visit www.abc-clio.com for details.

ISBN: 978-0-313-37209-4
EISBN: 978-0-313-37210-0

ABC-CLIO, LLC
130 Cremona Drive, P.O. Box 1911
Santa Barbara, California 93116-1911

This book is printed on acid-free paper ♾

Manufactured in the United States of America

Dedicated to my great-grandfather
MICHAEL FOLEY (1849–1897)
Born during the Irish Famine
Died from an experimental tuberculosis treatment

Contents

Preface

> It is my aspiration that health will finally be seen not as a blessing to be wished for, but as a human right to be fought for.
>
> —Attributed to Kofi Annan (1938–)

In the last week of April 2009, the manuscript for this book was nearly finished when reports of the Mexico City swine flu outbreak exploded onto computer and television screens worldwide. Two weeks later, the reported death toll of 150 had declined to 52—not because people came back to life, but because the cause of death was often hard to verify. It was an unusual flu strain that health departments had not seen before, but the news media treated it like the Apocalypse.

When revised numbers showed that the new strain was no more deadly than ordinary seasonal flu, the Monday-morning quarterbacking began. Was the public health response appropriate, or was it hype? Was it wrong for the Chinese to quarantine Mexican tourists, or for the Egyptians to slaughter their pigs? But just as public interest showed signs of fizzling, a new headline appeared: "SWINE FLU–HIV COULD DEVASTATE HUMAN RACE." Worse, the source was not the *Weekly World News*, but a major wire service[1] quoting the head of the World Health Organization.

The press release seemed to say that the new H1N1 swine flu strain could *combine* with the human immunodeficiency virus—the agent of AIDS—to create an airborne threat like nothing the world has ever seen. Many readers took this to mean that the two viruses might physically blend together and give rise to one ugly bug, but as far as we can tell, WHO said nothing of the sort. The original warning posted on the WHO website simply meant that the millions of people living with HIV would be at high risk for complications if they also caught the flu.

At the time this book went to press, the H1N1 swine flu outbreak had recently achieved pandemic status. In another year or two, the world will know whether the healthcare system has coped adequately with the result. At present, however, it seems unlikely that this is the long-awaited biological Big One. In May 2009, when the media breathlessly reported that N95 face masks were flying off the shelves as a result of high demand for flu protection, the author visited three hardware stores in a medium-sized city near San Diego. The shelves were well stocked with

1. United Press International, 4 May 2009.

N95 particulate respirators as usual, and nobody else seemed interested in them. In theaters and shopping malls and college campuses and buses, no one was wearing a mask. The next day, the top news story was no longer the swine flu, but the fact that a well-known actress had gained weight. These observations corroborated a Gallup poll, which showed that only 8 percent of Americans were worried about the swine flu. The virus was silent, but the people had spoken.

And then the people spoke again, in June 2009, this time in the form of a distraught father in upstate New York whose son caught the flu. Blaming the school district for exposing his child to what he perceived as a serious health threat, he stormed the district office and threatened the superintendent with a gun. Fortunately, no one was hurt and the son recovered quickly, but the incident suggested that not everyone was interpreting media reports in the same way. Some people took the exaggerated warnings seriously, while many others ignored the outbreak altogether. Information overload can do that.

By late June 2009, an estimated 1 million Americans had contracted the new H1N1 flu strain. Yet only about 3,000 had actually been hospitalized, and 127 had died. A vaccine was in production and would be available by October, but it was unclear how many people would request it, or what direction the pandemic would take.

Most biological threats—potentially harmful things of biological origin—are infectious diseases, and influenza is just one of many such diseases that this book discusses in detail. After an introductory chapter that defines some terms and concepts, Chapters 2 and 3 describe ten major human diseases (or disease categories) and related conditions. Each section draws on sources ranging from popular culture and urban legends to the recent biomedical literature.

Since biological entities that threaten the food supply are indirect threats to man, Chapters 4 and 5 explore this topic. Chapter 6 examines some human activities that have increased the level of risk associated with certain diseases, and Chapter 7 revisits the war on infectious disease from the perspective of the legendary (perhaps apocryphal) Chinese military strategist Sun Tzu.

1

Introduction

All interest in disease and death is only another expression of interest in life.

—Thomas Mann, *The Magic Mountain*

Most people in the Chicago area (and many others worldwide) have heard about the Great Epidemic of 1885, which resulted from a rainstorm that swept raw sewage into Lake Michigan and contaminated the city's water supply. For decades, historians have reported how an estimated 90,000 people, or one-eighth of the population of Chicago at that time, died of cholera or dysentery in the days that followed. The event has become nearly as famous as the Great Chicago Fire of 1871, and the players are as familiar as Mrs. O'Leary's cow. The story of the Great Epidemic features a Citizens' Association that fought for modern sewage treatment; clueless politicians who failed to heed warnings; comic relief in the form of waiters who continued serving oysters in a flooded restaurant; and thousands of helpless victims shoveled into mass graves. It almost sounds like the synopsis of a Hollywood movie, except that there are no movies about diarrhea.

But the strangest thing about the Great Epidemic of 1885 is that it *never happened.* That's right—the story is an urban legend. (The brave historian who finally debunked the Great Epidemic in 2000 reports that she was met with anger and disbelief.)

There are many things that everyone seems to know. Did you hear about the tourist in the Middle East who caught syphilis from a spitting camel? Did you know that malaria is more common in Siberia than in the tropics, so it is not a tropical disease, and global warming will not affect its range? Did you know that Einstein predicted that everybody on Earth will die within four years after honeybees become extinct? And how about the woman who opened a mysterious blue envelope and caught the deadly Klingerman virus? Sorry—camels and llamas do not carry syphilis, malaria is far more common in the tropics than in Siberia, Einstein made no such prediction about bees, and the Klingerman virus doesn't exist. But the true stories are even better.

PUBLIC HEALTH: A SHORT HISTORY

For thousands of years, civil and religious leaders have formulated rules to help people avoid diseases associated with foods, bodily secretions, lifestyle choices, or resource mismanagement. Some of the oldest rules were evidence-based in the modern sense, whereas others were seemingly invented out of thin air.

For example, the Bible tells us how Moses stopped an epidemic by ordering the deaths of all foreign women who had "known man by lying with him"—from which one can infer the nature of the disease. The measure was effective, if harsh by modern standards. Moses also enforced a quarantine and ordered the sterilization of clothing and equipment.[1] But sometimes public health takes a giant step backwards. Doctors in ancient India treated cholera with good results, by simply rehydrating the patient with herbal potions.[2] Later, English colonial doctors brought a belief system that favored more dramatic intervention, such as bloodletting, and the cholera death rate soared. (The English eventually rediscovered the earlier method.)

Today, government health agencies have largely assumed the role of high priest in such matters. Public health policies are generally based on scientific evidence rather than divine revelation, but scientists can make mistakes, and individuals have rights too. As a result, governments have limited authority to prosecute those who pose a risk to public health—such as tuberculosis patients who cough in crowded airplanes, parents who refuse to vaccinate their children, people who have unprotected sex with multiple partners, or animal lovers who share their garbage-filled homes with hundreds of cats. Mass execution, in particular, is no longer an option.

Despite scientific progress, such topics as food safety, water treatment, and the germ theory of disease itself are hotly debated to this day. Faced with information overload and skyrocketing medical costs, people often must figure out how to take care of themselves and their families. Thus, public health has come full circle.

KOCH AND HIS POSTULATES

Many statements in this book reflect the premise that certain diseases result from the presence of infectious agents ("germs") in the body. Familiar examples of germs include bacteria, viruses, parasites, and fungi. Such a disease is called an infectious disease, and this cause-and-effect relationship is generally known as the germ theory of disease. Not everyone subscribes to it, even today; but in 1890 the theory got a big boost from a German doctor named Robert Koch (1843–1910), who figured out a way to prove it. The criteria that he defined are known today as Koch's Postulates:

1. The microorganism must be present in all organisms with a specific disease.
2. The microorganism must be isolated from a diseased organism and grown in pure culture.
3. The cultured microorganism, when introduced into a healthy organism, must cause the same disease seen in the original organism.
4. The microorganism must be reisolated in pure culture from the experimental host and proven identical to the original specific causative agent.

HAZARD, THREAT, AND RISK

A *biological hazard* (or *biohazard* or *BH*), as defined in this book, is anything *of biological origin* that can harm human beings (or their goods or environment). Examples include disease-

1. Numbers 31:17–24.
2. Sushruta Samhita III, verse II.

causing bacteria and viruses, the toxic chemicals that some plants produce, and the venoms that some insects and reptiles inject when they sting or bite. A *biological threat* (or *biothreat* or *BT*) is about the same thing as a biological hazard, but usually worse.

By this definition, a nuclear weapon is not a biological hazard or threat. It certainly qualifies as a hazard or threat, but the weapon itself is not of biological origin. Cancer is not considered a biological hazard either, although some forms of cancer result from exposure to biological hazards, such as certain viruses or tobacco smoke. Some other cancers result from exposure to non-biological hazards, such as asbestos.

Risk is a measure of the expected loss resulting from a given hazard, based on how severe the loss might be and how likely it is to occur. A banana peel on the surface of the moon is a biohazard, in the sense that it is biological in origin and someone might slip on it; but this outcome is quite unlikely, so the risk is near zero. As another example, consider the smallpox vaccine, which can kill about one in every million people who are injected with it. Is this an acceptable risk, or not? That depends on the perceived benefits of protection from a disease that no longer exists.

There is a remarkable disconnect between what people are afraid of and what is really dangerous. Surveys have shown, for example, that the ten most feared diseases in the United States include five infectious diseases or categories: "foreign" diseases as a group, "deadly" diseases released by bioterrorists, AIDS, influenza, and Lyme disease (Table 1.1). The first two choices imply that people would prefer to die at the hand of a friend rather than a stranger, which makes sense, sort of. The last two are even less likely causes of death, considering that the mortality rate for influenza is usually less than 0.01 percent and that for Lyme is near zero. What these four diseases have in common is that all have received extensive media publicity. Only one of the diseases in Table 1.1, AIDS, is actually one of the world's most deadly infectious diseases (Table 1.2).

Among the leading causes of death in the United States and other high-income nations (Table 1.3), only one infectious disease, lower respiratory infection—essentially the same thing as pneumonia—even makes the cut. Pneumonia can result from various bacterial or

Table 1.1 Most Feared Diseases, United States, 2003

Disease or Condition	Percent "Very Worried"
Cancer (all forms)	16
Foreign disease such as SARS	15
Deadly disease released by terrorists	15
Heart disease	13
Diabetes	11
Arthritis	11
Stroke	9
High blood pressure	9
AIDS	7
Vision loss	7
Alzheimer's	6
Hearing loss	5
Emphysema	5
Influenza	4
Parkinson's	3
Lyme disease	3

Sources: Gallup Organization (survey GO 138154).

Table 1.2 Top Ten Infectious Diseases, 2002

Disease	Estimated Deaths (Worldwide)
Respiratory infections	3,871,000
HIV/AIDS	2,866,000
Diarrheal diseases	2,001,000
Tuberculosis	1,644,000
Malaria	1,224,000
Measles	645,000
Pertussis	285,000
Tetanus	282,000
Meningitis	173,000
Syphilis	167,000

Source: World Health Organization (WHO), World Health Report 2004.

viral infections, or it can occur as a complication of influenza or other diseases. The other causes of death on this list are noninfectious conditions such as heart disease and cancer. In low-income countries, the annual mortality picture (Table 1.4) is quite different, with five infectious diseases or disease categories ranking among the top ten killers: pneumonia, AIDS, diarrhea, malaria, and tuberculosis. In middle-income countries, two infectious diseases make the list (Table 1.5).

In terms of total healthcare costs (Table 1.6), the top ten medical conditions in the United States as of 2008 are heart disease, trauma, cancer, mental disorders, asthma and chronic obstructive pulmonary disease (COPD), high blood pressure, type 2 diabetes, osteoarthritis and other joint diseases, back problems, and normal childbirth. Many of these conditions are related to obesity, lifestyle choices, and the wear and tear of daily living. But notice that there are *no* infectious diseases on the list—unless, of course, most forms of heart disease and cancer turn out to be infectious after all.

Table 1.3 Top Ten Causes of Death, High-Income Countries, 2002

Disease or Condition	Annual Deaths
Coronary heart disease	1,340,000
Stroke and other cerebrovascular diseases	770,000
Cancers of the trachea, bronchus, or lung	460,000
Lower respiratory infections	340,000
Chronic obstructive pulmonary disease (COPD)	300,000
Cancers of the colon or rectum	260,000
Alzheimer's disease and other dementias	220,000
Diabetes mellitus	220,000
Cancer of the breast	150,000
Cancer of the stomach	140,000

Source: World Health Organization (WHO).

Table 1.4 Top Ten Causes of Death, Low-Income Countries, 2002

Disease or Condition	Annual Deaths
Coronary heart disease	3,100,000
Lower respiratory infections	2,860,000
HIV/AIDS	2,140,000
Perinatal conditions	1,830,000
Stroke and other cerebrovascular diseases	1,720,000
Diarrheal diseases	1,540,000
Malaria	1,240,000
Tuberculosis	1,100,000
Chronic obstructive pulmonary disease	880,000
Road traffic accidents	530,000

Source: World Health Organization (WHO).

Table 1.5 Top Ten Causes of Death, Middle-Income Countries, 2002

Disease or Condition	Annual Deaths
Stroke and other cerebrovascular diseases	3,020,000
Coronary heart disease	2,770,000
Chronic obstructive pulmonary disease	1,570,000
Lower respiratory infection	690,000
HIV/AIDS	620,000
Perinatal conditions	600,000
Cancer of the stomach	580,000
Cancer of the trachea, bronchus, or lung	570,000
Road traffic accidents	550,000
Hypertensive heart disease	540,000

Source: World Health Organization (WHO).

Table 1.6 Medical Conditions Ranked by Cost, United States, 2002

Disease or Condition	Annual Cost, in Billions of U.S. Dollars
Heart disease	76
Trauma	72
Cancer	70
Mental disorders	56
Asthma and COPD	54
High blood pressure	42
Type 2 diabetes	34
Osteoarthritis and other joint diseases	34
Back problems	32
Normal childbirth	32

Source: U.S. Agency for Healthcare Research and Quality.

OUTBREAKS, EPIDEMICS, AND PANDEMICS

About 400 million people in Third World nations contract malaria every year, and about 1 million of those people die of malaria every year. That is a mind-boggling tragedy, but not an epidemic. An *epidemic* is a sudden increase in the number of cases over the expected number; and the expected number is usually the average number recorded in recent years. An *outbreak* is similar to an epidemic, except that the affected population is usually smaller and more localized. A *pandemic* is essentially a worldwide epidemic, but there are other criteria, such as the number of regions affected and the extent to which the disease spreads within each region. (A pandemic is not necessarily dangerous; it depends on what the disease is.)

WHAT IS POPULAR CULTURE?

Every major biological threat described in this book has played a role in popular culture, and every section includes examples to illustrate this point. But what is popular culture? There are so many competing definitions that the term itself must have arisen as a product of popular culture.

As used here, "popular culture" encompasses not only the arts and entertainment media, but also widely publicized events or beliefs, such as urban legends, Internet myths, hoaxes, rumors, and folk traditions, whether ultimately validated or not.

MORE DEFINITIONS

Prions, the smallest known infectious agents, are not alive; they are proteins that seem to cause disease by inducing other proteins to change shape. *Viruses* are extremely small, probably nonliving entities that can reproduce only inside living cells. A *bacterium* (plural *bacteria*) is a small, one-celled organism that has no nucleus and differs in other ways from the cells of animals and plants. *Rickettsiae, spirochetes,* and *chlamydiae* are specific types (or relatives) of bacteria. *Protozoa* are not exactly single-celled animals, as their name might suggest, but close to it. A prion, virus, or bacterium that causes disease is called a *pathogen.* Protozoa or larger creatures (such as worms) that cause disease are usually called *parasites.*

Infectious diseases are diseases that result from the presence of a pathogen or parasite in the body. A *communicable disease* is a disease that one host can transmit to another. A *host* is a human or other organism that a pathogen or parasite uses as a source of food or shelter. If the host acts as a source of infection for other species, it is called a *reservoir host.*

Vectors are insects or other organisms that transmit pathogens between hosts. A *zoonosis* (plural, *zoonoses*) is a disease that infects both humans and other vertebrate species.

SO HOW BAD IS IT?

The Four Horsemen of the Apocalypse and other images in the *Book of Revelation* have become powerful nonsectarian symbols for many people. Even the swine flu pandemic of 2009 prompted some millennial thinking, with at least one TV celebrity advising viewers to isolate themselves and to stock up on canned food and bottled water. Is Pestilence (epidemic disease) really the one on the white horse? And if so, can diseases or other biological threats really bring about the end of the world? We can't comment on the first issue, and as for the second, it all depends on what you mean by "end" and "world."

If the question is whether any real or imagined biothreat could shatter the Earth's core, annihilate all living things, and hurl frozen bits of primordial soup screaming into outer space, to begin the long interstellar journey to seed some distant world with the victor's DNA—no, we don't see that happening in the near future. Nor do most scientists consider it likely that any single pathogen could destroy all life on Earth, or even drive the human species to extinction. But some diseases and contributing factors do have the potential to end hundreds of millions of individual lives, to transform human societies and economies, and to cause unimaginable hardship to our children and grandchildren.

A biothreat, then, is what Lewis Carroll called a portmanteau word—two words packed into one, like a portmanteau suitcase with two compartments. But what is inside? Whatever you take with you. Reading about biothreats really means confronting your worst nightmare, and every reader has a different one. Whether you fear paralysis, asphyxiation, pain, bleeding, disfigurement, infertility, starvation, or just plain death, rest assured that it's somewhere in this book.

References and Recommended Reading

Ackerman, G. A., and J. Giroux. "A History of Biological Disasters of Animal Origin in North America." *Revue Scientifique et Technique*, Vol. 25, 2006, pp. 83–92.

Ackerman, G. A. "It Is Hard to Predict the Future: The Evolving Nature of Threats and Vulnerabilities." *Revue Scientifique et Technique*, Vol. 25, 2006, pp. 353–360.

Beran, G. W. (Ed.) 1994. *Handbook of Zoonoses*. Boca Raton, FL: CRC Press.

Booker, C., and R. North. 2008. *Scared to Death: From BSE to Global Warming—How Scares Are Costing Us the Earth*. London: Continuum.

Brown, L. R., and B. Halwell. "Breaking Out or Breaking Down." *World Watch*, Vol. 12, 1999, pp. 20–29.

Callahan, J. R. 2002. *Biological Hazards: An Oryx Sourcebook*. Westport, CT: Oryx Press (imprint of Greenwood Publishing Group).

Delamothe, T. "Several Horsemen of the Apocalypse." *British Medical Journal*, Vol. 337, 2008, p. a1365.

Fielding, J. E. "Public Health in the Twentieth Century: Advances and Challenges." *Annual Review of Public Health*, Vol. 20, 1999, pp. xiii–xxx.

Gillett, J. D. "The Behaviour of *Homo sapiens*, the Forgotten Factor in the Transmission of Tropical Disease." *Transactions of the Royal Society of Tropical Medicine and Hygiene*, Vol. 79, 1985, pp. 12–20.

Goleman, D. "Hidden Rules Often Distort Ideas of Risk." *New York Times*, 1 February 1994.

Guadalupe, P., and L. Saad. "Public Perceptions of Worldwide Malaria and TB Risks Haven't Risen." Gallup News Service, 26 June 2007.

Hiiemäe, R. "Handling Collective Fear in Folklore." *Folklore*, Vol. 26, 2004, pp. 65–80.

Hill, L. 2000. *The Chicago River: A Natural and Unnatural History*. Chicago: Lake Claremont Press.

Hill, L. "The Chicago Epidemic of 1885: An Urban Legend?" *Journal of Illinois History*, Vol. 9, 2006, pp. 154–174.

Jackson, J. W. "Bioterrorist Attacks on Food Would Cause More Panic than Actual Damage." Knight Ridder/Tribune Business News, 6 March 2003.

Jones, K. E., et al. "Global Trends in Emerging Infectious Diseases." *Nature*, Vol. 451, 2008, pp. 990–993.

Kumate, J. "Infectious Diseases in the 21st Century." *Archives of Medical Research*, Vol. 28, 1997, pp. 55–61.

Lane, J. M., et al. "Deaths Attributable to Smallpox Vaccination, 1959 to 1966, and 1968." *Journal of the American Medical Association*, Vol. 212, 1970, pp. 441–444.

Lopez, A. D., et al. "Global and Regional Burden of Disease and Risk Factors, 2001: Systematic Analysis of Population Health Data." *Lancet*, Vol. 367, 2006, pp. 1747–1757.

Lorber, B. "Are All Diseases Infectious?" *Annals of Internal Medicine*, Vol. 125, 1996, pp. 844–851.

Maugh, T. H. "Worldwide Study Finds Big Shift in Causes of Death." *Los Angeles Times*, 16 September 1996.

McMichael, A. J. "Environmental and Social Influences on Emerging Infectious Diseases: Past, Present and Future." *Philosophical Transactions of the Royal Society of London B*, Vol. 359, 2004, pp. 1049–1058.

Moeller, S. D. 1999. *Compassion Fatigue: How the Media Sell Disease, Famine, War, and Death.* New York: Routledge.

Ong, A. K., and D. L. Heymann. "Microbes and Humans: The Long Dance." *Bulletin of the World Health Organization*, Vol. 85, 2007, pp. 422–423.

Patz, J. A., et al. "Unhealthy Landscapes: Policy Recommendations on Land Use Change and Infectious Disease Emergence." *Environmental Health Perspectives*, Vol. 112, 2004, pp. 1092–1098.

Sanders, J. W., et al. "The Epidemiological Transition: The Current Status of Infectious Diseases in the Developed World versus the Developing World." *Science Progress*, Vol. 91, 2008, pp. 1–37.

Slingenbergh, J., et al. "Ecological Sources of Zoonotic Diseases." *Revue Scientifique et Technique*, Vol. 23, 2004, pp. 467–484.

Smith, D. F. "Food Panics in History: Corned Beef, Typhoid, and 'Risk Society.'" *Journal of Epidemiology and Community Health*, Vol. 61, 2007, pp. 566–570.

Snowden, F. M. "Emerging and Reemerging Diseases: A Historical Perspective." *Immunological Reviews*, Vol. 225, 2008, pp. 9–26.

Strange, R. N., and P. R. Scott. "Plant Disease: A Threat to Global Food Security." *Annual Review of Phytopathology*, Vol. 43, 2005, pp. 83–116.

Trust for America's Health. "New Report Finds Rising Risk of Infectious Diseases in America." Press release, 29 October 2008.

"UPI Poll: Bioterrorism Seen as Top Threat." United Press International, 23 February 2007.

Weiss, R. A. "The Leeuwenhoek Lecture 2001: Animal Origins of Human Infectious Disease." *Philosophical Transactions of the Royal Society of London B*, Vol. 356, 2001, pp. 957–977.

Weiss, R. A., and A. J. McMichael. "Social and Environmental Risk Factors in the Emergence of Infectious Diseases." *Nature Medicine*, Vol. 10, 2004, pp. S70–S76.

Whipple, D. "Microbes Replace Wolves in Culling Herds." *UPI Science News*, 1 September 2003.

2

Five Big Ones

> We could easily be made to believe that nothing has happened, and yet we have changed, as a house changes into which a guest has entered.
>
> —R. M. Rilke, *Letters to a Young Poet* (1904)

This chapter discusses the five infectious diseases that (arguably) pose the greatest threat to humans at present. The first three are the ones that top almost everyone's list: HIV/AIDS, malaria, and tuberculosis. Goal 6 of the United Nations Millennium Project (Table 2.1) is to combat these three diseases. Targeted milestones include halting their spread, and perhaps reducing the numbers of new cases, by 2015.

Each of these diseases kills over 1 million people every year worldwide. As of 2009, there is no effective vaccine for HIV/AIDS. There are experimental vaccines for malaria, and an unsatisfactory 90-year-old vaccine for tuberculosis. But HIV is now manageable as a chronic condition, and most cases of TB and malaria are potentially curable—in those patients who have access to expensive and prolonged treatment. As a result, the worst diseases in the world are no longer the ones that the public fears the most. In a 2007 Gallup survey, only 24 percent of U.S. respondents felt that either tuberculosis or malaria was a very serious global health problem (Table 2.2), despite the high death tolls of these diseases.

In addition to the "Big Three," this chapter covers two other diseases that claim between 500,000 and 1 million human lives in a typical year: influenza (all types and subtypes) and hepatitis (types B and C).

HIV DISEASE AND AIDS

Summary of Threat

The human immunodeficiency virus (HIV) attacks the immune system and makes the body unable to fight infection. Transmission is by fluid exchange. Without treatment, most (not all) people develop full-blown acquired immunodeficiency syndrome (AIDS) and die from opportunistic infections or cancer within a few years. As of 2009, no effective vaccine or cure is available, but antiretroviral drugs can prolong life and help prevent transmission.

Other Names

In the early 1980s, HIV was known as human T-cell lymphotropic virus type 3 (HTLV-III or HTLV-3) or lymphadenopathy-associated virus (LAV). The name HTLV-3 reflected an early hypothesis that the virus was related to the HTLV-1 and HTLV-2 viruses, which cause certain forms of human leukemia. Two labs identified the new virus in 1983 and named it human immunodeficiency virus (HIV). The U.S. Centers for Disease Control and Prevention (CDC) first referred to the disease as acquired immunodeficiency syndrome (AIDS) in 1982. Other names in that era included chronic symptomatic HIV infection, Kaposi's sarcoma and opportunistic infections (KSOI), and community-acquired immune dysfunction. The term AIDS-related complex

Table 2.1 United Nations Millennium Project, Goal 6

Goal 6: Combat HIV/AIDS, Malaria, and Other Diseases

Target 7. Have halted by 2015, and begun to reverse, the spread of HIV/AIDS

Indicators

18. HIV prevalence among pregnant women aged 15–24 years (UNAIDS-WHO-UNICEF)
19. Condom use rate of the contraceptive prevalence rate (UN Population Division)
19a. Condom use at last high-risk sex (UNICEF-WHO)
19b. Percentage of population aged 15–24 years with comprehensive correct knowledge of HIV/AIDS (UNICEF-WHO)
19c. Contraceptive prevalence rate (UN Population Division)
20. Ratio of school attendance of orphans to school attendance of non-orphans aged 10–14 years (UNICEF-UNAIDS-WHO)

Target 8. Have halted by 2015, and begun to reverse, the incidence of malaria and other major diseases

Indicators

21. Prevalence and death rates associated with malaria (WHO)
22. Proportion of population in malaria-risk areas using effective malaria prevention and treatment measures (UNICEF-WHO)
23. Prevalence and death rates associated with tuberculosis (WHO)
24. Proportion of tuberculosis cases detected and cured under DOTS (internationally recommended TB control strategy) (WHO)

Source: United Nations.

Table 2.2 Perceptions of Global Health Issues, United States, 2007

	Very Serious %	Somewhat Serious %	Not Serious %
HIV/AIDS	82	16	2
Cancer	79	20	1
Poor nutrition	75	22	3
Tuberculosis	24	51	23
Malaria	24	50	22

Source: Gallup News Service.

(ARC) refers to milder symptoms of HIV infection, such as weight loss and enlarged lymph nodes. In French- and Spanish-speaking countries and communities, the translation of AIDS is abbreviated SIDA.

In the early 1980s, the media coined several pejorative or inaccurate names, such as gay-related immunodeficiency disease (GRID), gay cancer, gay compromise syndrome, gay plague, and—most offensive of all—the so-called 4-H Club (Haitians, hemophiliacs, heroin users, and homosexuals). Other slang terms for AIDS that appear to assign blame include "women's disease" and "men's disease," both used in Tanzania. But Africans also call HIV/AIDS by names that are highly descriptive in translation, such as "slim disease" (Kenya), "that which came" (Zimbabwe), "lack of guard in the body" (Tanzania), "a thing that sucks life out" (Uganda), "God is tracking you" (South Africa), "those that suffer from the germ" (Zambia), "stepped on a landmine" (Angola), or simply "death" (Nigeria).

Description

HIV (Figure 2.1) is one of the retroviruses—a group of RNA viruses that can copy their genetic material into DNA, which then becomes part of a host cell chromosome. HIV exists in

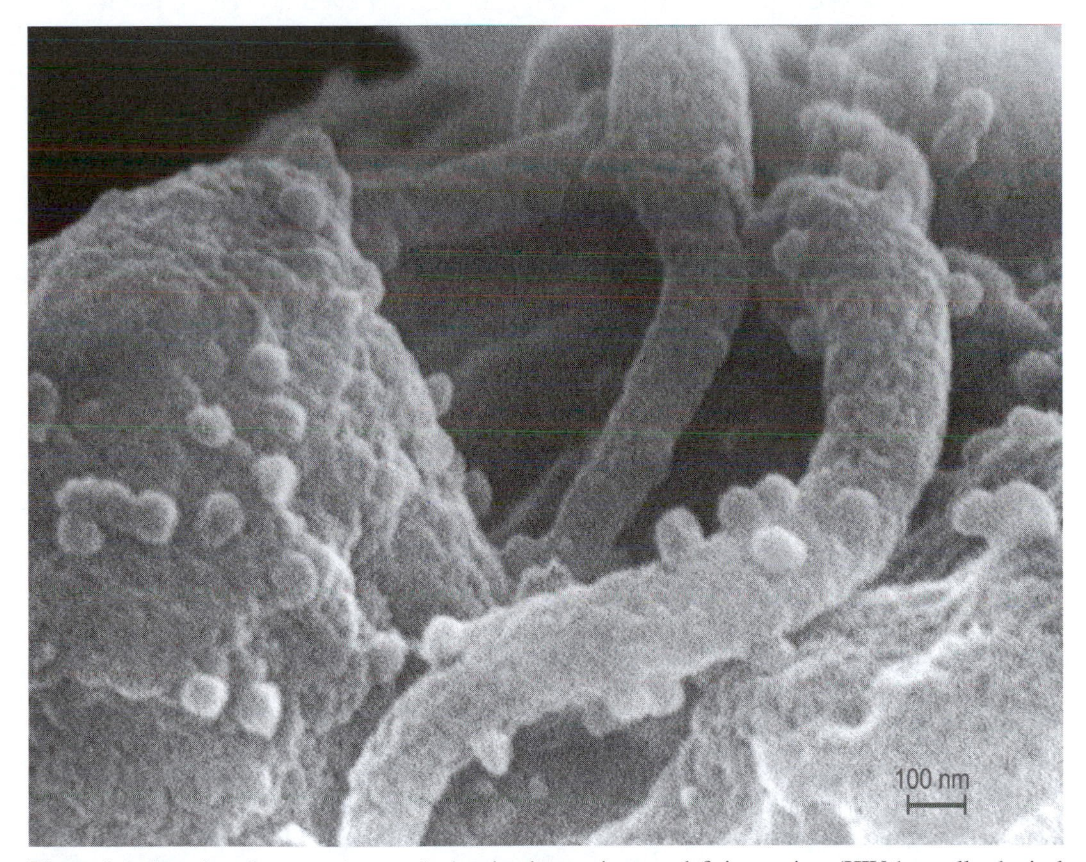

Figure 2.1 Scanning electron micrograph showing human immunodeficiency virus (HIV-1, small spherical objects) and human lymphocytes.

Source: U.S. Centers for Disease Control and Prevention, Public Health Image Library.

at least two forms, known as HIV-1 and HIV-2. The predominant form in most of the world is HIV-1, which includes three groups designated as M, N, and O. In 2009, French researchers identified a fourth group (tentatively designated as HIV-1 group P) that is closely related to a virus found in gorillas. Group M, the cause of most HIV-1 infections, comprises several genetic subtypes known as clades.

Thanks to its devastating effects and extensive media coverage, HIV/AIDS is perhaps better known to the American public than any other infectious disease. Yet the chief modes of HIV transmission are worth repeating: sexual contact (either homosexual or heterosexual) with exchange of body fluids; needle sharing by intravenous drug users; blood transfusions, organ transplants, or other medical procedures; laboratory transfer, such as needlestick injuries or splashing of contaminated body fluids on a skin lesion; or transfer during pregnancy or birth. The virus cannot replicate itself in mosquitoes, bedbugs, or ticks, so it is highly unlikely that these insects can serve as vectors. The virus can, however, survive for at least eight days in a bedbug or ten days in a tick.

Within the first few months after infection with HIV, many people develop a mild illness that lasts for a week or two, with symptoms such as fever, fatigue, and swollen lymph nodes. Otherwise, the person may be free of symptoms for several years while the virus silently destroys his or her immune system. Diagnosis requires blood testing for HIV antibodies (which are undetectable for at least a month after infection) or circulating HIV antigen. By the time HIV-related cancers or opportunistic infections (Figure 2.2) appear, it may be too late for antiretroviral drugs to be fully effective, and partners or other contacts may have contracted the disease. The actual cause of AIDS-related death is often an infection such as tuberculosis or *Pneumocystis* pneumonia.

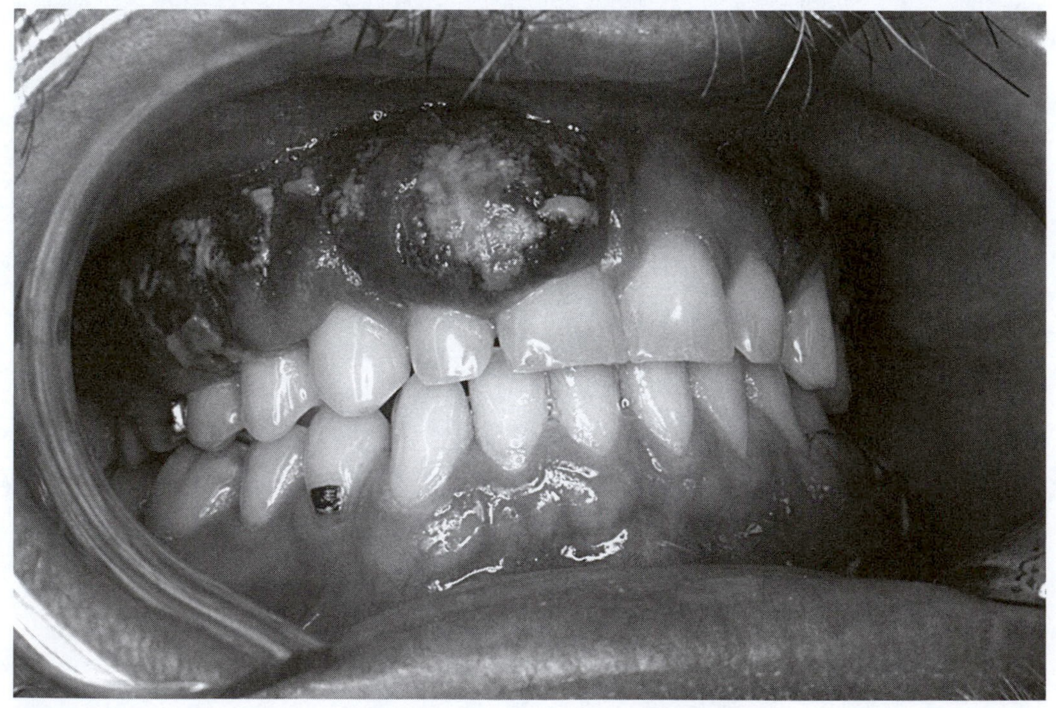

Figure 2.2 Oral Kaposi's sarcoma lesion and candidiasis (thrush) in HIV-positive patient.
Source: U.S. Centers for Disease Control and Prevention, Public Health Image Library.

Who Is at Risk?

Unprotected sex (without a condom) and contaminated needles account for most cases. Risk factors include coinfection with syphilis or other diseases that cause open sores through which the virus can enter. Circumcision may reduce HIV risk, at least in heterosexual men. Herpesvirus-6, which causes a childhood disease called roseola, appears to speed up the progress of HIV. Persons of European descent with a gene called Delta32 are resistant to HIV. Another gene, called DARC (Duffy antigen receptor for chemokines), common among West Africans, increases the risk of HIV while conferring some protection against malaria. About two-thirds of all HIV cases, and three-quarters of all AIDS-related deaths, are in sub-Saharan Africa.

Infection with HIV also increases the risk of contracting other infectious diseases, such as tuberculosis, visceral leishmaniasis, malaria, and many opportunistic infections that do not normally occur in healthy persons, including certain forms of pneumonia.

The Numbers

As of 2008, men who have sex with men (MSM) account for about 44 percent of AIDS cases in the United States, 65 percent in Canada, and 64 percent in Australia. These numbers do not imply that homosexuals are at disproportionate risk, but only that the first infected persons in North America and Australia were homosexual, and they transmitted the disease primarily to other homosexuals. In Africa, AIDS occurs mainly in heterosexuals, and about 60 percent of reported cases are in women.

Between 1981 and 2008, an estimated 32 million people died of AIDS, including 566,000 in the United States alone. The number of HIV-related deaths peaked in the United States in 1995, while worldwide numbers continued to rise. Recent projections are all over the map, but a 2008 WHO report states that annual HIV/AIDS deaths will peak at 2.4 million in 2012, up from 2.2 million in 2008. By a previous estimate, total deaths would peak at 6.5 million in 2030. These estimates are based on the best available data and assumptions, which change continually. The numbers do not include deaths from "new variant famine," a term recently applied to hunger that affects African households when one or more family members have AIDS and cannot work.

As of 2008, an estimated 33 million people worldwide were HIV-positive, including about 1 million in the United States. About half of all infected U.S. residents were African American, although this group represented only about 12 percent of the total population. By 2008, AIDS had become the leading cause of death among young African American women.

History

Researchers believe that retroviruses similar to HIV have existed for millions of years and were present in the earliest primates. HIV-1 probably originated in central Africa in about 1930 and spread to Haiti and the United States during the 1960s. Rumors of a new disease circulated in American prison populations in the 1970s, but the general public did not become aware of the epidemic until 1981, when doctors reported an unusual outbreak of *Pneumocystis* pneumonia among gay males in California. By then, HIV was established on every continent. The French researchers who identified the virus in 1983, Françoise Barre-Sinoussi and Luc Montagnier, shared the 2008 Nobel Prize in Physiology or Medicine. Since then, research has focused on the development of antiretroviral drugs and vaccines.

The AIDS era has been a time of great discoveries as well as great frustration regarding human rights and priorities. One example will suffice here: In July 2008, President George W. Bush reauthorized the 2003 PEPFAR (President's Emergency Plan for AIDS Relief) by signing into law the Tom Lantos and Henry J. Hyde United States Global Leadership against HIV/AIDS, Tuberculosis, and Malaria Reauthorization Act of 2008. In theory, this law ends a longstanding ban on HIV-positive travelers entering the United States; however, U.S. immigration law already prohibits foreigners with any communicable disease of public health importance from entering the country. As of 2008, it is not clear how much of a difference the new law will make, except that persons with HIV will no longer be singled out for exclusion.

Prevention and Treatment

The Bush administration (2001–2009) reportedly spent $6 billion per year to fight AIDS in Africa, but the program focused on antiretroviral drugs rather than prevention. The only known ways to avoid HIV are to avoid sex, needles, contact with injured people, and other potential sources of exposure, or to wear a condom or other protection whenever exposure might occur.

Since some belief systems or personal preferences discourage men from using condoms, recent effort has focused on the development of protective barriers for women, such as antiviral gels. As of 2009, these products are still on the drawing board. Female condoms are available, but their high cost and complexity may interfere with acceptance. Improved male condoms would also be beneficial; for example, the spermicide coatings may cause irritation that increases the probability of HIV transmission if the condom breaks.

Unfortunately, HIV vaccine trials to date have been unsuccessful or inconclusive. In 2003, a study of several thousand volunteers in North America, Europe, and Thailand concluded that a VaxGen vaccine was ineffective. In 2009, however, shortly before this book went to press, preliminary reports indicated that a VaxGen vaccine used in combination with a second vaccine appeared to reduce new HIV infections by about 30 percent. In 2007, a Merck vaccine failed in a study of 3,000 volunteers in nine countries. Clinical trials of a vaccine called Remune® yielded inconclusive results, and the manufacturer filed bankruptcy in 2008.

As of 2009, HIV treatment requires a combination of drugs: reverse transcriptase inhibitors to stop the virus from building new DNA, protease inhibitors to interfere with the action of viral enzymes, and fusion inhibitors to stop HIV from entering the host cell membrane. This treatment (called highly active antiretroviral therapy or HAART) has prolonged and improved many lives, but it often has side effects, such as nausea, vomiting, and diarrhea. Less frequent outcomes include diabetes, liver failure, pancreatitis, and enlargement of the breast or neck. Antiretroviral drugs can also cause immune restoration inflammatory syndrome (IRIS), in which opportunistic infections temporarily return. Some of the same drugs are used for short-term post-exposure prophylaxis (PEP), to treat people who have been exposed to HIV. Unfortunately, drug-resistant HIV strains have already appeared (Chapter 3).

There are recent reports of individuals who have apparently been cured of HIV by means of bone marrow transplants. These findings are encouraging, but even if this method works, it would be impractical on a large scale. In the near term, management of this disease will continue to focus on antiretroviral drugs, despite the side effects and the staggering cost of treatment.

Popular Culture

Any disease as deadly as AIDS becomes a nexus for exaggerated fears and rumors. Many of these misconceptions originated before science established that HIV is not transmitted by casual contact. Under the circumstances, it was reasonable to fear people with an unfamiliar disease that was known to be contagious.

The fear of HIV transmitted by accidental needlestick injury—which really happens on occasion—has given us the urban legend of Needle Boy, who supposedly hides HIV-infected needles in movie theater seats, coin return slots of public phones, and gas pump handles to stick the unwary. Typhoid Mary (Chapter 3) is the prototype for the urban legend of AIDS Mary, who picks up men in bars and infects them with HIV. Ironically, the real Typhoid Mary meant no harm; she simply refused to believe that typhoid is contagious, just as some people nowadays are in denial about HIV. (For examples of HIV denial, see Case Studies 2-1 and 2-2.)

In a 2007 Pew opinion survey, a startling 23 percent of Americans agreed with the statement "AIDS might be God's punishment for immoral sexual behavior." In 1987, 43 percent of Americans agreed with the same statement. But the wording of the question might be at fault, because almost anything *might* be true.

Some journalists and clergymen have attributed the AIDS epidemic to another authority figure, namely the United States government. One theory holds that doctors administering polio vaccine in Africa accidentally transferred the virus from chimpanzees to humans. Another claims that government agents created HIV as a genocidal weapon against Africans. In a third version, the government used gay American males as guinea pigs for an experimental hepatitis B vaccine that was contaminated with HIV and human herpesvirus-8 (the agent of Kaposi's sarcoma). These frightening scenarios have not held up to scrutiny, but the rumors continue.

Hollywood depictions of AIDS have varied in accuracy and sensitivity. Some critically acclaimed examples are *Longtime Companion* (1990), *Philadelphia* (1993), *And the Band Played On* (1993), and *Jeffrey* (1995). On a lighter note, the 1987 motion picture *Return to Salem's Lot* features a group of vampires who raise cows as a source of blood because of the danger of contracting AIDS or other diseases from human blood.

Although the world has been aware of HIV/AIDS for less than 30 years, alleged folk cures already exist, particularly in Africa and Latin America. Perhaps worst of all is the belief that having sex with a virgin will cure AIDS.

Case Study 2-1: HIV and the Down-Low

There is no way to address this controversial topic without offending someone, but it's too important to skip. The "down-low" is a slang term for a phenomenon that has plagued the African American community for decades and has probably caused many unnecessary deaths. The term refers to closeted sexual contact between men who do not regard themselves as homosexual and often do not use protection or seek HIV testing. Because of the difficulty of obtaining accurate data, it is unclear whether such behavior has been a major factor in the spread of HIV. But any source of denial that discourages people from checking their HIV status is one more barrier to defeating the epidemic. Anyone who engages in high-risk behavior must be tested, for the sake of their families and communities.

Case Study 2-2: HIV in Kyrgyzstan

Many people apparently still believe that AIDS is 100 percent avoidable by virtuous living, despite evidence to the contrary. In 2008, the news media reported a tragic story that took place in the central Asian republic of Kyrgyzstan. A seven-month-old baby boy developed a high fever, and his mother took him to the local hospital for treatment. At the hospital, the boy contracted HIV from a contaminated intravenous needle, but the hospital did not discover the problem immediately. Meanwhile, the mother took the baby home and continued breastfeeding. Later, both mother and child became ill and tested positive for HIV. It is not clear whether the virus entered through the mammary duct or through a small cut on her breast, but in either case, the mother caught HIV from her baby. Yet the woman's husband allegedly responded by accusing her of adultery, beating her up, and throwing her out of the house along with their son. One such incident would be dreadful enough, but the media reported that at least 72 children and 16 mothers were infected as a result of tainted blood transfusions or reused needles in two Kyrgyz hospitals.

Obviously, this won't work—it will more likely transmit HIV to the virgin, thus threatening two lives instead of one. This belief is not unique to AIDS; a century ago, European men reportedly sought out virgins to cure syphilis. According to another theory, sex with an animal will cure AIDS or other sexually transmitted diseases, because animals (like virgins) are said to be innocent. Other questionable AIDS cures include bee venom, apple cider vinegar, beet root with garlic, rattlesnake meat, and "AIDS nosode" (plain water mixed with a microscopic quantity of blood drawn from a deceased AIDS patient).

The Future

Someday scientists will create an effective HIV vaccine, and the world will rejoice. But will everyone who needs this vaccine choose to accept it? History has taught many people to fear new vaccines. In the 1950s, the first polio vaccines caused polio in hundreds of children, because the attenuated virus was not attenuated enough; and in 1942, some 50,000 U.S. military personnel received a yellow fever vaccine contaminated with hepatitis B. Even the smallpox vaccine that freed the world from that disease did so at the cost of many lives.

Also, requesting HIV vaccination or testing might be seen by some as tantamount to an admission of risk. By analogy, an effective vaccine for hepatitis B—a virus with a similar transmission pattern—has been available for nearly 30 years, yet many people have rejected it or have never heard of it. As a result, the hepatitis B epidemic continues (Chapter 3).

Until recently, it seemed unlikely that HIV could be eradicated without effective and mandatory vaccination. In 2008, however, WHO researchers published a mathematical simulation that appears to show that HIV transmission could be eliminated in as little as ten years, even without a vaccine. The plan is to test people every year and immediately give them antiretroviral drugs if they test positive. Implementation would pose some major practical problems, but the concept is intriguing (Case Study 7-16, Chapter 7).

Because of its effect on the human immune system, HIV poses at least one unique threat that may outlive the virus itself. Before the HIV epidemic, opportunistic infections such as *Pneumocystis* pneumonia were rare. But now that these pathogens can multiply and undergo mutation in millions of immunosuppressed human hosts, they may evolve into forms that can also infect healthy people. There is no proof that this has happened yet, but it is possible in theory. For that matter, HIV itself could eventually mutate to a form that is transmissible by casual contact, if it continues to circulate unchecked.

References and Recommended Reading

"AIDS Rates Among U.S. Blacks Rival Africa." *Science Online*, 29 July 2008.

Alexaki, A., et al. "Cellular Reservoirs of HIV-1 and Their Role in Viral Persistence." *Current HIV Research*, Vol. 6, 2008, pp. 388–400.

Altman, L. K. "Rare Cancer Seen in 41 Homosexuals." *New York Times*, 3 July 1981.

Bartelsman, M., and H. Veeken. "The HIV Pandemic in the Year 2007, an Overview." *Nederlands Tijdschrift voor Geneeskunde*, Vol. 151, 2007, pp. 2655–2660. [Dutch]

Bhavan, K. P., et al. "The Aging of the HIV Epidemic." *Current HIV/AIDS Reports,* Vol. 5, 2008, pp. 150–158.

Bockarie, M. J., and R. Paru. "Can Mosquitoes Transmit AIDS?" *PNG Medical Journal*, Vol. 39, 1996, pp. 205–207.

Boykin, K., and E. L. Harris. 2006. *Sex, Lies, and Denial in Black America*. New York: Carroll and Graf.

"California Prisons Giving Inmates Free Condoms." KNBC.com, 31 July 2008.

"CDC: More than 1 Million Have HIV in U.S." United Press International, 2 October 2008.

Cohan, G. R. "Another Bioterrorist." *The Advocate*, 4 December 2001.

Cohen, M. S., et al. "The Spread, Treatment, and Prevention of HIV-1: Evolution of a Global Pandemic." *Journal of Clinical Investigation,* Vol. 118, 2008, pp. 1244–1254.

Cohn, S. K., and L. T. Weaver. "The Black Death and AIDS: CCR5-Delta32 in Genetics and History." *QJM*, Vol. 99, 2006, pp. 497–503.

Correll, T. C. "'You Know About Needle Boy, Right?' Variation in Rumors and Legends about Attacks with HIV-Infected Needles." *Western Folklore*, Winter 2008, pp. 59–100. De Waal, A., and Whiteside, A. "New Variant Famine: AIDS and Food Crisis in Southern Africa." *Lancet,* Vol. 362, 2003, pp. 1234–1237.

Duran, D., et al. "Persons Tested for HIV—United States, 2006. *Morbidity and Mortality Weekly Report*, Vol. 57, 2008, pp. 845–849.

European Centre for Disease Prevention and Control. "HIV Prevention in Europe: Action, Needs and Challenges." Meeting Report, Stockholm, 2–3 October 2006.

Gaym, A. "Microbicides—Emerging Essential Pillars of Comprehensive HIV/AIDS Prevention." *Ethiopian Medical Journal*, Vol. 44, 2006, pp. 405–415.

Gilbert, M., et al. "The Emergence of HIV/AIDS in the Americas and Beyond." *Proceedings of the National Academy of Sciences*, Vol. 104, 2007, pp. 18566–18570.

Gottlieb, G. J., et al. "A Preliminary Communication on Extensively Disseminated Kaposi's Sarcoma in Young Homosexual Men." *American Journal of Dermatopathology*, Vol. 3, 1981, pp. 111–114.

Hall, H. I., et al. "Estimation of HIV Incidence in the United States." *Journal of the American Medical Association*, Vol. 300, 2008, pp. 520–529.

Hammer, S. M., et al. "Antiretroviral Treatment of Adult HIV Infections: 2008 Recommendations of the International AIDS Society—USA Panel." *Journal of the American Medical Association*, Vol. 300, 2008, pp. 555–570.

Humphery-Smith, I., et al. "Evaluation of Mechanical Transmission of HIV by the African Soft Tick, *Ornithodoros moubata*." *AIDS*, Vol. 7, 1993, pp. 341–347.

Iqbal, M. M. "Can We Get AIDS from Mosquito Bites?" *Journal of the Louisiana State Medical Society*, Vol. 151, 1999, pp. 429–433.

Jaffe, H. W., et al. "The Reemerging HIV/AIDS Epidemic in Men Who Have Sex with Men." *Journal of the American Medical Association*, Vol. 298, 2007, pp. 2412–2414.

Lacey, M. "Back from the US and Spreading HIV in Mexico." *International Herald Tribune*, 16 July 2007.

Laurencin, C. T., et al. "HIV/AIDS and the African-American Community: A State of Emergency." *Journal of the American Medical Association*, Vol. 100, 2008, pp. 35–43.

Leigh, J. P., et al. "Costs of Needlestick Injuries and Subsequent Hepatitis and HIV Infection." *Current Medical Research and Opinion*, Vol. 23, 2007, pp. 2093–2105.

Marsden, M. D., and J. A. Zack. "Eradication of HIV: Current Challenges and New Directions." *Journal of Antimicrobial Chemotherapy*, 4 November 2008.

"Math Model: HIV Can Be Eliminated in a Decade." Associated Press, 25 November 2008.

McGroarty, P. "Doctors Say Marrow Transplant May Have Cured AIDS." Associated Press, 13 November 2008.

Merson, M. H., et al. 2008. "The History and Challenge of HIV Prevention." *Lancet*, Vol. 372, 2008, pp. 475-488.

"New Hope on AIDS in Africa." Associated Press, 1 December 2008.

Russell, S. "Unsettling Re-emergence of 'Gay Cancer.'" *San Francisco Chronicle*, 12 October 2007.

"Scientists Lose Hope Over AIDS Vaccine." *Science Online*, 24 April 2008.

Shafer, R. W., and J. M. Schapiro. "HIV-1 Drug Resistance Mutations: An Updated Framework for the Second Decade of HAART." *AIDS Reviews*, Vol. 10, 2008, pp. 67–84.

Shin, L. Y., and R. Kaul. "Stay it with Flora: Maintaining Vaginal Health as a Possible Avenue for Prevention of Human Immunodeficiency Virus Acquisition." *Journal of Infectious Diseases*, Vol. 197, 2008, pp. 361–368.

Singh, M. "No Vaccine Against HIV Yet—Are We Not Perfectly Equipped?" *Virology Journal*, Vol. 3, 2006, p. 60.

"U.N. Agency Lowers AIDS Estimates." *Science Online*, 20 November 2007.

U.S. Centers for Disease Control and Prevention. 1981. "Kaposi's Sarcoma and *Pneumocystis* Pneumonia among Homosexual Men—New York City and California." *Morbidity and Mortality Weekly Report*, Vol. 30, 1981, pp. 305–308.

Walker, B. D., and D. R. Burton. "Toward an AIDS Vaccine." *Science*, Vol. 320, 2008, pp. 760–764.

Weiss, R. A. "The Leeuwenhoek Lecture 2001: Animal Origins of Human Infectious Disease." *Philosophical Transactions of the Royal Society of London B*, Vol. 356, 2001, pp. 957–977.
Weiss, R. A. "The Discovery of Endogenous Retroviruses." *Retrovirology*, 3 October 2006.
Willyard, C. "Circumcision Strategy Against HIV Continues to Prove Divisive." *Nature Medicine*, Vol. 14, 2008, p. 895.
Worobey, M., et al. "Direct Evidence of Extensive Diversity of HIV-1 in Kinshasa by 1960." *Nature*, Vol. 455, 2008, pp. 661–664.

MALARIA

Summary of Threat

Malaria parasites reproduce in the body and invade blood cells, causing severe illness that may recur. As of 2009, there is no commercially available vaccine, and over 1 million people (mostly children in Africa) die from this disease each year. Antimalarial drugs and hospital treatment could save most of them. Since mosquitoes transmit malaria, mosquito control is also essential.

Other Names

The parasite *Plasmodium falciparum* causes the most severe form of malaria, also called falciparum malaria or malignant tertian malaria. Milder forms called benign tertian malaria, quartan malaria, and ovale malaria result from infection with the related species *Plasmodium vivax*, *P. malariae*, and *P. ovale*, respectively. The simian malaria parasite (*P. knowlesi*) also infects humans in southeast Asia. Mixed infections with more than one parasite species are fairly common. Double tertian malaria or quotidian fever means either infection with two *Plasmodium* species or two distinct generations of parasites that mature asynchronously, producing daily attacks of fever and chills.

Malaria originally meant "bad air" in Italian. In the old days, doctors believed that poisonous vapors rising from swamps caused disease. Malaria was known as swamp fever, marsh fever, marsh miasma, paludism, paludeen fever, paludal fever, Roman fever, Chagres fever, Panama fever, West African fever, malarial fever, jungle fever, congestive fever, remitting fever, biduoterian fever, aestivoautumnal fever, blackwater fever, or ague. "Blackwater" refers to the presence of hemoglobin in the urine, whereas ague is a less specific term for the chills that are often a prominent feature of this disease.

Description

The infectious agents of malaria are not bacteria or viruses, but single-celled parasites called protozoans. Malaria does not spread directly from one person to another; in most cases, mosquitoes act as vectors. Some cases also result from contaminated needles or blood transfusions.

The malaria parasite (Figure 2.3) cannot live independently without a host. It spends part of its life cycle inside a female *Anopheles* mosquito, where the sexual stages unite, giving rise to cells called sporozoites that migrate to the mosquito's salivary glands. When the mosquito bites a human to obtain blood, it sprays the bite area with infected saliva. The sporozoites then enter the human bloodstream and find their way to the liver, where they pass through a series of developmental stages and finally reproduce in red blood cells (Figure 2.4). The next time a mosquito bites the infected person, it ingests blood that contains sexual stages of the parasite, and the cycle starts over.

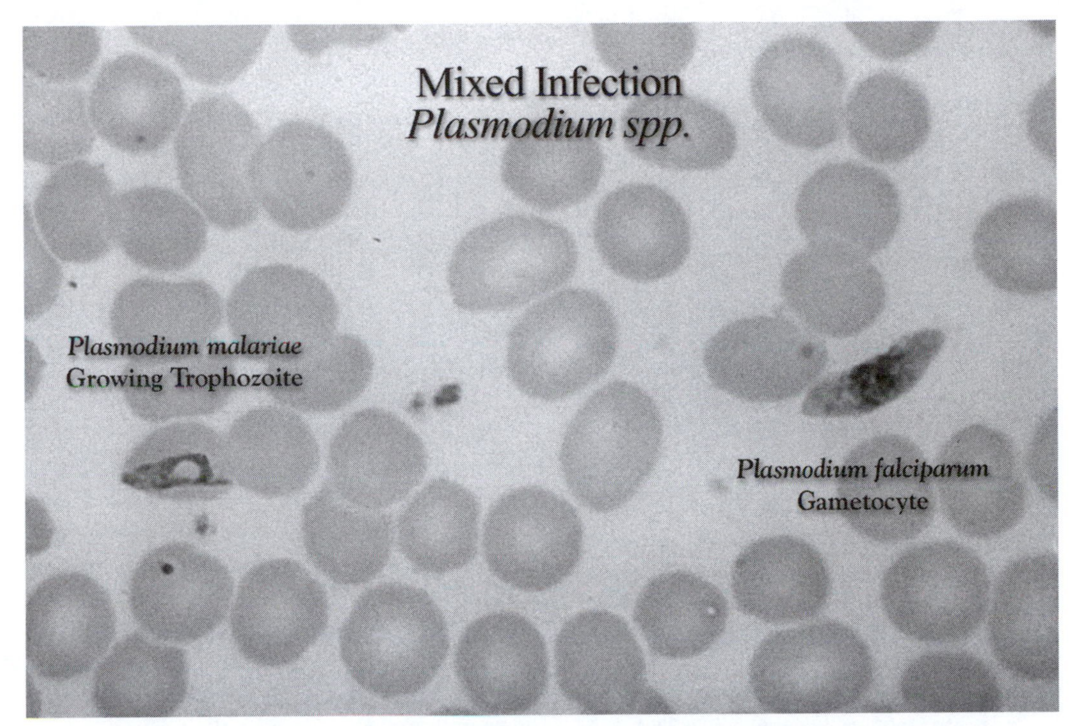

Figure 2.3 Micrograph showing mixed *Plasmodium falciparum* and *P. malariae* parasitic infection.
Source: U.S. Centers for Disease Control and Prevention, Public Health Image Library.

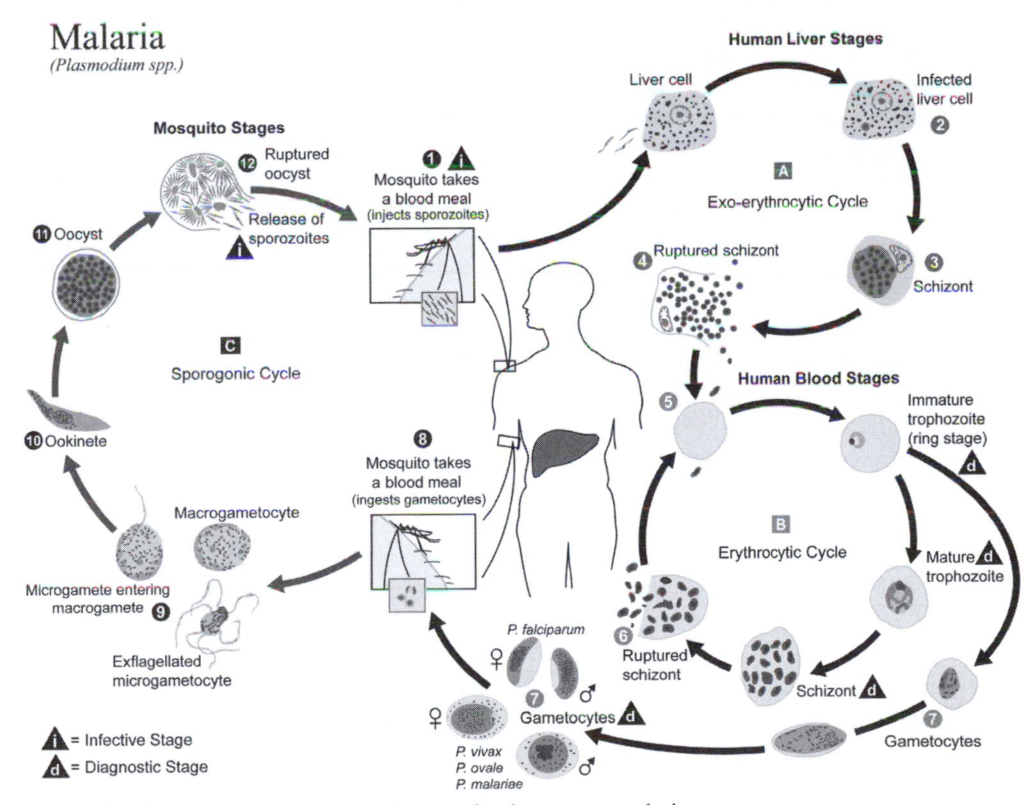

Figure 2.4 Life cycle of *Plasmodium,* the parasite that causes malaria.
Source: U.S. Centers for Disease Control and Prevention, Public Health Image Library.

Case Study 2-3: A Death in Palm Springs

In the spring of 2004, wildlife biologists in California were shocked to learn that malaria had claimed the life of a respected colleague. Although malaria kills many people in Third World nations, death is a rare outcome for malaria patients in American hospitals. The biologist had recently returned to California from a visit to Uganda, where he worked outdoors without using antimalarial drugs. A two-day history of fever and chills brought him to the hospital ER, where he was first diagnosed with *Plasmodium malariae* (quartan malaria). But after four days of unsuccessful treatment with chloroquine and primaquine, he developed respiratory distress and was transferred to the ICU. A second blood test revealed a mixed infection with *P. malariae* and *P. falciparum,* and the staff administered oral quinine and doxycycline. Unfortunately, it was too late. Twelve days after admission, he died of multiorgan system failure and cardiac arrest at age 43. Mixed-species malaria infections are common but often not recognized.

As each generation of parasites matures and red blood cells rupture, the person develops a high fever and chills. In some cases, the malarial parasites also invade the brain or other organs. Cerebral malaria is fatal in about 20 percent of children. Those who survive repeated malarial infections acquire natural immunity by about age 10, but many suffer permanent brain damage, impaired vision, ruptured spleen, or damage to the heart, kidneys, or liver.

Once nearly worldwide, malaria has been eliminated or controlled in most developed countries. It remains a major problem in the tropics and some subtropical areas, including sub-Saharan Africa, southeast Asia, Central America, and the Amazon region of South America. Malaria has essentially been eliminated from the United States, despite a few endemic cases as recently as 2003. Travelers to endemic areas may return with malaria (Case Study 2-3).

Who Is at Risk?

People are not at significant risk if they stay in a region where malaria does not occur. Global climate change and migration may change the present geographic limits of malaria, thus increasing the exposed population (Chapter 6).

At present, about one-third of the global population is at risk. Even those with protective genes (such as sickle cell trait or thalassemia) can contract malaria, but the number of parasites in the blood usually stays low and the symptoms do not become severe. Africans with a gene called DARC have partial resistance to malaria but are at increased risk for HIV. Blood group O may confer some protection against malaria, as does a common genetic disorder called glucose-6-phosphate dehydrogenase deficiency (G6PD). People with acquired resistance to malaria can tolerate the infection, but they are not fully immune either.

An iron-rich diet or iron supplementation in children may increase susceptibility to malaria, but general malnutrition is also said to be a risk factor. Malaria is more likely to be severe or fatal in people who have other diseases, such as HIV, tuberculosis, dengue fever, or yellow fever. In some countries, hospital patients and transfusion recipients are at high risk for malaria. Accidental transmission by blood transfusion is fairly common where malaria is endemic and blood bank screening is inadequate.

The Numbers

Every year, 400 to 500 million people contract malaria, and 1 to 2 million die as a result. An estimated 90 percent of these cases are in sub-Saharan Africa. Worldwide, about 20 percent of all childhood deaths result from malaria. On average, a child dies from this disease every 30 seconds. Sources vary regarding the actual number of deaths, because many malaria victims also have other diseases such as yellow fever or tuberculosis. Also, children who die from malaria often succumb to one of its effects (such as anemia), which might be recorded as the proximate cause of death instead.

Critics of global warming projections recently claimed that the largest malaria epidemic in history took place in Siberia during the 1920s and 1930s, with an estimated total of 13 million cases and 600,000 deaths over a 10-year period. These people concluded that malaria is not associated with hot climates and therefore will not expand its range if the Earth grows warmer. But this reasoning reflects a misunderstanding of the word "epidemic." Even if the Siberian epidemic were the largest in history—which it wasn't—malaria would still be a tropical disease, for the simple reason that the vast majority of cases occur in the tropics. An epidemic does not mean a large number of cases of a disease; it means a larger number *than usual*. Ethiopia, for example, normally has about 6 million cases of malaria every year. Those cases do not represent an epidemic, because they are normal for the region. But in 2003, the number of cases in Ethiopia jumped to about 12 million, and *that* was an epidemic.

History

Scientists estimate that malaria has killed more people than any other disease in the history of the world—perhaps as many as half of all humans who have ever lived. Today, malaria is not only a deadly disease in its own right, but also a factor that has contributed to the AIDS epidemic in Africa, as discussed above.

One of the oldest malaria treatments was a crude form of quinine obtained from the bark of South American evergreen trees in the genus *Cinchona*. This drug was called "the Jesuit powder" because missionary priests were among the first Europeans to recognize its value. Not everyone accepted this cure; Oliver Cromwell, Lord Protector of England, Scotland, and Ireland, contracted malaria in the marshes of Kent in his native England, but he refused treatment on religious grounds and died of the disease in 1658.

George Washington had his first malaria attack in Virginia in 1749, with recurrences in 1752, 1761, 1784, and 1798. Theodore Roosevelt caught malaria in Brazil in 1914, and John F. Kennedy had it while stationed in the South Pacific during World War II. Other U.S. presidents who survived malaria include James Monroe, Andrew Jackson, Abraham Lincoln, Ulysses S. Grant, and James A. Garfield.

Although malaria is primarily a tropical disease nowadays, the United States had millions of cases during the 1930s. The incidence diminished sharply by the 1940s, possibly as a result of New Deal malaria control projects or the depopulation of the rural South, where malaria was once prevalent. But malaria was already on the decline by 1928, before the New Deal or the use of DDT, thanks to a higher living standard that enabled people to live in houses with window screens. Without an infected host population, the disease tends to disappear.

Prevention and Treatment

In 2009, researchers announced a major milestone in the age-old battle against malaria. An experimental vaccine developed by GlaxoSmithKline was more than 50 percent effective in preventing malaria in infants and toddlers in Kenya and Tanzania. If further testing is successful, the company may seek marketing approval in 2011, thus bringing the world one step closer to the United Nations' goal of eliminating malaria (reducing the incidence to near zero) by 2015 and eradicating it altogether in the long term, perhaps by 2030.

Meanwhile, until a vaccine becomes available, the most effective preventive measures are also some of the oldest: wear protective clothing and mosquito repellent, avoid visiting endemic areas during the malaria season—unless, of course, you live there—and stay inside a screened building at night, preferably with a bed net.

Taking antimalarial drugs (such as chloroquine or mefloquine) can reduce the probability of contracting malaria. But these drugs often have unpleasant side effects, such as nausea and depression, and they are not 100 percent effective. Since protozoan cells have a nucleus and chromosomes similar to those of higher animals, the drugs that kill protozoans tend to make people sick too. Also, some malarial parasites have become drug-resistant, like many other pathogens. Vendors of diluted antimalarial drugs have contributed to this problem by promoting the evolution of resistance without curing the disease. Antimalarial drugs are also expensive, and the cost of treatment is a major problem in most countries where malaria is endemic.

In 2008, researchers reported progress in developing a method for inexpensive mass production of a phosphonate antimalarial compound. This process will help governments and individuals afford these drugs, but hopefully the new vaccine will soon render them redundant.

The other way to prevent malaria is by vector control, usually involving elimination or management of standing water combined with pesticide spraying (Case Study 2-4). As of 2009, the controversy regarding the pesticide DDT has raged for more than 50 years. In 1972, the U.S. government outlawed DDT because of its harmful effects on wildlife, and many other nations followed suit. Some people now claim that the removal of DDT from the arsenal sacrificed millions of human lives in the name of environmental protection; others point out that DDT was not working that well anyway, because some mosquito populations had already acquired resistance before the chemical was banned. This debate will most likely continue when malaria is only a bad memory. At present, DDT is used on a limited basis for indoor spraying in some areas.

Case Study 2-4: Malaria in Mauritius

The island nation of Mauritius is located in the southern Indian Ocean east of Madagascar, where conditions are favorable for malaria, yet the incidence of malaria is very low. The reasons are rooted in the history of the island. Neither humans nor malaria vectors (anopheline mosquitoes) were present on Mauritius until the mid-1700s, because cyclones made the island unsuitable for colonization. After Great Britain gained control of Mauritius in 1810, sugar cane plantations and irrigation canals replaced most of the original forest. The human population increased to about 300,000, malaria vectors arrived on ships from Africa, and a series of outbreaks followed. Between 1866 and 1868, about 10 percent of the population died of malaria. Survivors had partial immunity, but malaria was endemic on Mauritius for the next century. During World War II, British forces on the island started a vector control program that continued after the war. In 1973, the World Health Organization (WHO) declared Mauritius free of malaria, but four factors led to a resurgence between 1975 and 1984: standing water from a cyclone, new houses with flat roofs, infected migrant workers, and general complacence. The government redoubled its efforts, and the island has been free of indigenous malaria since 1996.

Popular Culture

According to folklorists, people in Ghana believe that malaria is the result of working too hard in hot weather. In Gambia, there is a widespread belief that malaria results from drinking sour milk during the rainy season. In Togo, malaria is said to result from eating too much red palm oil. Others attribute the disease to witchcraft. In a 2005 study of mothers in Nigeria, 85 percent insisted that there was no connection between malaria and mosquitoes. These traditions are no more foolish than the nineteenth-century European belief that malaria was the result of bad air from swamps. But humans and malaria have coexisted in Africa for hundreds of thousands of years, and most Africans must have noticed the association with mosquitoes. Perhaps they know, but prefer not to tell; more than one folklorist has failed to recognize when someone was pulling his or her leg. Or perhaps it's simply easier to attribute a disease to something controllable, such as food or work habits. Whatever the explanation, traditional beliefs persist in rural areas and may interfere with the acceptance of medical treatment.

The 1985 motion picture *Out of Africa* devotes exactly five words to the subject of malaria: "My water's gone black, Denys." (This is a reference to blackwater fever, a late stage of malaria in which hemoglobin appears in the urine.)

Mary Medearis's 1942 novel *Big Doc's Girl* is set in rural Arkansas during the Great Depression, where the title character takes on the state political establishment and a powerful real estate developer to fight a malaria epidemic.

In North America, crushed leaves of the American beautyberry plant (*Callicarpa americana*) are a traditional folk remedy to repel mosquitoes. Recent studies show that this plant actually works. Other alleged remedies include black oak tea, garlic, white bryony root, peach, cow urine, dogwood, willow bark, and poplar bark.

A popular malaria treatment in Thailand is a grass called *ya ka* (*Imperata cylindrica*) steeped in water. Known as cogon grass in English, it also grows in eastern and southern Africa, Indonesia, and Australia. In Togo, people reportedly make a tea called "malaria drink" from citronelle, ginger, and pineapple rinds to bring on sweating. West Africans reportedly use a bitter tea of neem leaves.

Traditional Chinese remedies for malaria include concoctions of tang-shen (*Campanumaea* sp.), skullcap (*Scutellaria baicalensis*), Chinese quinine (*Dichroa febrifuga*), Chinese knotweed (*Polygonum multiflorum*), or magnolia (*Magnolia officinalis*), with the addition of toasted tortoise shell and peach kernels for chronic cases. At least one of these plants, Chinese quinine, contains an effective antimalarial compound.

The side effects of antimalarial drugs are well documented in the medical literature. Depending on the drug, reported symptoms range from severe headache, upset stomach, and depression to nightmares, seizures, and violent behavior. Yet in every war since Vietnam, military personnel stationed in the tropics have been criticized for refusing to take their antimalarial medication due to unwarranted fears of side effects. In other words, there is an urban legend to the effect that the side effects of antimalarial drugs are nothing but an urban legend.

The Future

The complete eradication of malaria will rank among the greatest of human achievements. When that day comes, the world should place all wars on hold, have a huge party, and make sure every colonist on Mars has been vaccinated.

Chapter 6 discusses the potential effects of future climate change on malaria and several other diseases.

References and Recommended Reading

AlKadi, H. O. "Antimalarial Drug Toxicity: a Review." *Chemotherapy*, Vol. 53, 2007, pp. 385–391.

Altman, L. K. "Diagnosis Was Malaria, but Experts Disagreed on the Source." *New York Times*, 9 November 1999.

"Anti-malarial Tablets Often Fake in Asia." United Press International, 13 February 2008.

Black, J., et al. "Mixed Infections with *Plasmodium falciparum* and *P. malariae* and Fever in Malaria." *Lancet*, Vol. 343, 1994, p. 1095.

Bryson, D. "African Researchers Plan Malaria Vaccine Trial." Associated Press, 10 November 2008.

Campbell, C. C. "Malaria Control—Addressing Challenges to Ambitious Goals." *New England Journal of Medicine*, Vol. 361, 2009, pp. 522–523.

"CDC Allowed to Give Artesunate for Malaria." *Science Online*, 4 August 2007.

Chowdhury, K., and O. Bagasra. "An Edible Vaccine for Malaria Using Transgenic Tomatoes of Varying Sizes, Shapes and Colors to Carry Different Antigens." *Medical Hypotheses*, Vol. 68, 2007, pp. 22–30.

Cohen, S. "Immunity to Malaria." *Proceedings of the Royal Society of London, Series B, Biological Sciences*, Vol. 203, 1979, pp. 323–345.

Collins, W. E., and G. M. Jeffery. "*Plasmodium malariae*: Parasite and Disease." *Clinical Microbiology Reviews*, Vol. 20, 2007, pp. 579–592.

Connor, S., and M. Thomson. "Epidemic Malaria: Preparing for the Unexpected." *Science and Development Network*, 1 November 2005.

Cox-Singh, J., et al. 2008. "*Plasmodium knowlesi* Malaria in Humans Is Widely Distributed and Potentially Life Threatening." *Clinical Infectious Diseases*, Vol. 46, 2008, pp. 165–171.

Cox-Singh, J., and B. Singh. "Knowlesi Malaria: Newly Emergent and of Public Health Importance?" *Trends in Parasitology*, Vol. 24, 2008, pp. 406–410.

"DDT Resistance Protein Found in Mosquitoes." United Press International, 17 June 2008.

Egan, T. J., and C. H. Kaschula. "Strategies to Reverse Drug Resistance in Malaria." *Current Opinion in Infectious Disease*, Vol. 20, 2008, pp. 598–604.

Filler, S. J., et al. "Locally Acquired Mosquito-Transmitted Malaria: A Guide for Investigations in the United States." *Morbidity and Mortality Weekly Report*, 8 September 2006.

"Gates Unleashes Mosquitoes at Tech Conference." Associated Press, 6 February 2009.

Greenwood, B. M., et al. "Malaria: Progress, Perils, and Prospects for Eradication." *Journal of Clinical Investigation*, Vol. 188, 2008, pp. 1266–1276.

Humphreys, M. "Water Won't Run Uphill: The New Deal and Malaria Control in the American South, 1933–1940." *Parasitologia*, Vol. 40, 1998, pp. 183–191.

Hyde, J. E. "Drug-Resistant Malaria—An Insight." *FEBS Journal*, Vol. 274, 2007, pp. 4688–4698.

Ibidapo, C. A. "Perception of Causes of Malaria and Treatment-Seeking Behaviour of Nursing Mothers in a Rural Community." *Australian Journal of Public Health*, Vol. 13, 2005, pp. 214–218.

Jamieson, A., et al. 2006. *Malaria: A Traveller's Guide*. Cape Town: Struik Publishers.

Katz, T. M., et al. "Insect Repellents: Historical Perspectives and New Developments." *Journal of the American Academy of Dermatology*, Vol. 58, 2008, pp. 8865–8871.

Kitchen, A. D., and P. L. Chiodini. "Malaria and Blood Transfusion." *Vox Sanguinis*, Vol. 90, 2006, pp. 77–84.

Langhorne, J., et al. "Immunity to Malaria: More Questions than Answers." *Nature Immunology*, Vol. 9, 2008, pp. 725–732.

Lewison, G., and D. Srivastava. "Malaria Research, 1980–2004, and the Burden of Disease." *Acta Tropica*, Vol. 106, 2008, pp. 96–103.

MacArthur, J. R., et al. "Probable Locally Acquired Mosquito-Transmitted Malaria in Georgia, 1999." *Clinical Infectious Diseases*, Vol. 32, 2001, pp. E124–E128.

Mali, S., et al. "Malaria Surveillance—United States, 2006." *Morbidity and Mortality Weekly Report*, Vol. 57, 2008, pp. 24–39.

McKenzie, F. E., et al. "Strain Theory of Malaria: The First 50 Years." *Advances in Parasitology*, Vol. 66, 2008, pp. 1–46.

Mishra, S. K., and Mohanty, S. "Problems in Management of Severe Malaria." *Internet Journal of Tropical Medicine*, Vol. 1, 2003, No. 1.

Murdock, D. "DDT Key to Third World's Winning War on Malaria." *Human Events*, 14 May 2001.

Nano, S. "Malaria Vaccine Shows Promise in Africa Tests." Associated Press, 8 December 2008.

Oppenheimer, S. "Comments on Background Papers Related to Iron, Folic Acid, Malaria and other Infections." *Food and Nutrition Bulletin*, Vol. 28 (4 Suppl.), 2007, pp. S550–S559.

O'Shaughnessy, P. T. "Parachuting Cats and Crushed Eggs: the Controversy over the Use of DDT to Control Malaria." *American Journal of Public Health*, 17 September 2008.

RBM (Roll Back Malaria) Partnership. "Global Roadmap to End Malaria Launched at UN Summit." Press Release, 25 September 2008.

Ringwald, P. "Current Antimalarial Drugs: Resistance and New Strategies." *Bulletin de l'Académie Nationale de Médecin*, Vol. 191, 2007, pp. 1273–1284.
"Russia in Grip of Malaria Wave." Universal Service, 4 August 1923.
Skarbinski, J., et al. "Malaria Surveillance—United States, 2004." *Morbidity and Mortality Weekly Report*, Vol. 55, 2006, pp. 23–37.
"Study May Cut Malaria Treatment Costs." United Press International, 1 October 2008.
Thwing, J., et al. "Malaria Surveillance—United States, 2005." *Morbidity and Mortality Weekly Report*, Vol. 56, 2007, pp. 23–40.
Tripathy, V., and B. M. Reddy. 2007. "Present Status of Understanding on the G6PD Deficiency and Natural Selection." *Journal of Postgraduate Medicine*, Vol. 53, 2007, pp. 193–202.
United Nations. "World Leaders Commit Record Billions to Tackle Malaria." Press Release, 25 September 2008.
U.S. Centers for Disease Control and Prevention. "Treatment Guidelines: Treatment of Malaria (Guidelines for Clinicians)." March 2007.
Walther, B., and M. Walther. "What Does It Take to Control Malaria?" *Annals of Tropical Medicine and Parasitology*, Vol. 101, 2007, pp. 657–672.
White, N. J. 2008. "Qinghaosu (Artemisinin): The Price of Success." *Science*, Vol. 320, 2008, pp. 330–334.

TUBERCULOSIS

Summary of Threat

Tuberculosis (TB) is an airborne bacterial disease that affects the lungs or other organs. Although preventable and curable, TB kills about 2 million people every year worldwide—more than any other single infectious agent. As of 2009, available vaccines are only partially effective, and some TB strains have become resistant to antibiotic treatment. HIV infection and malnutrition are major risk factors for TB.

Other Names

Tuberculosis is commonly called TB. Older names include consumption ("con" for short), phthisis, bacillary phthisis, phthisis pulmonalis, king's evil, wasting disease, hectic fever, buck Irish, Pott's disease, Koch's disease, cachexia, and white plague. Specific forms of TB were called tabes mesenterica, lupus vulgaris, gibbus, fungous arthritis, or scrofula when the disease affected the abdominal lymph nodes, skin, spine, joints, or neck, respectively. The terms miliary tuberculosis and disseminated tuberculosis refer to TB that has invaded the circulatory system. A modern slang name for tuberculosis is "Victorian novel disease," an allusion to its popularity as a plot device among fiction writers of that era.

Primary tuberculosis is a person's first infection with TB, usually in the lung; *secondary tuberculosis* means reinfection or reactivation of a dormant infection. The word "lunger" formerly meant a person with tuberculosis, but in modern slang the same word means a large wad of spit. Inelegant derivatives (according to some sources) include the slang words "looger" and "loogie." TB has names in every language: *kekkaku* in Japanese, *shachepheth* in Hebrew, *eitinn* in Irish Gaelic, *gorley shymlee* in Manx Gaelic, and *lugnasjúka* in Faeroese.

Description

The word "tuberculosis" often conjures up an image of someone like the famous Doc Holliday, a thin person with interesting cheekbones who occasionally coughs blood into a handkerchief. In

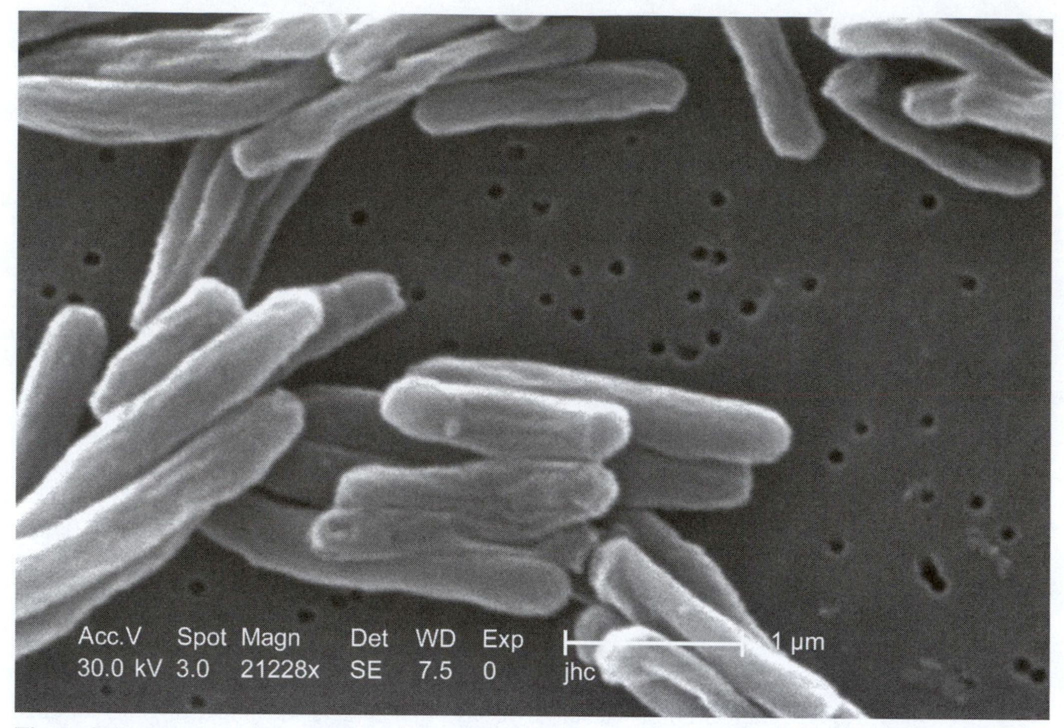

Figure 2.5 Scanning electron micrograph showing *Mycobacterium tuberculosis* bacteria, which cause tuberculosis in humans.

Source: U.S. Centers for Disease Control and Prevention, Public Health Image Library.

fact, tuberculosis takes many forms. Pulmonary (lung) tuberculosis is the most common, but TB can also affect the brain, lymph nodes, or other parts of the body (see Other Names).

Many infected people are unaware that they have tuberculosis, because the disease often becomes latent for years before symptoms develop. An estimated one-third of the human population is infected with TB, and about 10 percent of those with healthy immune systems will eventually develop clinical disease. People who are HIV-positive are at even higher risk.

In most of the world, the principal agent of human TB is a bacterium with the scientific name *Mycobacterium tuberculosis* (Figure 2.5). Several related species can also cause TB, particularly in persons with compromised immune systems. Transmission is usually by airborne droplet. Before the routine pasteurization of dairy products, people often developed scrofula or other forms of TB after drinking milk that was contaminated with the agent of bovine tuberculosis (*Mycobacterium bovis*).

Who Is at Risk?

Tuberculosis is a re-emerging disease in the industrialized world, where it was on the decline until about 1985. Its recent resurgence is related to the spread of HIV (a major risk factor) and to immigration from Third World countries with a high prevalence of TB. As of 2009, immigrants to the United States from Mexico are three times more likely to have tuberculosis than the general U.S. population. There is an urgent need for better treatment programs to serve the needs of this immigrant group and prevent the disease from spreading.

In addition to HIV, known risk factors for TB include alcoholism, malnutrition, smoking, air pollution, diabetes, and low serum vitamin D levels. Crowding also increases risk, because the disease is airborne and spread by coughing. Studies suggest that African American, Hispanic, and Native American populations are more likely than others in the United States to develop TB. Certain occupational groups are at risk, such as hospital employees who work with TB patients or HIV-positive patients. The reverse can also happen; in 2008, a hospital worker with TB exposed an estimated 960 infants to the disease. Employees and residents of prisons, long-term care facilities, and homeless shelters also are at risk for TB.

In at least two reported cases, a person with tuberculosis transmitted the disease to a group of friends while smoking marijuana in a shared bong or by "hotboxing" (smoking and exhaling inside a closed car).

The Numbers

Tuberculosis kills more people worldwide each year than any other single infectious agent. Recent estimates range from 1.5 million to 3 million deaths per year, depending on who is counting. People with AIDS often contract TB (and other diseases such as malaria), and the recorded cause of death may not be accurate.

During the first decade of the new millennium, about 9 million new cases of active TB appeared every year. About 95 percent of such patients are curable if they have access to treatment and are willing to cooperate, but about 5 percent have extensively drug-resistant tuberculosis (XDR-TB), which is essentially untreatable. In the United States, the cost of hospitalization for one XDR-TB patient averages about $483,000. Since XDR-TB does not respond to available antibiotics, these patients eventually die of the disease.

History

Scientists have found traces of the TB bacterium in the remains of a South American woman who died some 500 years before Columbus reached the New World, a fact suggesting that the disease already had a worldwide distribution. Studies of ancient Egyptian mummies found TB in about one-third of them—the same as the prevalence in today's human population. But tuberculosis appears to be even older than that. In 2007, anthropologists found evidence of a form of TB called tubercular meningitis in a 500,000-year-old human (*Homo erectus*) fossil from Turkey. It is possible that TB was originally a zoonosis found in wild ancestors of modern cattle and became a human disease soon after our ancestors began hunting and dismembering these animals.

The public health impact of tuberculosis has long challenged human ingenuity. In the Old West, for example, bartenders deployed the latest medical advances to limit the spread of TB. Since it would be impolite to ask customers to refrain from spitting indoors, public spittoons were partially filled with an antiseptic such as carbolic acid.

U.S. President Theodore Roosevelt recognized the association between tuberculosis and the conditions of poverty and crowding that many new immigrants encountered on their arrival in this country, but his proposed solution was harsh by today's standards. In 1916, he wrote:

> If I could I would have the kind of restriction which would not allow any immigrant to come here unless I was content that his grandchildren would be fellow-citizens of my grandchildren. They will not be so if he lives in a boarding house at $2.50 per month with ten other boarders and contracts tuberculosis and contributes to the next generation a body of citizens inferior not only morally and spiritually but also physically.[1]

1. "A Roosevelt Idea Made in Germany" (*New York Times*, 2 February 1916).

No sentient being nowadays would call immigrants morally and spiritually inferior as a result of poverty or tuberculosis. Yet it is a statistical fact that people arriving in the United States from other countries account for the majority of active TB cases, and a solution better than rhetoric must be found. Recovering from this disease requires more than pluck and determination. It requires medical treatment, an increasingly unaffordable luxury for longtime residents and immigrants alike.

During the first half of the twentieth century, the incidence of tuberculosis in the United States fell steadily, and many doctors were confident that this ancient enemy was on the run. Yet a cautious 1948 editorial predicted that TB might not yield so easily:

> On numerous occasions in the past, persons became over-enthusiastic and launched slogans such as "No Tuberculosis by 1920"; "No Tuberculosis by 1960"; "No Tuberculosis by the Year 2000"; "No Tuberculosis In Our Time," etc., etc. Obviously those who prepared and publicized such slogans had not carefully analyzed the situation. . . . Even if all infection were to stop today, tuberculosis could not possibly be eradicated by the year 2000.[2]

Prevention and Treatment

Many employers require regular tuberculin skin tests or chest X-rays. If more people would voluntarily request TB testing, many lives might be saved. Treatment of latent tuberculosis infection with drugs such as isoniazid helps to prevent progression to active disease. Nor is active TB an automatic death sentence. If it is detected early, and if the patient cooperates fully with treatment, full recovery is possible in about 95 percent of cases. Treatment usually lasts several months and requires a combination of several antimicrobial drugs (usually isoniazid, rifampin, and pyrazinamide) with frequent testing of sputum. A 2009 study found that these drugs are more effective at higher doses.

Like many other diseases, however, tuberculosis has given rise to drug-resistant strains that are difficult to treat: the increasingly prevalent multidrug-resistant TB (MDR-TB), found in up to 36 percent of previously treated TB patients, and extensively drug-resistant TB (XDR-TB), which accounts for about 5 percent of cases. Combinations of second-line drugs such as amikacin, kanamycin, capreomycin, or the fluoroquinolone wide-spectrum antibiotics, often are effective against resistant strains. In 2008, scientists at Rutgers University announced the discovery of a new class of antibiotics that might help combat drug-resistant TB and other diseases.

The BCG (Bacille Calmette-Guérin) tuberculosis vaccine has been available since 1921, but it is seldom used in the United States because of concerns regarding its efficacy. It can prevent disseminated TB and TB meningitis in children, but otherwise the results of testing have been inconsistent and appear to vary by population and latitude, with little or no protective effect near the equator. As of 2009, research was in progress to develop new TB vaccines or adjuvants.

In some cases, recovery from TB requires surgery as well as antibiotic treatment. In 2008, doctors in Europe successfully replaced a former TB patient's damaged windpipe by growing a new one from her own stem cells. Doctors may also use surgery to remove part of a damaged lung in cases of drug-resistant TB.

As a last resort, governments may isolate tuberculosis patients to protect the general population from infection. In 2008, the press reported that a South African hospital isolated TB patients by means of razor wire and armed guards. Although such policies might sound extreme, public health officials have limited options when dealing with noncompliant patients

2. "Tuberculosis Eradication Being Achieved" (*Chest*, Vol. 14, 1948, pp. 292–293).

or those with extensively drug-resistant tuberculosis. In 2007, more than one TB patient made headlines in the United States by ignoring doctors' orders and exposing others to the disease (Case Study 2-5).

Case Study 2-5: An Airplane Passenger

In the summer of 2007, the news media reported the case of a Georgia attorney with multidrug-resistant tuberculosis (MDR-TB) who allegedly ignored his doctor's advice and flew to Greece to get married. When he and his wife returned to the United States, the CDC thought he had a more dangerous form of the disease, known as extensively drug-resistant TB (XDR-TB) and took him into custody to protect the public. Soon after his release, he had surgery to remove diseased lung tissue. Public reaction seemed divided between (1) outrage at the government for restricting this man's freedom and (2) outrage at the government for allowing him to fly in the first place. A similar incident in 2006 received less publicity, when an Arizona TB patient was taken into custody. And in October 2007, the *Washington Times* reported that a Mexican national with XDR-TB had recently crossed the U.S. border 76 times and had taken multiple domestic flights. Given the apparent difficulty of enforcing public health laws or identifying passengers, the simplest solution might be to have all airline passengers wear protective masks, whether infected or not.

Popular Culture

In the old days before antibiotics, TB victims tended to be pale and thin, and they were often seen with blood on their faces. The explanation was obvious: they must be vampires! Perhaps the most famous example was Mercy Brown, who died of TB in Exeter, Rhode Island, in 1892 at age 20. Her mother and two sisters had previously died of the same disease, and her older brother had also contracted it. For some reason, the neighbors decided that one member of the Brown family must be a vampire who was attacking the others. So they opened Mercy's above-ground crypt, two months after her death, and found the body relatively fresh. Instead of attributing its condition to the cold winter weather, they identified her as the vampire, and insisted on removing and burning her heart. They even made her brother drink some of the ashes for good measure, but he died of TB a few months later. (This appears to be a true story, except for the part about Mercy Brown being a vampire.)

Tuberculosis is a theme in many works of art and literature. In the Puccini opera *La Bohème*, the heroine is a beautiful young seamstress named Mimi who dies of TB. In the 1936 motion picture *Camille* and the Verdi opera *La Traviata*, the heroine is a beautiful young courtesan who dies of TB. And then there was Fantine in *Les Miserables*, and Beth in *Little Women*, and Helen in *Jane Eyre*. Thomas Mann's 1924 novel *The Magic Mountain* takes place in a tuberculosis sanitarium. A more recent example is Nicole Kidman's character Satine in the 2001 motion picture *Moulin Rouge*.

A gritty treatment of TB appears in Upton Sinclair's 1906 novel *The Jungle*, in which meatpackers suffer most unromantically from this disease. John le Carré's 2001 novel *The Constant Gardener* deals with the testing of tuberculosis drugs on unaware subjects in Africa. And in the 1964 memoir *A Moveable Feast*, Ernest Hemingway notes that prostitutes in Kansas City believed that swallowing semen would cure tuberculosis.

North American and European folk remedies for TB include deer's tongue (a plant), elder, sagebrush, red clover, sanicle, cocklebur, creosote bush, or pine resin; a mixture of lemon, honey, and flaxseed; mistletoe, St. John's wort, blackberry, garlic, tobacco, eggs, ginseng, valerian, elecampane, logwood (*Haematoxylon campechianum*), yellow dock, whortle berries (*Vaccinium arboreum*), hellebore (*Veratrum viride*), wild lettuce (*Lactuca canadensis*), liverwort,

Case Study 2-6: Umckaloabo

Don't ask us how to pronounce this word, but it is a blend of two equally difficult Zulu words that mean "lung disease" and "breast pain." Umckaloabo is an herbal medicine that African healers have used for generations to treat tuberculosis. The principal ingredient is the root of *Pelargonium,* a relative of the garden geranium. Every culture has its traditional remedies, but this one has gained worldwide fame thanks to the marketing genius of an Englishman named Charles Henry Stevens (1880–1942). Diagnosed with TB in 1897, Stevens moved to South Africa on his doctor's advice and was allegedly cured by a local medicine man. After the Boer War, he founded a company to sell the wonderful root. Despite decades of insults from an unsympathetic medical establishment, Stevens became a wealthy man and eventually died from an infected scratch on his leg. The story of Umckaloabo did not end there; a Swiss physician heard about the remedy and became its next champion. He reported some anecdotal success in treating TB patients, but he was unable to conduct clinical trials because Stevens never disclosed the identity of the plant. That mystery was solved in 1977, when a laboratory identified a sample as *Pelargonium.* As of 2005, the product is popular in Germany and other European countries, but nobody has yet figured out whether it really works or not.

mullein, nettle leaf tea, juniper tea, buttercup, or dew shaken from the flowers of chamomile; warm milk from a red cow or from a goat of any color; dew from a manure pile; the fat of a turtle, alligator, or rattlesnake; pickled rattlesnake skin, small frogs, cooked raccoon, or blood from the tail of a cat without a white hair; bogbean or buckbean (*Menyanthes trifoliata*); soup made from a black dog, or the hind limb of any dog; tea made by boiling potato peelings; walking to a spot where four winds meet; being bitten by a rattlesnake; wearing a cat skin on the chest, or allowing a cat to sleep on the bed; inhaling fumes from a cattle shed or horse stall, or those of a skunk, or the air of pine woods; or applying a salve of twinleaf root.

Traditional Chinese herbal remedies for pulmonary tuberculosis include concoctions of fresh marlberry (*Ardisia japonica*), figwort (*Scrophularia* sp.), or fritillary (*Fritillaria verticillata*), or powdered roots of Indian madder (*Rubia cordifolia*), rhubarb (*Rheum rhabarbarum*), oriental arborvitae (*Thuja orientalis*), or Chinese ground orchid (*Bletilla* sp.) mixed with juice of crushed radish. See also Case Study 2-6.

The Future

As of 2009, WHO was midway through its Global Plan to Stop TB, with the goal of reducing new cases and deaths by 50 percent by the year 2015 and eliminating the disease (reducing its incidence to one case per million population) by 2050. Key elements of this plan include making treatment more accessible and affordable, and promoting research to develop new TB tests and drugs. Unfortunately, the consensus appears to be that these goals are not yet feasible. Elimination of TB will most likely require an effective vaccine, reduction of risk factors, and more rapid progress in improving healthcare for low-income people.

References and Recommended Reading

Al-Jahdali, H., et al. "Tuberculosis in Association with Travel." *International Journal of Antimicrobial Agents*, Vol. 21, 2003, pp. 125–130.

Bloom, B. R., and P. M. Small. "The Evolving Relation Between Humans and *Mycobacterium tuberculosis*." *New England Journal of Medicine*, Vol. 338, 1998, pp. 677–678.

Blue, L. "A New Class of Antibiotics Could Offer Hope Against TB." *Time*, 17 October 2008.

Brosch, R., and V. Vincent. "Cutting-Edge Science and the Future of Tuberculosis Control." *Bulletin of the World Health Organization*, Vol. 85, 2007, pp. 410–412.

Budha, N. R., et al. "Biopharmaceutics, Pharmacokinetics and Pharmacodynamics of Antituberculosis Drugs." *Current Medicinal Chemistry*, Vol. 15, 2008, pp. 809–825.

Burzynski, J., and W. Schluger. "The Epidemiology of Tuberculosis in the United States." *Seminars in Respiratory and Critical Care Medicine*, Vol. 29, 2008, pp. 492–498.

Castro, K. G. "Tuberculosis Elimination in the United States: Why, How, and What Will It Take?" *Kekkaku*, Vol. 83, 2008, pp. 93–100.

"Cincy Homeless Shelter Hit with TB." *Science Online*, 5 May 2007.

"Diabetes May Increase Risk of Developing TB." Reuters, 15 July 2008.

"Drug-Resistant TB Is 5 Percent of All TB." United Press International, 26 February 2008.

European Centre for Disease Prevention and Control. "Framework Action Plan to Fight Tuberculosis in the European Union." Stockholm, February 2008.

Fattorini, L., et al. "Extensively Drug-Resistant (XDR) Tuberculosis: An Old and New Threat." *Annali dell'Istituto Superiore di Sanità*, Vol. 43, 2007, pp. 317–319.

Fernandez, E. "960 Babies in TB Scare at Kaiser in S. F." *San Francisco Chronicle*, 27 August 2008.

Khan, K., et al. "Tuberculosis Infection in the United States: National Trends Over Three Decades." *American Journal of Respiratory and Critical Care Medicine*, Vol. 177, 2008, pp. 455–460.

Kilner, J. "Siberian Jail Is Champion in Fight against TB." Reuters, 4 July 2008.

Klein, A. "New Drug Targets May Fight Tuberculosis and Other Bacterial Infections in Novel Way." Weill Cornell Medical College, press release, 27 December 2007.

LoBue, P., et al. "Plan to Combat Extensively Drug-Resistant Tuberculosis: Recommendations of the Federal Tuberculosis Task Force." *Morbidity and Mortality Weekly Report*, 12 February 2009.

Lönnroth, K., and M. Raviglione. "Global Epidemiology of Tuberculosis: Prospects for Control." *Seminars in Respiratory and Critical Care Medicine*, Vol. 29, 2008, pp. 481–491.

Maâlej, S., et al. "Pulmonary Tuberculosis and Diabetes" [French]. *Presse Medicale*, 2 September 2008.

Marks, S. M., et al. "Knowledge, Attitudes and Risk Perceptions about Tuberculosis: U.S. National Health Interview Survey." *International Journal of Tuberculosis and Lung Disease*, Vol. 12, 2008, pp. 1261–1267.

"Most Ancient Case of Tuberculosis Found in 500,000-Year-Old Human; Points to Modern Health Issues." University of Texas at Austin, press release, 7 December 2007.

Naidoo, R. "Surgery for Pulmonary Tuberculosis." *Current Opinion in Pulmonary Medicine*, Vol. 14, 2008, pp. 254–259.

Nnoaham, K. E., and A. Clarke. "Low Serum Vitamin D Levels and Tuberculosis: A Systematic Review and Meta-Analysis." *International Journal of Epidemiology*, Vol. 37, 2008, pp. 113–119.

Oeltmann, J. E., et al. "Tuberculosis Outbreak in Marijuana Users, Seattle, Washington, 2004." *Emerging Infectious Diseases*, Vol. 12, 2006, pp. 1156–1159.

Pieters, J. "*Mycobacterium tuberculosis* and the Macrophage: Maintaining a Balance." *Cell Host and Microbe*, Vol. 3, 2008, pp. 399–407.

Raviglione, M. C. "The New Stop TB Strategy and the Global Plan to Stop TB, 2006–2015." *Bulletin of the World Health Organization*, Vol. 85, 2007, p. 327.

Stevenson, C. R., et al. "Diabetes and the Risk of Tuberculosis: A Neglected Threat to Public Health?" *Chronic Illness*, Vol. 3, 2007, pp. 228–245.

Storla, D. G., et al. "A Systematic Review of Delay in the Diagnosis and Treatment of Tuberculosis." *BMC Public Health*, Vol. 8, 2008, p. 15.

"TB Eradication in the U.S. by 2010 Unlikely: Survey." Reuters Health, 5 February 2008.

"TB on a Plane? Expect More of It, Experts Say." Associated Press, 31 May 2007.

"TB Patient Gets New Windpipe Made with Own Stem Cells." *Medical News Today*, 19 November 2008.

"Tuberculosis Found in Ancient *Homo erectus*." *Science Online*, 7 December 2007.

U.S. Centers for Disease Control and Prevention. "BCG Vaccine." Fact Sheet, April 2006.

Weinberg, E. D. "Iron Out-of-Balance: A Risk Factor for Acute and Chronic Diseases." *Hemoglobin*, Vol. 32, 2008, pp. 117–122.

Young, D. B., et al. "Confronting the Scientific Obstacles to Global Control of Tuberculosis." *Journal of Clinical Investigation*, Vol. 118, 2008, pp. 1255–1265.

INFLUENZA

Summary of Threat

Influenza (flu) is an airborne viral disease that causes fever and respiratory illness, sometimes with complications such as pneumonia. About 1 million people die from influenza in a typical year, but large epidemics of certain flu strains have caused many more deaths. As of 2009, flu vaccines and antiviral drugs are only partially effective. Influenza affects not only humans, but also birds, pigs, and other animals.

Other Names

The name "influenza" refers to the fact that medieval astronomers attributed disease epidemics to cosmic influences. English names for influenza include flu, grippe, and epidemic

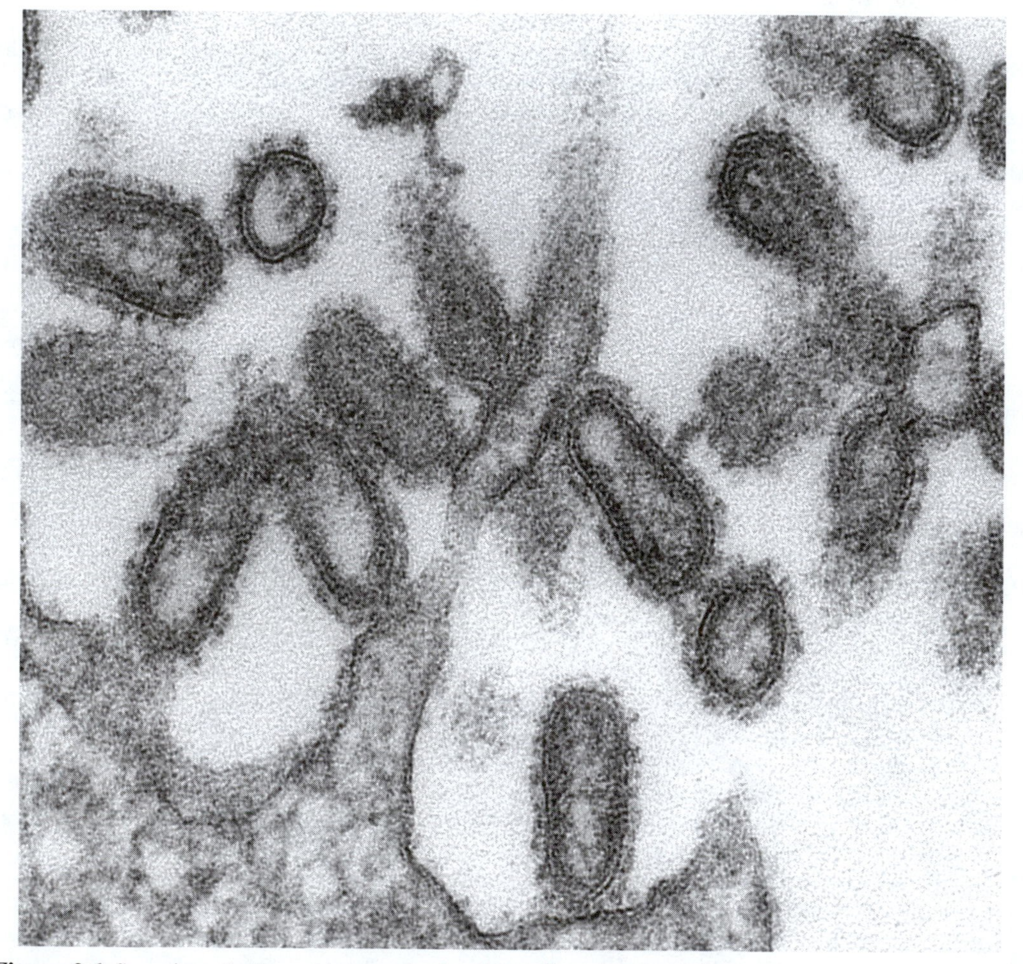

Figure 2.6 Scanning electron micrograph showing influenza virus of the type that caused the 1918 pandemic.

Source: U.S. Centers for Disease Control and Prevention, Public Health Image Library.

catarrhal fever. The nickname "panflu" for pandemic influenza also appears to be catching on. Words for flu in most other languages are cognates of the English words; for example, *floo* in Manx, *grip* in Turkish, and *imfluwenza* in Zulu.

Outbreaks and subtypes of influenza tend to acquire nicknames. The 1918 pandemic (source unknown) was called the Spanish Flu, Spanish Lady, Naples Soldier, La Grippe Espagnole (French for "Spanish grippe"), or La Pesadilla (Spanish, "the nightmare"). Other names may refer to a host (swine flu, bird flu, chicken flu, avian influenza) or point of origin (Asian flu, London flu, Hong Kong flu, Johannesburg flu, Guangdong flu, New Caledonian flu, Panama flu, Sichuan flu, Shanghai flu, London flu, Kamamoto flu); a full catalog would exceed the scope of this book. The so-called "stomach flu" (viral gastroenteritis) is not related to influenza, although some flu strains may cause nausea or diarrhea in addition to the usual flu symptoms. (Likewise, despite its name, the bacterium *Haemophilus influenzae* is not the agent of influenza.)

Antigenic subtypes of the influenza A virus (Figure 2.6) get their official names from the types of glycoproteins found on the surface of the virus (Figure 2.7). The letter H followed by a number identifies a glycoprotein called a hemagglutinin, and the letter N followed by a number identifies another glycoprotein called a neuraminidase. For example, H1N1 refers to a subtype

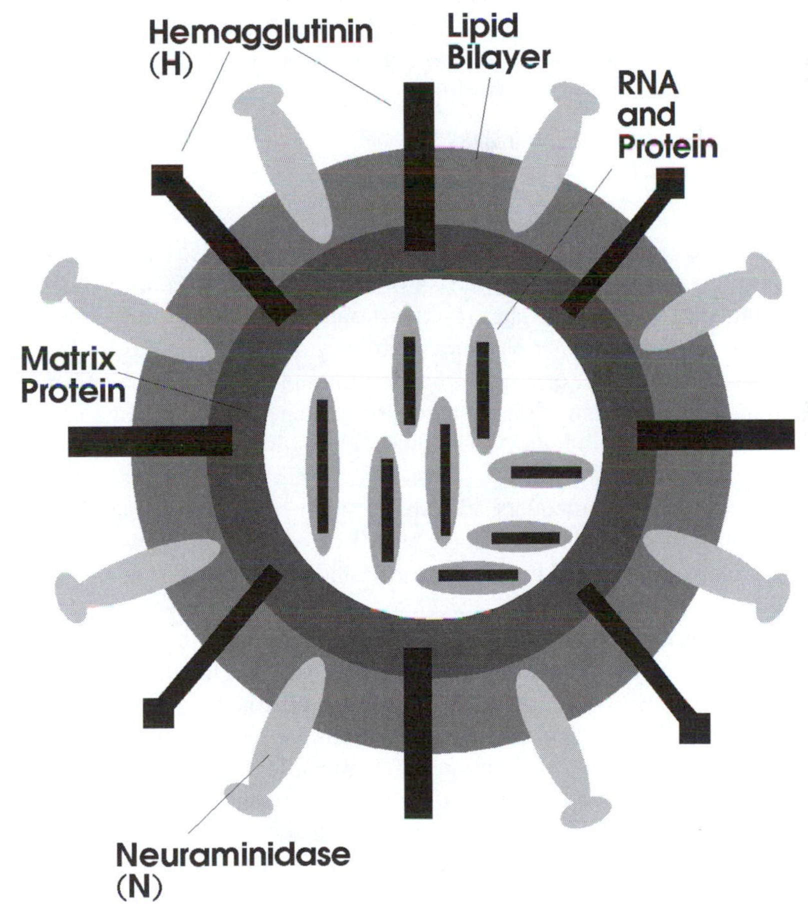

Figure 2.7 Diagram of Influenza A virion.

Source: J. R. Callahan, *Biological Hazards* (Oryx Press, 2002).

with hemagglutinin 1 and neuraminidase 1. Influenza is also characterized by outbreak pattern: seasonal, pandemic, or sporadic.

Description

Influenza A, a disease of humans and various domesticated and wild animals, is the type responsible for most human epidemics and all recorded pandemics. The media's favorite flu viruses, such as H1N1 and H5N1, are subtypes of influenza A. Influenza B has caused some regional epidemics, and influenza C is usually limited to sporadic cases and small outbreaks. This discussion will focus on influenza A.

Influenza usually starts abruptly, with a high fever, headache, and muscle pain. After a few days, these symptoms often subside, and the person develops a persistent dry cough, sore throat, and runny nose. The cough and general fatigue may last for weeks, and serious complications may develop, such as pneumonia or encephalitis (Chapter 3).

Influenza is an example of a zoonosis—a disease that infects both humans and other species, in this case primarily birds and pigs. In the tropics, influenza A occurs throughout the year, but in temperate regions of both hemispheres, epidemics occur mainly in winter. These seasonal epidemics usually start in Asia, apparently because of farming practices that bring humans into close contact with millions of ducks, chickens, and pigs, which serve as a reservoir for the disease and as a de facto laboratory for the evolution of new strains. Worldwide flu epidemics, known as pandemics, also occur periodically (see History).

Who Is at Risk?

Anyone can catch the flu, but for most people it is a mild disease with a very low death rate (usually less than 0.01 percent). The risk of bacterial pneumonia, or other serious complications of influenza, is usually higher if at least one of the following factors is present: age under 2 or over 65 , asthma or other chronic respiratory disease, diabetes, poor nutritional status, low-dose arsenic exposure, or a compromised immune system. Some forms of influenza are more virulent than others; in 2008, researchers estimated that the human case fatality rate of highly pathogenic avian influenza (Chapter 4) is somewhere between 14 percent and 33 percent.

Healthcare workers and bartenders are often at high risk for influenza because of indoor exposure to people and their sputum. People who live or work in crowded conditions, those exposed to secondhand smoke, and those who use public transportation are also at increased risk for airborne diseases in general.

Factors that reduce the probability of catching the flu include youth (for most strains), acquired immunity from previous exposure to the same subtype, having annual flu shots, wearing a protective mask and gloves in public, and avoiding all contact with other human beings or contaminated objects. Clearly, some of these precautions are more feasible than others.

These risk factors refer to recent subtypes and epidemics. In the 1918 pandemic, however, a high proportion of those who died were military recruits and other young, healthy people. That discrepancy is one of several reasons why scientists believe that secondary or concurrent infections, such as TB and pneumonia, were largely responsible for the high death rate. As of 2009, it is unclear if youth is a risk factor for complications resulting from the H1N1 swine flu, or if older people simply have partial resistance because of prior exposure to a similar strain.

The Numbers

In a typical year, 10 to 20 percent of U.S. residents contract influenza, and some 200,000 are hospitalized with complications of influenza. About 36,000 to 40,000 die, usually as the result of a secondary infection such as pneumonia. Worldwide, there are about 500,000 to 1 million flu-related deaths every year.

In a pandemic year, these numbers change. The 1918 pandemic killed at least 40 to 50 million people, with some recent estimates running as high as 100 million. These numbers tend to dominate every discussion of influenza. The global population in 1918 was about 1.8 billion; in 2009, there were about 6.8 billion, including hundreds of millions with compromised immune systems, diabetes, or other chronic medical conditions that represent risk factors for influenza. As a result, some (not most) sources predict that the next influenza pandemic might kill as many as 1 billion people.

History

Between the isolation of the influenza virus in 1933 and the advent of HIV in 1981, influenza was perhaps the focus of more intensive study than any other viral disease. There were two influenza pandemics during that interval, in 1957 and 1968. Sources do not agree on the exact number of past pandemics, since records are incomplete, but there have been at least nine since written records became available (Table 2.3). The H1N1 influenza outbreak of 2009 achieved pandemic status shortly before this book went to press. Note that the word "pandemic" refers to the numbers and distribution of cases, and does not necessarily imply a high death rate.

The 1918 influenza pandemic changed the world in many ways, not all of them immediately apparent. For example, in April 1919, President Woodrow Wilson contracted influenza (perhaps complicated by encephalitis) while representing the United States at the Paris Peace Conference. He survived, but was unable to participate effectively in the conference. Without his input, the resulting treaty created economic hardship in postwar Germany that would soon favor the rise of a ruthless dictator.

Prevention and Treatment

Preventive measures include avoidance and vaccination. Researchers prepare an influenza vaccine each year, based on the viral strains that appear to be circulating, but this approach

Table 2.3 Influenza Pandemics

Date	Name	Estimated Deaths	Subtype
1580	none (Europe)	Unknown	Unknown
1729	Russian Flu	Unknown	Unknown
1782	Blitz Catarrh	Unknown	Unknown
1830	China Flu	1 to 2 million?	Unknown
1847	none (Europe)	<1 million	H1N1?
1889	Russian Flu	1 million	H3N8 or H2N2
1918	Spanish Flu	50 million	H1N1
1957	Asian Flu	2 million	H2N2
1968	Hong Kong Flu	1 to 2 million	H3N2

requires some guesswork and is not uniformly successful. Many people—including the majority of healthcare workers in recent surveys—avoid getting their flu shots, either because they are worried about side effects or because they doubt that the vaccine will work. The 1976 swine flu outbreak and its aftermath may have reduced public confidence in vaccination programs (Case Study 2-7).

Case Study 2-7: The 1976 Swine Flu Controversy

In February 1976 a soldier at Fort Dix, New Jersey, somehow contracted a form of influenza that appeared similar to the strain that killed millions in the 1918 pandemic. (As it later turned out, the two strains were quite different; the soldier had swine flu, but the 1918 strain was eventually identified as bird flu.) The infected soldier died within 24 hours, and several others were hospitalized. Fearing a deadly epidemic, the Ford administration made the decision to immunize all Americans. By October 1976, the swine flu vaccine was ready. During the next two months, some 40 million people received the vaccine before unexpected side effects forced the government to suspend the program. Several hundred people developed neurological problems after vaccination, including the dreaded Guillain-Barré syndrome, and 32 died. Ironically, the swine flu epidemic never materialized, and no one ever figured out how the virus reached Fort Dix. The debate continues to the present day: Did the government do the right thing, based on available information? Or did it make a series of bad decisions that cost lives? The reader may decide.

Paradoxically, while public health agencies constantly urge people to get their flu shots, hospitals often report vaccine shortages. In 2004, for example, the news media reported that a critical shortage made the vaccine unavailable to tens of millions of Americans who were at risk for influenza. Either there are major distribution problems, or these shortages represent a marketing strategy to increase demand. In either case, the people who need the flu vaccine most are often the least likely to get it.

As of 2009, injectable influenza vaccines (flu shots) contain killed viruses that cannot cause flu. There is also a nasal spray vaccine (called FluMist) that uses weakened live viruses. Although potentially transmissible to other people, these weakened viruses do not cause disease symptoms either. (Since the vaccine is prepared using eggs, people who are allergic to eggs should check with their doctor before getting a flu shot.)

The ideal solution would be a "universal" influenza vaccine that works against all subtypes and strains of these viruses. In 2009, Japanese researchers reported substantial progress in that direction. Existing vaccines target proteins on the surface of the flu virus, but these proteins mutate so often that a new vaccine is needed every year. The new vaccine reportedly targets proteins inside the virus that rarely change. To date, it has been tested only on mice.

Several antiviral drugs have shown promise in reducing the symptoms or duration of influenza. The best known are Tamiflu® (oseltamivir) and Relenza® (zanamivir). When first introduced in the 1990s, both drugs were effective. Unfortunately, as of 2009, an estimated 98.5 percent of influenza A cases are resistant to Tamiflu®. Resistance has been less of a problem with Relenza®, apparently because it is less widely used. Two older antiviral drugs, amantadine and rimantadine, are also used to treat or prevent influenza, but again there are reports of increasing resistance. All these drugs have potentially serious side effects, but they are useful for treating high-risk patients.

Inevitably, some sources claim that megadoses of ascorbic acid (vitamin C) or other vitamins can cure or prevent influenza. People with actual deficiencies of vitamin C, vitamin D, or other nutrients are probably more susceptible to influenza and other diseases (Chapter 7). But if you are already getting enough of these vitamins, there is no evidence that taking more will cure anything. Elderberry extract may help relieve influenza symptoms or shorten the course of the disease; unfortunately, elderberry juice may contain toxins that cause severe illness, depending on how it is prepared.

The most virulent forms of influenza may require additional measures. For example, physicians in southeast Asia have reported some success using monoclonal antibodies or plasma from survivors to treat humans infected with highly virulent avian influenza H5N1.

Popular Culture

Although the media called the 1918 pandemic the "Spanish Flu"—because about 80 percent of the Spanish population caught the flu, and it was widely reported in the Spanish press—the actual source of the pandemic is unknown. In 1918, most countries other than Spain apparently tried to minimize disease statistics because of the need to keep up morale during the war, and rumors filled the information vacuum as usual. Some people claimed that the Germans spread the flu via contaminated Bayer aspirin tablets. Others, possibly closer to the mark, reported that strange-looking birds started the disease. And children played and sang throughout the crisis, as they have always done: "I opened the door, and in flew Enza."

A recent source claims that deadly bird flu pandemics happen every 84 years, when the planet Uranus is in the sign of Pisces. Uranus was in Pisces for seven years starting in 1918, and of course there was a pandemic. Uranus was not in Pisces in 1957 or 1968, the dates of the next two pandemics, but the circulating strains were not bird flu. Uranus entered Pisces again in 2002, and soon we had a small but scary outbreak of SARS and several hundred cases of bird flu. But by the time this book is published, Uranus will have moved out of Pisces. If the bird flu pandemic doesn't hurry up and happen, we will just need to wait another 84 years. (The 2009 swine flu pandemic doesn't count, although the new virus reportedly does have some avian components.)

Katherine Anne Porter's 1939 novella *Pale Horse, Pale Rider* depicts American society in the fall of 1918 during the influenza pandemic. Porter herself caught the disease and nearly died, so the story is partly autobiographical.

According to folklore books, burning orange rind, garlic, onions, or flowers of sulfur will relieve the symptoms of influenza (see also Case Study 2-8). In fact, any type of smoke tends to irritate the lungs, and if some family members already have the flu, it would probably make their cough worse. But if you burn something that really stinks, it might serve to keep other people out of your house during an epidemic, thus reducing the probability of catching the flu from visitors (or giving them the flu if you already have it).

Other North American or European folk remedies for flu include wearing a bag around the neck containing garlic, a spider, a woodlouse, or a foul-smelling plant resin called asafoetida; thrusting a live fish into the throat and returning it to the water; burying a piece of hair or toenail in a tree hole; being passed three times under the belly of a horse; inhaling the vapor of grated horseradish; or drinking any of the following: goat's milk, holly

Case Study 2-8: The Carbolic Smoke Ball

The influenza pandemic of 1889, like those before and since, served to showcase both the best and worst qualities of humanity. In 1890, the British manufacturer of a device called the Carbolic Smoke Ball offered a reward of 100 pounds sterling to any person who used the device as directed and yet developed influenza. Near the end of the pandemic in 1891—not long before Prince Albert (the one recently made famous by Bart Simpson) died of influenza complicated by pneumonia—a woman named Emily Carlill purchased a smoke ball, used it as directed, and promptly caught the flu. After making a full recovery, she requested payment from the manufacturer, only to be told that the advertisement was not intended to be taken literally. The resulting brouhaha ended in 1893 with one of the most famous legal precedents in the doctrine of contracts, *Carlill v. Carbolic Smoke Ball Company.* The court ruled in Ms. Carlill's favor.

leaf tea, elder flower infusion, elderberry wine, horehound tea, wild cherry gum dissolved in wine, infusions of almost anything (elder flower, yarrow, elecampane, thyme, juniper, mint, or mullein), or whiskey mixed with hot water, lemon, and honey. Or, of course, chicken soup. Traditional Chinese herbal remedies for influenza include packing one nostril with crushed coriander, or drinking a concoction of white onions, black beans, and fresh ginger.

Did the famous cross-dressing pirate Anne Bonny really die of influenza? In one version of the story, yes. In others, it was Anne's husband who died of influenza, or the husband of another pirate named Mary Read. In any case, they are all dead now.

The Future

There will be future influenza pandemics, but the consensus (despite media claims to the contrary) is that the death rate is unlikely to approach the catastrophic levels seen in 1918. Many of those deaths were from secondary bacterial infections that are now treatable. On the other hand, there are nearly four times as many people in the world today as there were in 1918, and many are already sick with HIV, diabetes, or other chronic diseases that make them harder to treat. Also, there is the problem of drug resistance (Chapter 3). Regardless of the actual death toll, the worldwide economic impact may cause great hardship. In 2008, the United Nations predicted that a flu pandemic could cost $3 trillion and cause a 5 percent decline in global production.

In the next pandemic, hospitals may quickly reach and exceed surge capacity. Those that run out of respirators and other equipment may turn to the Strategic National Stockpile. As always, it will be advisable to keep people out of hospitals to the extent possible. The pandemic may provide a boost for online education and shopping, telecommuting, and multiplayer computer games, which are already popular due to rising fuel costs. When the next pandemic is over, perhaps the majority of people will simply stay at home.

References and Recommended Reading

Albright, F. S., et al. "Evidence for a Heritable Predisposition to Death Due to Influenza." *Journal of Infectious Diseases*, Vol. 197, 2008, pp. 18–24.

Aoki, F. Y., et al. 'Influenza Virus Susceptibility and Resistance to Oseltamivir." *Antiviral Therapy*, Vol. 12, 2007, pp. 603–616.

"Bird Flu Survivors' Antibodies Effective." *Science Online*, 29 May 2007.

Blackburne, B. P., et al. "Changing Selective Pressure During Antigenic Changes in Human Influenza H3." *PLoS Pathogens*, 2 May 2008.

Boffey, P. M. "Editorial Observer; Ruminating on Smallpox Vaccine and the Notorious Swine Flu Fiasco." *New York Times*, 27 October 2002.

Cannell, J. J., et al. "On the Epidemiology of Influenza." *Virology Journal*, Vol. 5, 2008, p. 29.

Davis, M. 2005. *The Monster at Our Door*. New York: New Press.

Dowd, J. B., et al. "Early Origins of Health Disparities: Burden of Infection, Health, and Socioeconomic Status in U.S. Children." *Social Science and Medicine*, 17 January 2009.

Engel, M. "Researchers Tout Potential Flu-Fighting Tool." *Los Angeles Times*, 26 February 2009.

European Centre for Disease Prevention and Control. "Avian Influenza in Cats: ECDC Advice for Avoiding Exposure of Humans." *ECDC Scientific Advice*, March 2006.

Fiore, A. E., et al. "Prevention and Control of Influenza: Recommendations of the Advisory Committee on Immunization Practices (ACIP), 2008." *Morbidity and Mortality Weekly Report*, Vol. 57, 2008, pp. 1–60.

"The Flu." *New York Times*, 22 November 2008.

"Flu Deaths Linked to Staph Infections." Associated Press, 6 October 2008.

"Flu Not Responding to Tamiflu Treatment." United Press International, 3 March 2009.

"Flu Pandemic Tops Terrorism as U.K. threat." United Press International, 8 August 2008.

Gambotto, A., et al. "Human Infection with Highly Pathogenic H5N1 Influenza Virus." *Lancet*, Vol. 317, 2008, pp. 1464–1475.

Germann, T. C., et al. "Mitigation Strategies for Pandemic Influenza in the United States." *Proceedings of the National Academy of Sciences* (USA), Vol. 103, 2006, pp. 5935–5940.

Greenstreet, R. "Adjustment of Rates of Guillain-Barré Syndrome among Recipients of Swine Flu Vaccine, 1976–1977." *Journal of the Royal Society of Medicine*, Vol. 76, 1983, pp. 620–621.

Hsieh, Y. C., et al. "Influenza Pandemics: Past, Present and Future." *Journal of the Formosan Medical Association*, Vol. 105, 2006, pp. 1–6.

"Influenza A Virus Movement Is Tracked." United Press International, 20 September 2007.

Kandun, I. N., et al. "Factors Associated with Case Fatality of Human H5N1 Virus Infections in Indonesia: a Case Series." *Lancet*, Vol. 372, 2008, pp. 744–749.

Kozul, C. D., et al. "Low Dose Arsenic Compromises the Immune Response to Influenza A Infection *In Vivo*." *Environmental Health Perspectives*, 20 May 2009.

Li, F. C., et al. "Finding the Real Case-Fatality Rate of H5N1 Avian Influenza." *Journal of Epidemiology and Community Health*, Vol. 62, 2008, pp. 555–559.

Longman, R. E., and T. R. Johnson. "Viral Respiratory Disease in Pregnancy." *Current Opinions in Obstetrics and Gynecology*, Vol. 19, 2007, pp. 120–125.

McCullers, J. A. "Planning for an Influenza Pandemic: Thinking Beyond the Virus." *Journal of Infectious Diseases*, Vol. 198, 2008, pp. 945–947.

McCullers, J. A. "Preparing for the Next Influenza Epidemic." *Pediatric Infectious Disease Journal*, Vol. 27 (10 Suppl.), 2008, pp. S57–S59.

McGinnis, J. D. "*Carlill v. Carbolic Smoke Ball Company*: Influenza, Quackery, and the Unilateral Contract." *Canadian Bulletin of Medical History*, Vol. 5, 1988, pp. 121–141.

Morens, D. M., and A. S. Fauci. "The 1918 Influenza Pandemic: Insights for the 21st Century." *Journal of Infectious Disease*, 1 April 2007.

Morens, D. M., et al. "Predominant Role of Bacterial Pneumonia as a Cause of Death in Pandemic Influenza: Implications for Pandemic Influenza Preparedness." *Journal of Infectious Diseases*, Vol. 198, 2008, pp. 962–970.

Mumford, E. L., et al. 2007. "Avian Influenza H5N1: Risks at the Human-Animal Interface." *Food and Nutrition Bulletin*, Vol. 28 (Supplement), 2007, pp. S357–S363.

"Mutated Flu Virus Resists Medicine." United Press International, 1 February 2008.

O'Connor, A. "The Claim: Nasal-Spray Vaccine Can Make You Sick." *New York Times*, 18 November 2008.

Potter, C. W. "A History of Influenza." *Journal of Applied Microbiology*, Vol. 91, 2001, pp. 572–579.

Pyle, G. F. 1986. *The Diffusion of Influenza*. Totowa, NJ: Rowman & Littlefield.

"Report: Tamiflu May Increase Abnormal Acts." United Press International, 19 April 2009.

Rothberg, M. B., et al. "Complications of Viral Influenza." *American Journal of Medicine*, Vol. 121, 2008, pp. 258–264.

Specter, M. "Nature's Bioterrorist." *The New Yorker*, 5 February 2005.

"Students Wear Masks to Try to Stop Flu." Associated Press, 8 February 2007.

Taubenberger, J. K., and D. M. Morens. "1918 Influenza: The Mother of All Pandemics." *Emerging Infectious Diseases*, Vol. 12, 2006, pp. 15–22.

Taubenberger, J. K., and D. M. Morens. "The Pathology of Influenza Virus Infections." *Annual Review of Pathology*, Vol. 3, 2008, pp. 499–522.

U.S. Centers for Disease Control and Prevention. "Key Facts about Avian Influenza (Bird Flu) and Avian Influenza A (H5N1) Virus." Fact sheet, 30 June 2006.

"Universal Flu Vaccine Nears Development." Agence France-Presse, 30 January 2009.

Walsh, B. "Danger from the Bird-Flu Drug?" *Time*, 20 March 2007.

Wang, H., et al. "Probable Limited Person-to-Person Transmission of Highly Pathogenic Avian Influenza A (H5N1) Virus in China." *Lancet*, Vol. 371, 2008, pp. 1427–1434.

Zakay-Rones, Z., et al. "Randomized Study of the Efficacy and Safety of Oral Elderberry Extract in the Treatment of Influenza A and B Virus Infections." *Journal of International Medical Research*, Vol. 32, 2004, pp. 132–40.

HEPATITIS B AND C

Summary of Threat

Hepatitis B and C are viral diseases that cause inflammation of the liver and increase the probability of liver or pancreatic cancer. Both forms of hepatitis are transmitted by fluid exchange or household contact. Hepatitis B alone kills about 1 million people each year worldwide, although an effective vaccine is available. As of 2009, there is no hepatitis C vaccine, but at least one is being tested.

Other Names

Many different diseases are called hepatitis, a term that just means inflammation of the liver. Only two forms of hepatitis, called hepatitis B and hepatitis C, fall within the criteria for this book: many cases, many deaths. Their infectious agents are the hepatitis B virus (HBV) and hepatitis C virus (HCV), respectively.

Serum hepatitis and transfusion hepatitis are old names for hepatitis B. Before the agent of hepatitis C was identified in 1989, researchers called it non-A non-B infectious hepatitis or non-B transfusion-associated hepatitis. Any form of hepatitis caused by a virus may be referred to as viral hepatitis, infectious hepatitis, serum hepatitis, hepatitis epidemica, epidemic jaundice, or icterus epidemicus. Older names include Botkin's disease, icterus catarrhalis, icterus simplex, and hepatia. Jaundice, a yellow discoloration of the skin and eyes, is really a symptom of hepatitis and several other diseases, not identical with hepatitis itself.

In the old days, the colloquial expression "bilious fever" meant almost any disease that might involve the liver, such as hepatitis, malaria, or typhoid fever. "Soldier's disease" often referred to infectious hepatitis, but it was only one of several diseases that plagued military garrisons. The words for hepatitis in most other languages are cognates, such as *hépatite* (French), *epatite* (Italian), and *hepatit* (Turkish). The Czech term is *zánet jater* (liver inflammation), and the Manx call it *chingys aane* (liver disease).

Description

Like most human diseases, hepatitis B and C were probably zoonoses at one time. The hepatitis B virus (Figure 2.8) is similar to a virus that causes liver cancer in woodchucks. The hepatitis C virus is one of the flaviviruses—a group of viruses that cause many of the world's most serious human and animal diseases, such as yellow fever and West Nile encephalitis. Unlike most of its relatives, HCV is not transmitted by mosquitoes or other vectors (at least not yet).

Hepatitis B is one of the world's most neglected public health threats. It is about 100 times more infectious than HIV and 10 times more infectious than hepatitis C. An effective hepatitis B vaccine has existed since 1982, yet hepatitis B remains the most common chronic liver disease, and it kills about 1 million people every year worldwide.

An estimated 25 percent of hepatitis B cases are sexually transmitted. Other routes include needle sharing by intravenous drug users, wound infection by contaminated body fluids, vertical transmission (mother to child), and a poorly defined grab bag called "household contact." HBV can survive on environmental surfaces for at least 7 days, and there is some evidence that mosquitoes may transmit hepatitis B in Africa. Besides killing people directly, hepatitis B more than doubles the risk of liver cancer. The risk is even higher if the person also has hepatitis C.

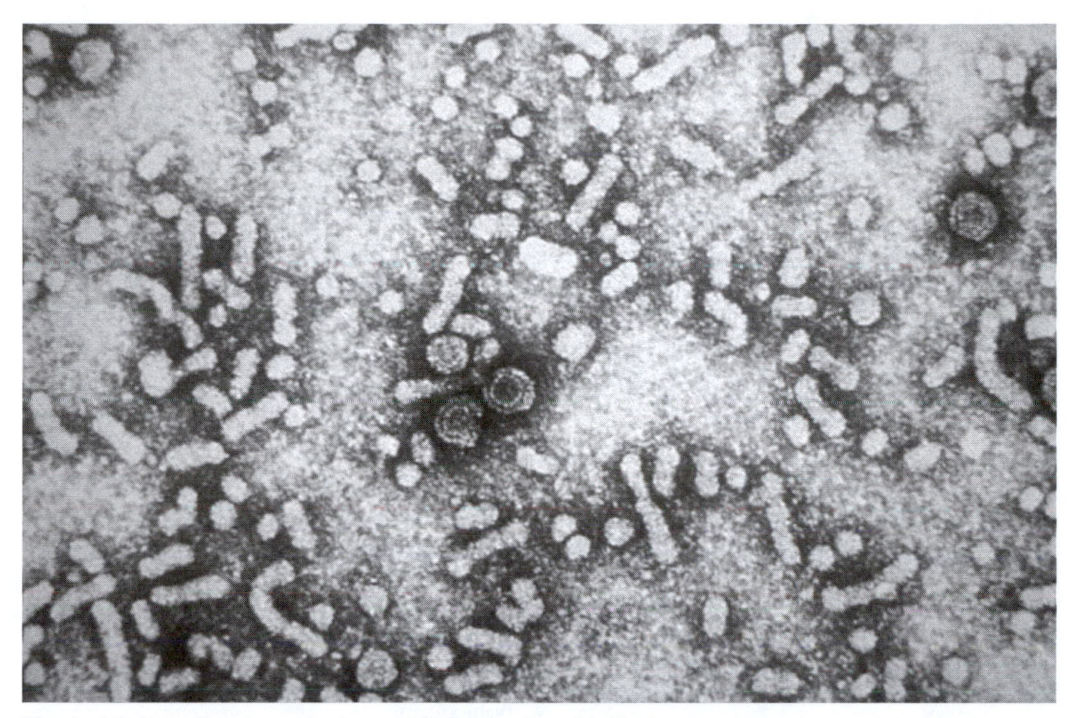

Figure 2.8 Scanning electron micrograph showing hepatitis B virus.
Source: U.S. Centers for Disease Control and Prevention, Public Health Image Library.

Hepatitis C spreads mainly by parenteral transmission, such as needle sharing, dental work, lab accidents, or the use of contaminated body piercing and tattooing equipment. People who share such articles as razor blades and toothbrushes may also share hepatitis C. It spreads among cocaine users sharing banknotes or straws, because cocaine damages the inside of the nose, causing abrasions where the virus can enter. Wrestlers have transmitted hepatitis C to other wrestlers through contact with sweat. Vertical transmission (mother to child) also occurs, but sexual transmission is rare.

Before about 1990, many people contracted hepatitis B and C from blood transfusions, hemodialysis, or organ transplants. Blood screening procedures have improved, but accidental transmission of hepatitis still occurs, often as a result of contaminated equipment or reuse of syringes. A study of hepatitis transmission in the United States between 1998 and 2008 revealed a total of 33 outbreaks (12 in outpatient clinics, 6 in hemodialysis centers, and 15 in long-term care facilities), in which a total of 448 persons became infected with hepatitis B or C. In December 2008, a Japanese pharmaceutical company settled with more than 1,000 hepatitis C patients who were infected by tainted blood products between 1970 and the early 1990s.

The symptoms of acute hepatitis B and C are similar, except that hepatitis C is often less severe: fatigue, loss of appetite, fever, nausea, vomiting, diarrhea, joint pain, and rash. The urine may become darker in color, and jaundice may appear. Jaundice results from high levels of the pigment bilirubin, a product of the breakdown of hemoglobin. When inflammation of the liver blocks the excretion of bilirubin into the intestine, the excess passes into the bloodstream instead.

Chronic infection, often resulting in cirrhosis or liver cancer, occurs in less than 20 percent of people with hepatitis B and about 85 percent of those with hepatitis C. HCV (unlike HBV) also causes encephalitis on occasion.

Case Study 2-9: Dr. Hep

In 2007 and 2008, a New York anesthesiologist with a history of negligent medical practice made world headlines (as "Dr. Hep" or "Dr. Disease") after the news media reported that he had infected at least 14 patients with hepatitis C by reusing syringes. Health officials reportedly contacted 4,500 patients and advised them to be tested for the virus. At latest word, the doctor was working as a car salesman, and several patients were suing him. Unhappily, this case was far from unique. In 2008, a Las Vegas endoscopy clinic discovered that some of its nurses had reused syringes, and notified some 50,000 patients that they might have been exposed to hepatitis B, hepatitis C, or HIV. After 22 patients were found to have contracted hepatitis C, the City of Las Vegas closed the clinic, and the owner reportedly had a stroke. As of 2009, the case was slowly making its way through the courts. Worse, a subsequent investigation revealed that nearly half of all outpatient surgical centers in Nevada had similar deficiencies.

Who Is at Risk?

Given the known modes of transmission for hepatitis B and C, intravenous drug abusers and sex workers are prime candidates. But many of the people at risk do not fit the popular image at all. Police, firefighters, and other first responders are at risk for hepatitis B, because they often come in contact with injured people. Sewer workers have caught hepatitis C from work-related contact with contaminated material. Young children have contracted hepatitis B or C from ordinary household contacts, such as unsanitary ear piercing, sharing a towel or chewing gum, walking barefoot on a contaminated surface, or handling used razor blades or dental floss discarded by an infected adult. Unsanitary medical equipment can expose anyone to hepatitis (Case Study 2-9). An estimated 1 in 200 U.S. adults has chronic HBV infection, and an estimated 1 in 80 has chronic HCV infection, often without symptoms.

Prison inmates are at very high risk for infectious hepatitis. As of 2008, the prevalence of hepatitis B among inmates in the United States was five times that of the general population. Many inmates have a history of intravenous drug use, and many more acquire unsanitary habits or engage in high-risk behavior while incarcerated. For the sake of the community at large, prison populations must receive better healthcare, including post-exposure immunization (if indicated), health education, and release planning.

About 80 percent of U.S. residents with hepatitis B are Asian American, whereas hepatitis C disproportionately affects the Hispanic population. About 35 percent of HIV-positive U.S. residents also have hepatitis C. Children who are shorter than average, non-Hispanic black children, and those whose parents have less education or lower income are at higher risk for hepatitis than taller children, non-Hispanic white children, or those whose parents have more education or more money. Excessive alcohol consumption after infection increases the risk that chronic hepatitis will progress to cirrhosis of the liver.

The Numbers

About 400 million people worldwide have hepatitis B, one-third of them in China. Recent estimates of annual hepatitis B deaths range from 500,000 to 1.2 million. About one-third of the global population either has hepatitis B or has had it in the past. New cases of hepatitis B in the United States declined sharply between 1995 and 2005, reaching the lowest level in the 40 years since the CDC started collecting data (Table 2.4). The vaccine was probably responsible, since the decline was greatest in children and adolescents under age 15.

An estimated 180 to 200 million people worldwide (3 percent of the population) have hepatitis C. There were only about 52,000 reported deaths in 2002, but underreporting is likely. Cir-

Table 2.4 Reported New Cases of Hepatitis B and C, United States

	Hepatitis B		Hepatitis C	
Year	**Number**	**Rate/100,000**	**Number**	**Rate/100,000**
1992	16,126	6.3	6,010	2.4
1993	13,361	5.2	4,786	1.9
1994	12,517	4.8	4,470	1.8
1995	10,805	4.1	4,576	1.7
1996	10,637	4.0	3,716	1.4
1997	10,416	3.9	3,816	1.4
1998	10,258	3.8	3,518	1.3
1999	7,694	2.8	3,111	1.1
2000	8,036	2.9	3,197	1.1
2001	7,844	2.8	1,640[1]	0.7[1]
2002	8,064	2.8	1,223[2]	0.5[2]
2003	7,526	2.6	891[2]	0.3[2]
2004	6,212	2.1	758	0.3
2005	5,494	1.8	671	0.2

Source: U.S. Centers for Disease Control and Prevention (CDC).
[1] Excludes cases from New Jersey and Missouri.
[2] Excludes cases from Missouri.

rhosis of the liver is responsible for about 1 percent of all deaths worldwide, and undetected hepatitis C may contribute to many of these cases. New reported cases of hepatitis C in the United States have declined, possibly as a result of risk-reduction practices among intravenous drug users. Again, this trend is hard to interpret, because many infected people have no symptoms. The CDC estimates that, for every reported case, there are between two and five unreported cases.

The prevalence of hepatitis C is highest in Africa and the eastern Mediterranean region, where more than 10 percent of people are infected. In the 1920s, Egypt started a mass injection program to control schistosomiasis. Despite good intentions, the use of inadequately sterilized equipment resulted in an epidemic of hepatitis C, which now infects an estimated 20 percent of the Egyptian population.

History

Intravenous drug abuse and its associated diseases may be modern phenomena, but ancient practices such as tattooing, circumcision, and the neglect of open wounds enabled parenteral infections to spread long before the invention of the hypodermic syringe. In 2007, scientists in South Korea isolated the hepatitis B virus from a 500-year-old mummified child.

One of the earliest known references to a contagious form of hepatitis appears in a letter written by Pope Zachary sometime before his death in A.D. 752, regarding his decision to quarantine patients with signs of jaundice in order to stop an epidemic of liver disease in Rome.[3] Some 1,100 years earlier, in the seminal work *Of Internal Affections*, Hippocrates (460–375 B.C.) described "epidemic jaundice, from obstruction induced by over-eating and drinking," but he was probably referring to hepatitis A (which does not result from overeating but is often food-borne).

3. Contrary to rumor, this Pope did not really excommunicate Virgil the Geometer for claiming that the Earth was round.

Case Study 2-10: Baby on the Run

In 2008, the public controversy over the hepatitis B vaccine had already raged for 15 years, but the battle lines were drawn when Australian health officials unwittingly played into the hands of opposing forces. According to the news release, a young married couple went into hiding to shield their infant son from mandatory hepatitis B vaccination—a procedure they believed would cause permanent neurological damage. Granted, the child's mother had hepatitis B, and the child's future health was in danger. Granted, the state has a responsibility to protect children. But the image of this loving couple fleeing with their child conjured up the most sacred images and served to heat up the debate regarding parental rights and the proper role of government. At latest word, the case was unresolved.

In 1967, Dr. Baruch S. Blumberg showed that hepatitis B was associated with a protein called the Australia antigen, which often turned up in the blood of people who had leukemia or had recently received transfusions. This discovery led to a number of medical advances, including blood screening and the discovery of other hepatitis viruses. In 1976, Dr. Blumberg shared the Nobel Prize in Physiology or Medicine with Dr. D. Carleton Gajdusek, who identified the agents of scrapie and kuru (Chapter 4).

Much of the recent history of hepatitis B has focused on controversy regarding the vaccine, which has been available since 1982 and is now recommended for all children (Case Study 2-10). In 2004, a group that opposes hepatitis B vaccination claimed that the vaccine is more dangerous than the disease, citing "Merck's package insert showing the frequency of vaccine damage at a rate of 10.4%."[4] This statement would be alarming, if true—but if the vaccine predictably "damaged" 10 percent of recipients, it would not be on the market. In fact, the Merck package insert states:

> Injection site reactions and systemic complaints were reported following 0.2% and 10.4% of the injections, respectively. The most frequently reported systemic adverse reactions (>1% injections), in decreasing order of frequency, were irritability, fever (>101°F oral equivalent), diarrhea, fatigue/weakness, diminished appetite, and rhinitis.[5]

These symptoms last for hours or days, not for the rest of the person's life, and they are not necessarily related to the vaccine. In clinical trials, a comparable percentage of subjects who take a placebo report similar symptoms.

This controversy regarding the hepatitis B vaccine apparently started in the early 1990s, with scattered reports of neurological symptoms that developed within a few weeks or months after HBV vaccination. In 2001, a French court ordered a vaccine manufacturer to pay damages to two women who developed multiple sclerosis. A higher court overturned that judgment in 2004, and further studies have shown no connection to MS, but the public relations damage was done. About 1 in 600,000 hepatitis B shots does cause an allergic reaction called anaphylaxis, and the long-term consequences (if any) are unknown; but no one on record has died from this complication, which is clearly preferable to liver cancer or other known sequelae of hepatitis B.

Prevention and Treatment

Obvious preventive measures include avoidance of dirty needles, unprotected sex, or contact between open wounds and contaminated surfaces. Women should be tested for hepatitis B and C before having children. Again, however, not all cases of hepatitis reflect a careless lifestyle. Many

4. Association of American Physicians and Surgeons, letter to Cook County Board of Commissioners, 10 March 2004.
5. Merck & Co., Inc., product information, Recombivax HB Hepatitis B Vaccine, December 2007.

people become infected as the result of medical or dental accidents, or other circumstances they cannot prevent.

The hepatitis B vaccine is safe for most people, unless they are allergic to baker's yeast. According to some sources, however, this vaccine may be unavailable to many low-income adults and older children. Many other people seem unaware of the vaccine or avoid it because of concerns about side effects. There is also some evidence that exposure to shortwave ultraviolet light reduces the effectiveness of HBV vaccination.

In 2005, researchers announced an experimental edible hepatitis B vaccine in a transgenic potato. This discovery may or may not improve the public image of the vaccination program, particularly if the potato must be eaten raw.

As of 2009, there is no hepatitis C vaccine. Doctors sometimes treat both hepatitis B and hepatitis C with antiviral drugs (usually interferon and ribavirin), but drug-resistant mutant viruses have already begun to appear. These antiviral drugs can cause psychiatric side effects such as depression and anxiety. A 2008 study concluded that drinking large amounts of coffee may slow the progression of hepatitis C, but this finding is hard to evaluate. An alternative explanation for the data is that people whose livers are in bad shape might tend to reduce or underreport their coffee intake.

Popular Culture

Some researchers claim that an Indian plant called keezhanali (*Phyllanthus amarus*) is an effective treatment for hepatitis B, but others regard the evidence as inconclusive. In a 2001 study, for example, this herbal treatment reportedly cured 30 percent of cases within a month. But even without treatment, the majority of patients recover.

Plutarch (A.D. 46–120) claimed that it was possible to cure hepatitis by having the patient look at a stone curlew, a European bird with large yellow eyes that supposedly drew out the disease. Other reported folk cures for hepatitis or jaundice include eating artichokes, spiders, head lice (live or boiled), sheep dung, tobacco, or extracts of at least 100 different wild plants, some of them poisonous; or wearing yellow clothes, copper coins, or a necklace of beets; or placing hard-boiled eggs in the armpits overnight, or urinating through a bored-out carrot. Traditional Chinese herbal remedies for infectious hepatitis include concoctions of wormwood (*Artemesia capillaris*), self-heal (*Prunella vulgaris*), azuki bean (*Phaseolus angularis*), crowdipper (*Pinella ternata*), three-leaf corydalis (*Corydalis ternata*), or nut grass (*Cyperes rotundus*).

In 2005, the American Liver Foundation condemned the motion picture *Bewitched* because of a scene in which a woman discourages romantic overtures by claiming she has hepatitis C. The incident is reminiscent of a 1970s "Mary Worth" cartoon in which a young woman told an unwanted admirer that she was a mumps carrier. Politically correct or not, this strategy is probably as old as dating.

In 1999, the TV show *20/20* did a segment "Who's Calling the Shots?" about children and adults who developed arthritis or multiple sclerosis after receiving the hepatitis B vaccine. The program was somewhat misleading, because these disorders occur at the same frequency in people who have not had the vaccine. Opponents argue that babies rarely catch hepatitis B, so what's the point of vaccinating them? Why not wait until they grow up, and let them decide? Aside from the obvious problems, children infected before age 6 are more likely to get chronic liver disease and liver cancer. Some children apparently get hepatitis B through ordinary household contacts with infected family members.

The Future

The world clearly needs a hepatitis C vaccine. Unless the healthcare establishment can gain the trust of the general public, however, this new vaccine (like the hepatitis B vaccine) may be underutilized, and many children and adults at risk may go unprotected. Further study of the hepatitis B vaccine, and further investigation of opponents' claims, might alleviate some concerns.

References and Recommended Reading

"40 Die in Hepatitis-B Outbreak in Western Indian State." Xinhua News Agency, 21 February 2009.

Baldo, V., et al. "Epidemiology of HCV Infection." *Current Pharmaceutical Design*, Vol. 14, 2008, pp. 1646–1654.

Bereket-Yücel, S. "Risk of Hepatitis B Infections in Olympic Wrestling." *British Journal of Sports Medicine*, Vol. 41, 2007, pp. 306–310.

Blackard, J. T., and K. E. Sherman. "HCV/HIV Co-infection: Time to Re-evaluate the Role of HIV in the Liver?" *Journal of Viral Hepatitis*, Vol. 15, 2008, pp. 323–330.

Brautbar, N., and N. Navizadeh. "Sewer Workers: Occupational Risk for Hepatitis C—Report of Two Cases and Review of Literature." *Archives of Environmental Health*, Vol. 54, 1999, pp. 328–330.

Brook, M. G. "Sexually Acquired Hepatitis." *Sexually Transmitted Infections*, Vol. 78, 2002, pp. 235–240.

Calabrese, L. H., et al. "Hepatitis B Virus (HBV) Reactivation with Immunosuppressive Therapy in Rheumatic Diseases: Assessment and Preventive Strategies." *Annals of the Rheumatic Diseases*, Vol. 65, 2006, pp. 983–989.

Cavalheiro, N. P. "Sexual Transmission of Hepatitis C." *Revista do Instituto de Medicina Tropical de São Paulo*, Vol. 49, 2007, pp. 271–277.

Craxi, A., et al. "Hepatitis C Virus (HCV) Infection: a Systemic Disease." *Molecular Aspects of Medicine*, Vol. 29, 2008, pp. 85–95.

de Carvalho, J. F., and Y. Shoenfield. "Status Epilepticus and Lymphocytic Pneumonitis Following Hepatitis B Vaccination." *European Journal of Internal Medicine*, Vol. 19, 2008, pp. 383–385.

De Sanjose, S., et al. "Hepatitis C and Non-Hodgkin Lymphoma among 4784 Cases and 6269 Controls from the International Lymphoma Epidemiology Consortium." *Clinical Gastroenterology and Hepatology*, Vol. 6, 2008, pp. 451–458.

Dowd, J. B., et al. "Early Origins of Health Disparities: Burden of Infection, Health, and Socioeconomic Status in U.S. Children." *Social Science and Medicine*, 17 January 2009.

Edelman, S., and J. Culora. "Doctor 'Disease.'" *New York Post*, 17 June 2007.

Ghoda, M. K., and R. A. Shah. "A Prospective Epidemiological Study to See if Mosquito Bite Could Be Responsible for Spread of Hepatitis B Virus Infection." *Tropical Gastroenterology*, Vol. 26, 2005, pp. 29–30.

Hatfield, G. 2004. *Encyclopedia of Folk Medicine*. Santa Barbara, CA: ABC-CLIO.

"Hepatitis B Vaccine Not Linked to MS." United Press International, 30 September 2008.

Hsu, E. K., and K. F. Murray. "Hepatitis B and C in Children." *Nature Clinical Practice Gastroenterology and Hepatology*, Vol. 5, 2008, pp. 311–320.

"Infection Control Flaws Found at Most Nevada Clinics." Associated Press, 10 March 2009.

Jancin, B. "Sweat of Infected Patients May Contain Hepatitis C: Inapparent Parenteral Transmission." *OB/GYN News*, 15 December 2003.

Kamal, S. M. "Acute Hepatitis C: a Systematic Review." *American Journal of Gastroenterology*, Vol. 103, 2008, pp. 1283–1297.

Marshall, E. "A Shadow Falls on Hepatitis B Vaccination Effort." *Science*, Vol. 281, 1998, pp. 630–631.

Mauss, S., and H. Wedemeyer. "Treatment of Chronic Hepatitis B and the Implications of Viral Resistance to Therapy." *Expert Review of Anti-Infective Therapy*, Vol. 6, 2008, pp. 191–199.

Mikaeloff, Y., et al. "Hepatitis B Vaccination and the Risk of Childhood-Onset Multiple Sclerosis." *Archives of Pediatric and Adolescent Medicine*, Vol. 161, 2007, pp. 1176–1182.

Mikaeloff, Y., et al. "Hepatitis B Vaccine and the Risk of CNS Inflammatory Demyelination in Childhood." *Neurology*, 8 October 2008.

Missiha, S. B., et al. "Disease Progression in Chronic Hepatitis C: Modifiable and Nonmodifiable Factors." *Gastroenterology*, Vol. 134, 2008, pp. 1699–1714.

Nadler, J. P. "Multiple Sclerosis and Hepatitis B Vaccination." *Clinical Infectious Diseases*, Vol. 17, 1993, pp. 928–929.

Nash, K. L., and G. J. Alexander. "The Case for Combination Antiviral Therapy for Chronic Hepatitis B Virus Infection." *Lancet Infectious Diseases*, Vol. 8, 2008, pp. 444–448.

"Ninth Las Vegas Hepatitis C Case Confirmed." United Press International, 24 July 2008.

Panella, H., et al. "Transmission of Hepatitis C Virus During Computed Tomography Scanning with Contrast." *Emerging Infectious Diseases*, Vol. 14, 2008, pp. 333–336.

Perez, C. M., et al. "Hepatitis C in Puerto Rico: a Time for Public Health Action." *Puerto Rico Health Sciences Journal*, Vol. 26, 2007, pp. 395–400.

"Police Hunt for Two-Day-Old Baby to Enforce Hepatitis Jab Order." Agence France Presse, 22 August 2008.

Ramirez, A. "Three NYC Hepatitis Cases Linked to M.D." *The New York Times,* 15 June 2007.

Rodriguez-Torres, M. "Latinos and Chronic Hepatitis C: A Singular Population." *Clinical Gastroenterology and Hepatology*, Vol. 6, 2008, pp. 484–490.

Rougé-Maillart, C. I., et al. "Recognition by French Courts of Compensation for Post-Vaccination Multiple Sclerosis: The Consequences with Regard to Expert Practice." *Medicine, Science and Law*, Vol. 47, 2007, pp. 185–190.

Saunders, J. C. "Neuropsychiatric Symptoms of Hepatitis C." *Issues in Mental Health Nursing*, Vol. 29, 2008, pp. 209–220.

Seifert, F., et al. "In Vivo Detection of Hepatitis C Virus (HCV) RNA in the Brain in a Case of Encephalitis: Evidence for HCV Neuroinvasion." *European Journal of Neurology*, Vol. 15, 2008, pp. 214–218.

Sharfstein, J. "Inadequate Hepatitis B Vaccination of Adolescents and Adults at an Urban Community Health Center." *Journal of the National Medical Association*, Vol. 89, 1997, pp. 86–92.

Sheikh, M. Y., et al. "Hepatitis C Virus Infection: Molecular Pathways to Metabolic Syndrome." *Hepatology*, Vol. 47, 2008, pp. 2127–2133.

Strickland, G. T., et al. "Hepatitis C Vaccine: Supply and Demand." *Lancet Infectious Diseases*, Vol. 8, 2008, pp. 379–386.

Thanavala, Y., et al. "Immunogenicity in Humans of an Edible Vaccine for Hepatitis B." *Proceedings of the National Academy of Sciences (USA),* Vol. 102, 2005, pp. 3378–3382.

Thompson, N. D., et al. "Nonhospital Health Care-Associated Hepatitis B and C Virus Transmission: United States, 1998–2008." *Annals of Internal Medicine*, Vol. 150, 2009, pp. 33–39.

Wasley, A., et al. "Surveillance for Acute Viral Hepatitis—United States, 2005." *Morbidity and Mortality Weekly Report Surveillance Summaries*, Vol. 56, 2007, pp. 1–28.

Weissenborn, K., et al. "Hepatitis C Virus Infection and the Brain." *Metabolic Brain Disease*, 7 January 2009.

Wise, M., et al. "Changing Trends in Hepatitis C–Related Mortality in the United States, 1995–2004." *Hepatology*, Vol. 47, 2008, pp. 1128–1135.

"Witnesses Quiet on Hepatitis C Outbreak." United Press International, 30 September 2008.

3

Five More (and Complications)

My name is Legion, for we are many.

—King James Bible, Mark 5:9

In 1924, when 16-year-old Calvin Coolidge Jr. (the president's son) had an infected blister on his foot, the antibiotics that we take for granted today were unknown. The infection spread to his bloodstream, and in a few days he was dead.

If the same thing happened in 1954 or 1974 instead of in 1924, an antibiotic such as penicillin might have quickly cured the problem. As early as the 1960s, however, doctors noticed that some infections were becoming unresponsive to antibiotics. By the 1980s, the window of opportunity for miracle cures had already begun to close. Pharmaceutical companies were under pressure to invent new drugs, while doctors found themselves losing a battle with natural selection. In 2009, an otherwise healthy 20-year-old Brazilian model made international headlines for the last time when she died of a blood infection caused by a common drug-resistant bacterium.

In the United States alone, about 100,000 people die from drug-resistant infections every year. The problem of resistance is just one (arguably the worst) of the five major biological threats discussed in this chapter. The others are diarrheal diseases and dengue fever, plus an assortment known as "emerging diseases"—and measles, a disease that few Americans take seriously. Potential complications of these diseases include pneumonia, meningitis, and encephalitis.

MEASLES

Summary of Threat

Measles is a highly contagious respiratory infection that can lead to life-threatening pneumonia, meningitis, or other complications. Although an effective vaccine is available, and the 2001 Measles Initiative has saved many lives, nearly 200,000 children still die from this disease every year, most of them in the Third World. Crowding, malnutrition, and HIV infection are major risk factors.

Other Names

Names for measles include rubeola, morbilli, ten-day measles, nine-day measles, eight-day measles, seven-day measles, red measles, big red measles, and hard measles. The hemorrhagic form is sometimes called black measles. Complications and symptoms associated with measles add more names to the list: Bosin's disease, Van Bogaert encephalitis, and Dawson's encephalitis, all rare forms of encephalitis; and Koplik's spots, lesions that occur inside the mouth before the measles rash appears.

The names Edmonston A, Edmonston B, Enders, Schwartz, AIK-C, Connaught, Philips, Beckenham, Moraten, Zagreb, and Belgrade refer to specific strains of the measles virus that researchers have used in developing vaccines. Measles is *sarampión* in Spanish, *griùrach* in Scottish Gaelic, *spalnièky* in Czech, and *campak* in Indonesian. The Zulu call it *isimungumungwane.*

Some sources confuse measles (rubeola) with other diseases that cause a red rash, such as rubella (German measles) or roseola. Sometimes the word "measles" refers to tapeworm infestation of beef or pork, and the meat thus infested is said to be "measly."

Description

Measles is an airborne respiratory disease that was once a major scourge of childhood. The infectious agent is a virus (Figure 3.1) that infects only humans, but it is closely related to the agents of several animal diseases. In 1941, there were 894,134 reported cases and 2,279 deaths from measles in the United States alone. The true number of cases in 1941 was probably closer to 2 million, as virtually all children contracted measles before the vaccine was invented, and many cases were not reported. Although now infrequent in the United States—there were only 55 reported cases in 2006—measles remains a far more serious disease than many Americans realize.

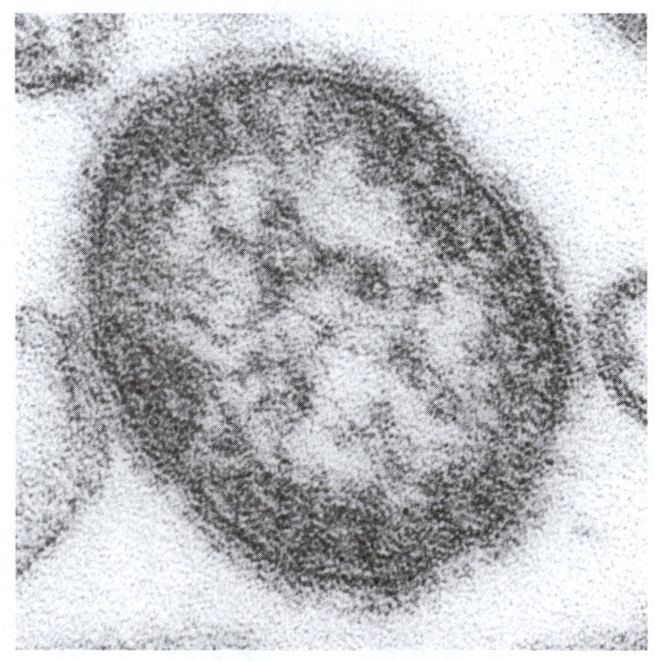

Figure 3.1 Scanning electron micrograph showing measles (rubeola) virus.
Source: U.S. Centers for Disease Control and Prevention, Public Health Image Library.

Table 3.1 Examples of Measles Outbreaks

Year	Location	Estimated Cases	Estimated Deaths
1846	Faeroes	6,000	102
1875	Fiji	135,000	36,000
1941–1942	New Haven, CT	3,200	0
1975	Greensville, Ontario	47	0
1977–1978	Shetland, UK	1,032	0
1989–1991	United States	55,000	135
1991	Israel	1,036	0
1997	São Paulo, Brazil	42,055	42
2000	Japan	200,000	88
2001–2002	Ukraine	25,000	14
2005	Chad	21,812	870
2006	Catalonia, Spain	381	0

Sources: World Health Organization (WHO), U.S. Centers for Disease Control and Prevention (CDC).

Measles is highly infectious, spreading both by airborne droplet inhalation and by direct or indirect contact. Symptoms begin with a slight fever and cough, usually ten to twelve days after exposure. The fever gradually rises, and small whitish spots with red borders may appear inside the mouth. The measles rash starts on the face and spreads downward, eventually covering the body. Other symptoms may include eye irritation, diarrhea, and swollen lymph nodes. After recovery, most people have permanent immunity.

About one-third of measles cases involve at least one complication, such as severe diarrhea, ear infections, pneumonia, or seizures. More serious problems, such as meningitis or deafness, occur in about one out of every 1,000 cases. The death rate is usually low, but it varies from one region and outbreak to another (Table 3.1). A rare complication called subacute sclerosing panencephalitis (SSP) can cause death or permanent neurological damage. In a pregnant woman, measles—like several other diseases, including the unrelated German measles—can cause miscarriage and possibly birth defects. Measles can also activate an existing tuberculosis infection.

Case Study 3-1: Measles and the Yanomami

The 2000 book *Darkness in El Dorado* claims that American scientists who provided measles vaccination to the South American Yanomami people in 1968 were trying to start an epidemic to test a eugenics theory. It is hard to determine exactly what really happened, except that the Yanomami may have been treated and interviewed without their informed consent, and that many died of measles as a result of contact with outsiders. Some 20 years after this fiasco, a second invasion force visited the Yanomami, this time in the form of gold prospectors who reportedly poisoned the rivers with mercury and introduced malaria, tuberculosis, and other diseases. Regardless of how this story began, it is unlikely to have a happy ending.

Who Is at Risk?

Measles is so highly infectious that the only people not at risk during an outbreak are those who have already had measles. Those who were vaccinated at least twice after age 1 are about 90 to 95 percent safe.

Complications are least likely to occur in children over age 5 and adolescents. Elderly or malnourished people, babies, and those with weakened immune systems are at highest risk. In underdeveloped nations with high rates of malnutrition and poor healthcare, and in populations with no prior exposure to measles, the case fatality rate may exceed 10 percent (Case Study 3-1).

The Numbers

Before 1963, when the first measles vaccine became available, 500 to 2,000 children died of this disease in a typical year in the United States. Many more children developed encephalitis or other complications that damaged the brain, heart, or eyes. The MMR (measles-mumps-rubella) vaccine greatly reduced the incidence of measles, but not everyone wants this vaccine or has access to it. In 2000, measles caused 757,000 deaths worldwide, most of them in Africa and India.

By 2004, however, the global death toll from measles was down to 410,000, and by 2008 it fell to 197,000. The world owes this stunning achievement to worldwide vaccination and health education campaigns sponsored by the 2001 Measles Initiative, a partnership of the American Red Cross, CDC, UNICEF, WHO, and the United Nations. The Initiative's goal is to reduce measles deaths to 10 percent of the 2000 level by 2010. But even that 10 percent will keep the measles virus alive and able to re-emerge wherever public health vigilance or public cooperation falters.

History

Hippocrates (460–375 B.C.) described many diseases, including mumps and scarlet fever, but nothing in his surviving work sounds like measles. Either it had not yet reached Greece, or he did not distinguish it from other common childhood rashes. The Persian physician Razes or Rhazes (Abu Bakr Razi) published a description of measles in the tenth century A.D. He cited earlier sources who believed that measles arrived in the Middle East from Africa during the siege of Mecca in the sixth century A.D., but this claim is hard to evaluate. In 1552, 15-year-old Edward VI of England (son of Henry VIII) caught measles and smallpox in rapid succession, followed by tuberculosis, which ended his life in 1553.

American microbiologist Maurice Ralph Hilleman (1919–2005) invented more than 40 vaccines, including the MMR vaccine, the first flu shots, the hepatitis A and B vaccines, the chickenpox vaccine, and many others that are still in use today. When the first attenuated measles vaccine released in 1963 had unacceptable side effects, he developed a safe version that became available in 1965. Dr. Hilleman's discoveries have saved hundreds of millions of lives, and it is ironic that anti-vaccination crusaders and conspiracy theorists have vilified his work.

In 1989–1991, the United States had a measles epidemic, with about 55,000 cases and 135 deaths. Many were children or college students who had never been vaccinated or whose childhood vaccinations were given too early. Although this outbreak represented a small fraction of the annual toll before 1963, opponents of immunization point to the 1989–1991 measles epidemic as proof of the vaccine's failure. On the contrary, the numbers show that the vaccine is about 95 percent effective.

Japan repealed its mandatory vaccination laws in 1994, and many Japanese children have not had a second booster shot. As a result, the country still has large measles epidemics. In 2000, Japan had about 200,000 cases of measles and 88 deaths. In 2008, health officials reported that measles was once again endemic in Britain and showing "continuous spread," thanks to scare campaigns about autism and MMR vaccine (Case Study 3-2).

Prevention and Treatment

Vaccination is recommended for anyone who was born after 1957 and has never had measles. All states require children to be vaccinated twice (after age one and before starting school), but most states offer religious or personal belief exemptions.

Measles vaccines have been available since 1963, but they are not perfect. Both the standard subcutaneous vaccine and a recently developed aerosol vaccine are 90 to 95 percent effective in preventing infection. Unimmunized contacts of measles patients may be vaccinated within 72 hours after exposure. Oral or intravenous ribavirin is sometimes used to treat measles, either alone or in combination with immune globulin.

In 2009, animal studies of a new measles DNA vaccine were in progress. If approved, this vaccine might be more effective than those presently in use, particularly in younger children. Since it does not contain live virus, it might also alleviate some safety concerns, whether justified or not.

Case Study 3-2: The 1998 Autism Scare

In 1998, the journal *Lancet* published what appeared to be evidence of a link between autism and the MMR vaccine. It was every parent's worst nightmare come true. The authors of the 1998 study later retracted their findings, however, and several other studies have exonerated the vaccine. As of 2009, the consensus is that neither thimerosal (a mercury-containing preservative) nor the MMR vaccine itself causes autism. Yet the controversy has continued, with some parents blaming the vaccine not only for autism but also for epilepsy, arthritis, fatigue, attention deficit disorder, and other conditions.

Popular Culture

In the 1995 motion picture *Apollo 13*, NASA replaced the command module pilot at the last minute because he was exposed to measles. Since he had no history of measles, and thus no immunity, there was concern that he might become ill during the mission. Accounts vary, but apparently the real-life astronaut was exposed to rubella (German measles), not measles.

In the 2007 remake of the motion picture *I Am Legend*, doctors use a genetically engineered measles virus to treat cancer. The virus mutates and escapes, causing a deadly pandemic that somehow turns its victims into the walking dead, who then devote their lives (or deaths or whatever) to the task of infecting the few remaining healthy people on Earth. According to at least one source, the story symbolizes the ongoing struggle between the allegedly indoctrinated zombie-like masses who accept vaccination and the enlightened holdouts who oppose it.

In Mark Twain's 1876 novel *Tom Sawyer*, Tom contracts measles and is sick for two weeks. A few days after his recovery, Tom has a relapse and is bedridden for another three weeks. The early medical literature reported a few cases of relapsing measles, but modern sources do not recognize this phenomenon. It would appear that Tom developed a secondary infection as a complication of measles. The fact that he felt depressed just before the relapse, and interpreted a thunderstorm as divine retribution for his sins, might suggest encephalitis.

In the 1950s and early 1960s, parents often held "measles parties" when one neighborhood child contracted the disease, so all of them could catch it and get it over with. This practice reflected the popular belief that measles was not dangerous; nobody, to the author's knowledge, ever held diphtheria or polio parties. In the twenty-first century, measles parties reportedly are making a comeback.

In North America and northern Europe, folk remedies for measles once included cow dung, roast mouse, mistletoe growing on hawthorn, mare's milk, bread or tea made from corn shucks, holly leaf tea, chamomile tea, coneflower (*Echinacea*), the blood of a black hen, or sheep dung mixed with porter ale, sulfur, and water. Traditional Chinese herbal remedies include drinking a concoction of chopped pumpkin vine or Japanese holly fern (*Cyrtomium fortunei*).

The Future

The complete eradication of measles, although possible, remains an elusive dream. The cattle disease rinderpest (Chapter 4) is on the brink of eradication, and it is closely related to measles; but herd immunity is hard to achieve without vaccination, and cows, unlike humans, can be vaccinated without their consent. Although there is little evidence that the measles vaccine causes autism, persistent rumors interfere with public cooperation. Also, since the measles vaccine is less effective in persons with weakened immune systems, the HIV epidemic may complicate the eradication of measles.

References and Recommended Reading

Barnard, D. L. "Inhibitors of Measles Virus." *Antiviral Chemistry and Chemotherapy*, Vol. 15, 2004, pp. 111–119.

Baird, G., et al. "Measles Vaccination and Antibody Response in Autism Spectrum Disorders." *Archives of Disease in Childhood*, Vol. 93, 2008, pp. 832–837.

Baur, M. P., et al. "International Genetic Epidemiology Society: Commentary on *Darkness in El Dorado* by Patrick Tierney." *Genetic Epidemiology*, Vol. 21, 2001, pp. 81–104.

Burgess, D. C., et al. "The MMR Vaccination and Autism Controversy in United Kingdom 1998–2005: Inevitable Community Outrage or a Failure of Risk Communication?" *Vaccine*, Vol. 24, 2006, pp. 3921–3928.

Dabbagh, A., et al. "Progress in Global Measles Control and Mortality Reduction, 2000–2006." *Morbidity and Mortality Weekly Report*, Vol. 56, 2007, pp. 1237–1241.

Dales, L. G., et al. "Measles Epidemic from Failure to Immunize." *Western Journal of Medicine*, Vol. 159, 1993, pp. 455–464.

Deer, B. "MMR Doctor Andrew Wakefield Fixed Data on Autism." *The Sunday Times*, 8 February 2009.

de Quadros, C. A., et al. "Feasibility of Global Measles Eradication After Interruption of Transmission in the Americas." *Expert Review of Vaccines*, Vol. 7, 2008, pp. 355–362.

Dugger, C. W. "Mothers of Nepal Vanquish a Killer of Children." *New York Times*, 30 April 2006.

Grais, R. F., et al. "Unacceptably High Mortality Related to Measles Epidemics in Niger, Nigeria, and Chad." *PLoS Medicine*, January 2007.

Hiremath, G. S., and S. B. Omer. "A Meta-analysis of Studies Comparing the Respiratory Route with the Subcutaneous Route of Measles Vaccine Administration." *Human Vaccines*, Vol. 1, 2005, pp. 30–36.

"Inquiry Planned into MMR Scare." United Press International, 23 February 2004.

Kumar, V. "Measles Outbreak in Gibraltar, August–October 2008: A Preliminary Report." *Eurosurveillance*, 6 November 2008.

"Men Against Measles." *Time*, 8 August 1960.

"New Book, Article Accuses Scientists of Disrupting Yanomami Tribes." CNN, 2 October 2000.

Norrie, J. "Japanese Measles Epidemic Brings Campuses to Standstill." *Sydney Morning Herald*, 27 May 2007.

Olsson, K. "An Ethics Firestorm in the Amazon." *U.S. News and World Report*, 2 October 2000, p. 51.

Orenstein, W. A., et al. "Measles Eradication: Is It in Our Future?" *American Journal of Public Health*, Vol. 90, 2000, pp. 1521–1525.

Roosevelt, M. "Yanomami." *Time*, 2 October 2000.

"Scientists Retract Earlier MMR–Autism Tie." United Press International, 4 March 2004.

Singh, V. K., and R. L. Jensen. "Elevated Levels of Measles Antibodies in Children with Autism." *Pediatric Neurology*, Vol. 28, 2003, pp. 292–294.

Smith, M. J., et al. "Media Coverage of the Measles-Mumps-Rubella Vaccine and Autism Controversy and Its Relationship to MMR Immunization Rates in the United States." *Pediatrics*, Vol. 121, 2008, pp. e836–e843.

Station, E., and M. Guran. "The Smoke That Kills." *Links*, Vol. 9, 1992, pp. 11–12.

Talley, L., and P. Salama. "Short Report: Assessing Field Vaccine Efficacy for Measles in Famine-Affected Rural Ethiopia." *American Journal of Tropical Medicine and Hygiene*, Vol. 68, 2003, pp. 545–546.

U.S. Centers for Disease Control and Prevention. "Measles—United States, January 1–April 25, 2008." *Morbidity and Mortality Weekly Report*, Vol. 57, 2008, pp. 494–498.

Wakefield, A. J., et al. "Ileal-Lymphoid-Nodule Hyperplasia, Non-Specific Colitis, and Pervasive Developmental Disorder in Children." *Lancet*, Vol. 351, 1998, pp. 637–641.
World Health Organization, et al. "Global Goal to Reduce Measles Deaths in Children Surpassed: Measles Deaths Fall by 60 Per Cent." Press release, 19 January 2007.
"World Measles Deaths Halved in 6 Years." United Press International, 10 March 2006.

DYSENTERIES AND ENTERIC FEVERS

Summary of Threat

As a group, waterborne and foodborne diseases that target the human gastrointestinal tract kill more people every year than HIV. Most of these diseases cause diarrhea and dehydration, whereas others can damage the intestine and other organs. Examples include cholera, shigellosis, salmonellosis, typhoid fever, and *E. coli*. Clean drinking water would prevent most of these deaths.

Other Names

Diarrhea or dysentery in general is *rhyddni* ("looseness") in Welsh, *Dünnschiss* ("thin feces") in German, and *ishal* ("the trots") in Turkish. Forms of diarrhea that afflict travelers have acquired colorful nicknames, such as the "Rangoon runs," "Bali belly," and the "Tijuana two-step." (These names reflect no discredit on the host nations; visitors to the United States often contract a similar malady here.) English names for the specific diseases in this section include the following.

Cholera: Asiatic cholera, Indian cholera, epidemic cholera, blue cholera, spasmodic cholera, cholera asphyxia, or vibriosis.

Shigellosis: Bacillary dysentery, bloody flux, *Shigella* enteritis, Flexner's dysentery, Japanese dysentery, or Schmidt's bacillus.

Salmonellosis: *Salmonella* food poisoning, *Salmonella* enterocolitis, or nontyphoidal salmonellosis.

Typhoid and paratyphoid fever together are known as enteric fever or enteromesenteric fever. Specific names for typhoid include typhus abdominalis, cesspool fever, Lent fever, nervous fever, night-soil fever, and Peyerian fever. Specific names for paratyphoid include Schottmüller's disease and Brion-Kayser disease.

Escherichia coli (*E. coli*): The word "escherichiasis" is hard to pronounce, so most sources refer to the strain instead, such as *E. coli* O157:H7, diarrheogenic *E. coli* (DEC), enterohemorrhagic *E. coli* (EHEC), enteroinvasive *E. coli* (EIEC), enteropathogenic *E. coli* (EPEC), enterotoxigenic *E. coli* (ETEC), Shiga toxin–producing *E. coli* (STEC), Shiga toxigenic *E. coli* (STEC), or verotoxin-producing *E. coli* (VTEC).

Description

In the United States, most highly publicized diarrhea outbreaks are traceable to bacteria such as toxigenic *E. coli* and nontyphoidal *Salmonella*. In the Third World, shigellosis and cholera cause hundreds of thousands of deaths every year (Table 3.2). Typhoid and paratyphoid fever often damage the intestines, but they do not always cause diarrhea. The fact that Westerners tend to dismiss diarrhea as a social inconvenience should not distract readers from the seriousness of these diseases.

Cholera (agent *Vibrio cholerae*, serogroup O1 or O139): This disease can cause profuse watery diarrhea, severe dehydration, and death within a few hours after onset. Humans appear to be the only animal reservoir, but the bacteria (Figure 3.2) can persist in coastal waters, often in

Table 3.2 Some Outbreaks of Dysentery and Similar Diseases

Disease	Year	Location	Estimated Cases	Estimated Deaths
Cholera	1849	Chicago, IL	NR	678
Cholera	1849	Canada	NR	6,000
Cholera	1854	Chicago, IL	NR	1,424
Cholera	1991–1999	South America	1 million	10,000
Cholera	2008–2009	Zimbabwe	100,000+	4,300+
E. coli O157	1994	Washington	501	3
E. coli O157	1999	Albany, NY	1000+	2
E. coli O157	2002	Pennsylvania	51	0
Salmonellosis	1965	Riverside, CA	15,000	3
Salmonellosis	2007–2008	United States	1,442	2
Salmonellosis	2008	Estonia	94	0
Salmonellosis	2008–2009	United States	600+	11+
Shigellosis	1997	Lubbock, TX	480	0
Shigellosis	1998	Minnesota	83	0
Shigellosis	2005	Spain	196	0
Typhoid	1964	Scotland	500+	0
Typhoid	1996–1998	Tajikistan	24,000	NR
Typhoid	2004	D.R. Congo	42,564	214
Typhoid	2005	South Africa	400+	3

Sources: World Health Organization (WHO), U.S. Centers for Disease Control and Prevention (CDC).

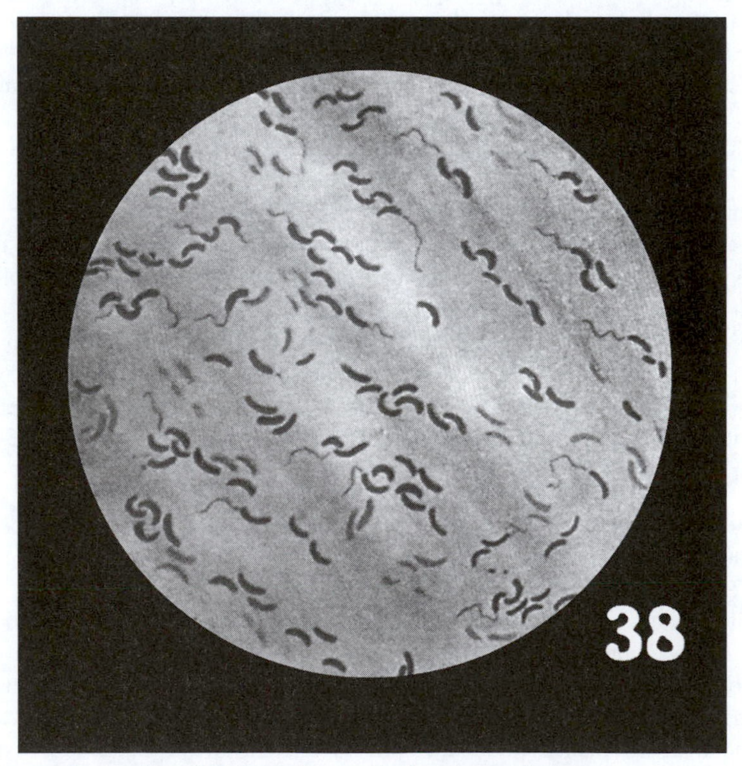

Figure 3.2 Micrograph showing a *Vibrio cholera* strain that causes cholera.

Source: U.S. Centers for Disease Control and Prevention, Public Health Image Library.

association with small crustaceans called copepods. Cholera is usually curable with prompt treatment, but in Third World countries where cholera is prevalent, clean rehydration fluids are often unavailable.

Shigellosis (agents *Shigella dysenteriae*, *S. flexneri*, *S. boydii*, and *S. sonnei*): This disease causes bloody diarrhea (dysentery) with a fever and abdominal cramps. Typical sources are contaminated water or vegetables fertilized with raw sewage. The bacteria produce the Shiga toxin, which causes these symptoms by inhibiting protein synthesis. This toxin (but not the *Shigella* bacterium itself) is classified as a biosecurity threat under the Bioterrorism Protection Act of 2002. Reactive arthritis may develop months or years after recovery from shigellosis.

Salmonellosis (agent *Salmonella enterica*, several serovars): This disease is usually foodborne, but there are other ways to catch it—for example, by touching a live turtle or a raw chicken, and then putting the same hand in your mouth. The disease causes profuse watery diarrhea, sometimes with vomiting, headache, and/or abdominal pain. Reactive arthritis may develop months or years later. The case fatality rate, however, is less than 1 percent.

Typhoid (agent *Salmonella typhi*): Symptoms of this waterborne and foodborne disease include high fever, headache, a rash of flat pink spots, and an enlarged spleen. Either diarrhea or constipation may occur, and the intestine may bleed or become perforated. About 10 to 20 percent of untreated patients die (Case Study 3-3). Paratyphoid (agent *Salmonella paratyphi*) is similar but less severe. Both diseases affect only humans, who can become lifelong carriers, like the famous Typhoid Mary.

E. coli (agent, *Escherichia coli* O157:H7 and others): Most strains of this bacterium (Figure 3.3) are normal, harmless occupants of the mammalian intestine, but some produce a dangerous toxin. People have contracted this disease from unpasteurized milk or juice, contaminated ground beef, lake water, airborne sawdust in a dirty building, and therapy dogs fed on raw meat. Severe cases can cause kidney failure and death. Most cases are relatively mild, but survivors may develop kidney damage or pancreatitis years after infection.

Case Study 3-3: A Death in Florence

Leland Stanford Jr. died of typhoid fever in Florence, Italy, on 13 March 1884, two months before his sixteenth birthday, while he and his parents were on a Grand Tour of Europe. He caught the disease in Constantinople, where they had gone because the sultan wanted Leland Sr.'s advice about building a railroad. The family returned to Florence for treatment, but all efforts failed. The night after his death, the young man appeared to his father in a dream and said: "Do not say that you have nothing to live for . . . Father, serve humanity." Mr. Stanford woke from his dream and told his wife, "The children of California shall be our children." The couple founded Stanford University to honor their son's memory.

Who Is at Risk?

Children, malnourished persons, those with low incomes or limited education, those who live under crowded conditions, and those with compromised immune systems are all at increased risk from diarrheal diseases in general. The following paragraphs describe additional risk factors for specific diseases.

Persons with blood group O, and those with below-average levels of stomach acid, are at increased risk for cholera. Disasters contribute to cholera outbreaks by contaminating water sources and disrupting sanitation services. Breastfed infants have some protection, either because of maternal antibodies or because they do not drink contaminated water. Survivors have some resistance to reinfection with the same strain.

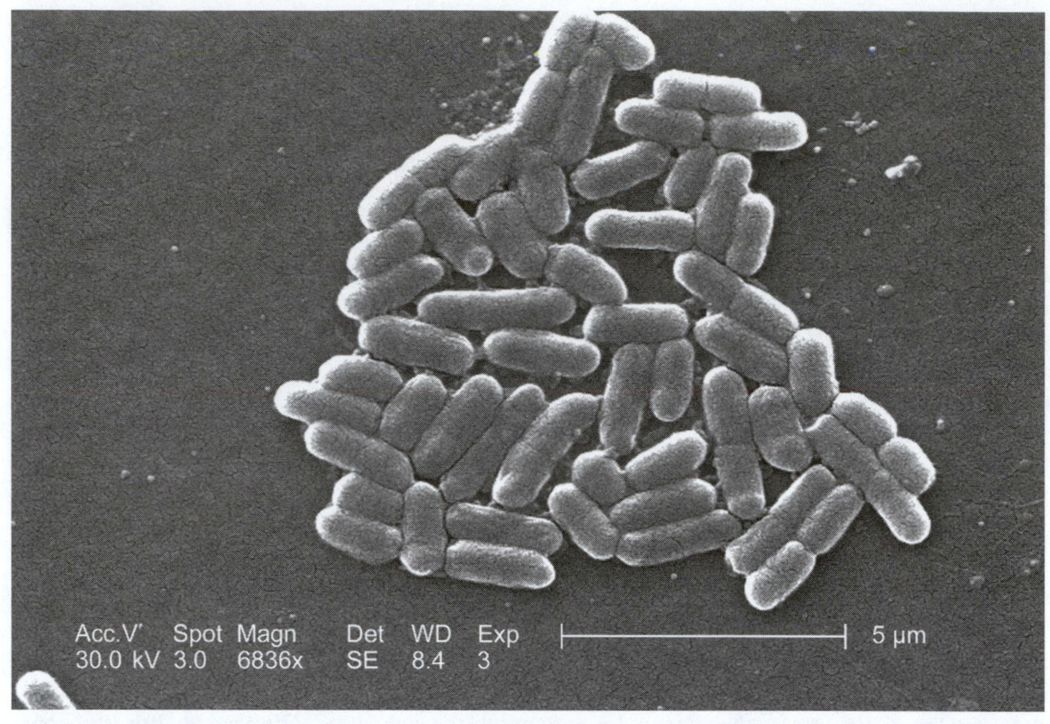

Figure 3.3 Scanning electron micrograph showing *Escherichia coli* bacteria of the strain O157:H7, which produces a toxin that can cause severe bloody diarrhea in humans.

Source: U.S. Centers for Disease Control and Prevention, Public Health Image Library.

A 2007 study of shigellosis in Vietnam showed that households located near a river or a hospital, and those that practiced ancestor worship, were at highest risk. The first two factors may be attributable to water pollution or improper waste disposal; the third is unclear, but may reflect socioeconomic status.

Nontyphoidal salmonellosis is hard to avoid, because almost any food, drink, or pet might be a source. In 1996, some 224,000 people in the United States contracted salmonellosis from one contaminated batch of ice cream, while 50 others caught it by touching a fence surrounding a zoo exhibit of Komodo dragons.

Persons with below-average levels of stomach acid are at increased risk for typhoid fever. In endemic areas, children and adolescents are most likely to develop severe illness. Persons with cystic fibrosis may be less susceptible to typhoid.

Elderly people appear to be at highest risk of contracting *E. coli*, possibly as a result of low stomach acid secretion. Children are at highest risk for a complication called hemolytic uremic syndrome (HUS), in which red blood cells rupture and capillaries become blocked, often resulting in kidney failure.

The Numbers

In 2004, WHO reported 101,383 cases of cholera worldwide, with 2,345 deaths. In 2009, one outbreak in Zimbabwe alone caused 4,127 deaths. The incidence varies from year to year, and many cases go unreported.

There are about 25,000 reported cases of shigellosis each year in the United States, but the actual total is probably much higher. Worldwide, there are an estimated 165 million cases each year, with about 1 million related deaths.

There are about 50,000 to 100,000 reported cases of salmonellosis in the United States in a typical year. There are an estimated 38 unreported mild cases for every one reported. About 6 percent of cases result from handling amphibians or reptiles. Many others result from eating raw eggs (Case Study 3-4).

Typhoid causes about 13 to 21 million cases worldwide each year, with 300,000 to 500,000 deaths. The United States has only 400 to 500 typhoid cases in a typical year.

The infamous O157:H7 strain of *E. coli* caused about 73,000 illnesses and 60 deaths in the United States in 2002. For example, in 1999, at the Washington County Fair, near Albany, New York, more than 1,000 people became ill and two died from *E. coli* infections after drinking contaminated well water.

Case Study 3-4: Egg Wars

In 1988, British author and former Member of Parliament Edwina Currie was forced to resign her position as Junior Health Minister after angering the poultry industry with a statement that "most of the egg production in this county, sadly, is now affected with salmonella." Whether or not the statement was defensible at the time, it caused egg sales to plummet, forcing the government to pay millions of pounds in compensation to poultry farmers. But a 1989 report concluded that Currie was right. The British government had, in fact, covered up a recent salmonellosis epidemic, during which salmonella could be recovered from any cooked egg with liquid yolk.

History

The first recorded cholera pandemic was in Asia in 1816. Others followed in 1829, 1852, 1863, 1881, and 1899. The seventh pandemic began in Indonesia in 1961, and a closely related strain that turned up in South America in 1991 is sometimes called the eighth pandemic. The disease might have reached South America in ballast water from a ship, but the reasons for its rapid spread and high mortality have become controversial (see Chapter 7, page 254).

In 2008–2009, the collapse of Zimbabwe's healthcare system and infrastructure resulted in a cholera outbreak that grew rapidly into a major epidemic. Several thousand people died (Table 3.2), although cholera is normally curable with simple rehydration.

During World War II, before antibiotics were available, German bacteriologists found that infusions of *Bacillus subtilis* fed to troops in northern Africa helped control shigellosis. The so-called hay bacillus has remained a popular component of alternative medicine and biological control.

Contrary to rumor, the bacterial genus *Salmonella* was not named for a spoiled fish loaf in the back of someone's refrigerator. The name is a patronymic to honor American veterinary pathologist Daniel Elmer Salmon (1850–1914), the administrator of a USDA program that discovered a form of salmonellosis in swine.

In 2009, the U.S. Food and Drug Administration (FDA) traced a *Salmonella* outbreak to a peanut butter processing plant in Georgia. Inspectors reported that the firm knowingly shipped contaminated peanut products. There were large leaks in the roof directly above open product containers; the same sink reportedly was used to wash hands, utensils, and mops; and a storage area was coated with "a slimy, black-brown residue."

No discussion of typhoid would be complete without the story of Typhoid Mary, aka Mary Mallon (1869–1938), a New York cook who insisted that germs did not cause disease. If Mary were alive today, she would probably have a website to showcase her beliefs and protest her treatment at the hands of the public health establishment. But an autopsy revealed that Mary's gallbladder contained a thriving colony of typhoid bacteria, and during her career as a cook, she infected at least 47 people and caused three deaths. She refused to change occupations and died in quarantine.

The first known *E. coli* O157:H7 outbreaks were in 1982. Some scientists believe the strain originated in Central America in the 1970s, when a bacterial virus transferred a toxin-producing gene from *Shigella* to the intestinal bacteria of cattle. Others have implicated the cholera bacterium as the source of the gene. Whatever its origin, O157:H7 does not harm infected cattle, but in humans it is sometimes lethal.

Prevention and Treatment

Severe diarrhea sometimes clears the gut of normal bacteria that aid in digestion, produce certain vitamins, and provide other useful services. Researchers have proposed that the role of the human appendix may be to store a backup supply of these bacteria.

Since most diarrheal diseases are waterborne or foodborne, the best preventive measures include water treatment and sanitary food preparation. In 2008, the FDA allowed irradiation of vegetables to kill bacteria. A zinc supplement may also help prevent and treat diarrhea in children. A study in Bangladesh showed that filtering water through old sari cloth reduced the number of cholera cases by about half.

The existing live attenuated single-dose cholera vaccine is only about 70 percent effective, but cholera is fairly simple to treat where rehydration fluids and other basic medical supplies are available. Antimicrobial drugs such as tetracycline or trimethoprim-sulfamethoxazole may shorten the course of the disease.

As of 2009, no shigellosis vaccine was licensed for use outside China, but candidate vaccines were in clinical trials. Treatment consists of rehydration, electrolyte replacement, and oral antibiotics such as ciprofloxacin in adults or trimethoprim-sulfamethoxazole in children.

Salmonellosis usually requires no treatment other than rehydration and electrolyte replacement. The bathroom will also need cleaning, in this or any other acute diarrheal disease; studies have shown that *Salmonella* can survive for weeks in biofilms inside toilet bowls. Antibiotic therapy is not recommended except for high-risk groups.

In a 2002 study of a salmonella outbreak, 100 percent of exposed people who drank only nonalcoholic beverages became sick, but a mere 78 percent of those who had one or two alcoholic drinks became sick. Alcohol (particularly white wine) stimulates gastric secretion of acid, thus creating a lethal environment for bacteria. Another possibility is that some of the people were simply too drunk to realize they were sick.

Typhoid vaccination (available since about 1899) is recommended for those at high risk of exposure to typhoid fever. Most cases are treatable with antimicrobial drugs such as ciprofloxacin, chloramphenicol, ampicillin, or trimethoprim-sulfamethoxazole.

In 2009, doctors in Michigan were testing a vaccine to protect against enterotoxigenic *E. coli.* The usual treatment is rehydration and electrolyte replacement, but severe cases or complications may require hospital treatment. It is unclear whether antimicrobial drugs such as fluoroquinolones are helpful or not. Another new approach is vaccination of cattle to prevent shedding of *E. coli* O157:H7.

Popular Culture

An ever-popular urban legend claims that an unknown evildoer is mailing sponges contaminated with "the Klingerman virus" to random victims, who then contract severe or fatal dysentery.

There is no such virus, but anyone who receives an unexpected sponge should probably discard it anyway.

Cocoa has been a popular diarrhea remedy in South America and Europe for centuries. It might actually work, since cocoa contains flavonoids that can inhibit diarrhea. Other sources recommend eating fenugreek seeds, cheese, or nuts. People who feel well enough to eat these foods will probably recover with or without them.

Chinese herbal remedies for shigellosis include concoctions of siler (*Siler divaricatum*), purple giant hyssop (*Agastache rugosa*), snow rose (*Serissa foetida*), Chinese buttonbush (*Adina rubella*), or huang-qi (*Astragalus henryi*). For typhoid, Chinese herbalists recommended black cardamom (*Amomum costatum*), sweet flag (*Acorus* sp.), bai zhu (*Atractylis ovata*), and other herbs, with the addition of rhinoceros or buffalo horn in the event of intestinal bleeding.

In nineteenth-century Europe and North America, one standard treatment for enteric fever (typhoid) was to cover the patient with a poultice of horse dung. Another was to tie fish to the patient's feet, or tie cabbage leaves around his neck, or wrap him in a fresh, warm sheepskin, immediately after removing it from the sheep. It's amazing that anyone survived medical treatment in those days.

In the 1985 novel and 2007 motion picture *Love in the Time of Cholera*, the main character's husband is a doctor who treats cholera. In the 1997 movie *Contagious*, an epidemiologist investigates a cholera epidemic and traces it to airline food.

One of the most startling T-shirts ever designed features a covered wagon and the words: "You have died of dysentery." This slogan refers to a classic computer game called *Oregon Trail*, in which players try to reach their simulated destination without succumbing to the diseases that ended the lives of many settlers.

Since *E. coli* first made headlines in 1982, the disease has had little opportunity to find its way into popular culture. In the 1998 motion picture *Urban Legend*, one character says to another: "If we ever have another *E. coli* crisis in the cafeteria, I want you to have the biggest, juiciest burger. My treat."

The following urban legend appeared on the Web in 2007: "It has been scientifically proven that if we drink 1 liter of water each day, at the end of the year we would have absorbed more than 1 kilogram of *Escherichia coli*." In fact, to achieve this level of consumption it would be necessary to drink untreated sewage. (This claim refers to coliform bacteria in general, not the toxigenic strains that cause dysentery.)

The Future

In 2009, about 1 billion people lacked access to safe drinking water, and about 80 percent of all illnesses in the Third World were related to this problem. While the literature on this subject grows, the water supply shrinks. Water is a renewable resource, but that fact does not make it potable. The reasons for this problem include overpopulation, crumbling infrastructure, political instability, industrial waste, climate change, and local custom. More than 3 million people die every year from the effects of diarrhea and related diseases, and the cost of treating survivors and supporting their economies may exceed the cost of preventive measures, such as water filters and health education.

References and Recommended Reading

"Anti-Typhoid Vaccine." *New York Times*, 9 February 1913.

Barker, J., and S.F. Bloomfield. "Survival of *Salmonella* in Bathrooms and Toilets in Domestic Homes Following Salmonellosis." *Journal of Applied Microbiology*, Vol. 89, 2000, pp. 137–144.

Bercu, T. E., et al. "Amebic Colitis: New Insights into Pathogenesis and Treatment." *Current Gastroenterology Reports*, Vol. 9, 2007, pp. 429–433.

Bhattacharya, S. K. "Progress in the Prevention and Control of Diarrhoeal Diseases since Independence." *National Medical Journal of India*, Vol. 2003, pp. 15–19.

Bruce, M. G., et al. "Lake-Associated Outbreak of *Escherichia coli* O157:H7 in Clark County, Washington, August 1999." *Archives of Pediatrics and Adolescent Medicine*, Vol. 157, 2003, pp. 1016–1021.

Chen, X., et al. "Differences in Perception of Dysentery and Enteric Fever and Willingness to Receive Vaccines among Rural Residents in China." *Vaccine*, Vol. 24, 2006, pp. 561–571.

"Dark Chocolate Helps Diarrhea." Press Release, Children's Hospital and Research Center at Oakland, California, 29 September 2005.

"Evolution of Typhoid Bacteria Traced." United Press International, 28 November 2006.

"FDA: OK to Zap Spinach, Lettuce with Radiation." Associated Press, 21 August 2008.

Forsberg, B. C., et al. "Diarrhoea Case Management in Low- and Middle-Income Countries—An Unfinished Agenda." *Bulletin of the World Health Organization*, Vol. 85, 2007, pp. 42–49.

Gaffga, N. H., et al. "Cholera: a New Homeland in Africa?" *American Journal of Tropical Medicine and Hygiene*, Vol. 77, 2007, pp. 705–713.

Green, L. R., et al. "Beliefs about Meals Eaten Outside the Home as Sources of Gastrointestinal Illness." *Journal of Food Protection*, Vol. 68, 2005, pp. 2184–2189.

Griffith, D. C., et al. "Review of Reported Cholera Outbreaks Worldwide, 1995–2005." *American Journal of Tropical Medicine*, Vol. 75, 2006, pp. 973–977.

Holt, K. E., et al. "High-Throughput Sequencing Provides Insights into Genome Variation and Evolution in *Salmonella typhi*." *Nature Genetics*, Vol. 40, 2008, pp. 987–993.

Kato, Y., et al. "Multidrug-Resistant Typhoid Fever Outbreak in Travelers Returning from Bangladesh." *Emerging Infectious Diseases*, Vol. 13, 2007, pp. 1954–1955.

Kim, D. R., et al. "Geographic Analysis of Shigellosis in Vietnam." *Health & Place*, Vol. 14, 2008, pp. 755–767.

King, A. A., et al. "Inapparent Infections and Cholera Dynamics." *Nature*, Vol. 454, 2008, pp. 877–880.

Kweon, M. N. "Shigellosis: The Current Status of Vaccine Development." *Current Opinion in Infectious Diseases*, Vol. 21, 2008, pp. 313–318.

Lockary, V. M., et al. "Shiga Toxin-Producing *Escherichia coli*, Idaho." *Emerging Infectious Diseases*, Vol. 13, 2007, pp. 1262–1264.

Lucas, R. "Clinical Significance of the Redefinition of the Agent of Amoebiasis." *Revista Latinoamericana de Microbiología*, Vol. 43, 2001, pp. 183–187.

Millward, D. "Currie 'Was Right' on Salmonella." *Telegraph*, 26 December 2001.

Muniesa, M., et al. "Occurrence of *Escherichia coli* O157:H7 and Other Enterohemorrhagic *Escherichia coli* in the Environment." *Environmental Science and Technology*, Vol. 40, 2006, pp. 7141–7149.

Neergard, L. 2008. "Food Poisoning's Legacy." Associated Press, 22 Jan 2008.

Nicolas, X., et al. "Shigellosis or Bacillary Dysentery." *Presse Médicale*, Vol. 36, 2007, pp. 1606–1618. [French]

Nullis, C. "Zimbabwe's New Export: Cholera." Associated Press, 26 November 2008.

Ochiai, R. L., et al. "A Study of Typhoid Fever in Five Asian Countries: Disease Burden and Implications for Controls." *Bulletin of the World Health Organization*, Vol. 86, 2008, pp. 260–268.

Petri, W., et al. "Enteric Infections, Diarrhea, and Their Impact on Function and Development." *Journal of Clinical Investigation*, Vol. 118, 2008, pp. 1277–1290.

"Purpose of Appendix Believed Found." Associated Press, 5 October 2007.

Sanders, J. W., et al. "Military Importance of Diarrhea: Lessons from the Middle East." *Current Opinion in Gastroenterology*, Vol. 21, 2005, pp. 9–14.

Santamaria, J., and G. A. Toranzos. "Enteric Pathogens and Soil: A Short Review." *International Microbiology*, Vol. 6, 2003, pp. 5–9.

Schneider, J., et al. "*Escherichia coli* O157:H7 Infections in Children Associated with Raw Milk and Raw Colostrums from Cows—California, 2006." *Morbidity and Mortality Weekly Report,* Vol. 57, 2008, pp. 625–628.

Shukla, V. K., et al. "Carcinoma of the Gallbladder—Is It a Sequel of Typhoid?" *Digestive Diseases and Sciences*, Vol. 45, 2000, pp. 900–903.

Snow, M., et al. "Differences in Complement-Mediated Killing of *Entamoeba histolytica* Between Men and Women—an Explanation for the Increased Susceptibility of Men to Invasive Amebiasis?" *American Journal of Tropical Medicine and Hygiene*, Vol. 78, 2008, pp. 922–923.

Sur D., et al. "Shigellosis: Challenges and Management Issues." *Indian Journal of Medical Research*, Vol. 120, 2004, pp. 454–462.
Taylor, R. H. "Cholera and the Royal Navy 1817–1867." *Journal of the Royal Naval Medical Service*, Vol. 83, 1997, pp. 147–156.
U.S. Centers for Disease Control and Prevention. 2008. Investigation of Outbreak of Infections Caused by *Salmonella saintpaul*. Updated 18 June 2008.
"Vaccine for *E. coli* Developed." United Press International, 16 April 2009.
Varma, J. K., et al. "An Outbreak of *Escherichia coli* O157 Infection Following Exposure to a Contaminated Building." *Journal of the American Medical Association*, Vol. 290, 2003, pp. 2709–2712.
"WHO Calls for Aid for Cholera Outbreak." United Press International, 6 February 2009.
World Health Organization. "Prevention and Control of Cholera Outbreaks: WHO Policy and Recommendations." 25 November 2008.

DENGUE AND DENGUE HEMORRHAGIC FEVER

Summary of Threat

Dengue is a mosquito-borne viral disease that causes pain and fever. About 100 million people contract dengue every year, but until recently the death rate was very low. A more dangerous form called dengue hemorrhagic fever (DHF) is becoming more common, possibly as a result of climatic or demographic changes that favor repeated exposures to multiple strains. As of 2009, no vaccine or specific treatment is available.

Other Names

The correct pronunciation is "DENG-gay," not "DENG-yoo" or "DENG-goo." Some people omit the second syllable and pronounce it "deng," by analogy to words such as "tongue" and "meringue."

Other names include breakbone fever, bone-break fever, stiffneck fever, bonecrushers' disease, dandy fever, three-day fever, five-day fever, seven-day fever, duengero, broken wing, African fever, and saddleback fever. ("Saddleback" refers to the fact that a graph of the fever sometimes reaches two peaks separated by a dip.) The first four names may also refer to influenza, malaria, or any disease that causes chills, fever, and pain.

Less common names for dengue are denguis, tootia, Aden fever, bouquet fever, date fever, polka fever, solar fever, scarlatina rheumatica, febris exanthematica articularis, and exanthesis arthrosia. A 1901 medical text lists nearly 100 names for dengue. Severe forms are known as dengue hemorrhagic fever (DHF) and dengue shock syndrome (DSS).

Some English dictionaries claim that we owe the name of this disease to the Swahili phrase *ka dinga pepo*, meaning "cramps brought on by an evil spirit." (According to a Swahili dictionary, *dinga* means a car, but *dege* means cramps.) Other sources claim that dengue is an Indian, Spanish, or Arabic word. Names for dengue in other languages include *trópusi náthaláz* ("tropical fever") in Hungarian, *calentura roja* ("red fever") in Spanish, *Fuenftage-Fieber* ("five-day fever") in German, and *fièvre épidémique inflammatoire* ("epidemic inflammatory fever") in French.

Description

We debated whether dengue belongs in a book on biological threats, because until recently the death rate was very low—except in rare cases when it progressed to dengue hemorrhagic fever (DHF), which can kill 10 to 20 percent of untreated patients. A few years ago, however, doctors made the unsettling discovery that DHF is no longer rare. In Mexico, its reported incidence increased by 600 percent between 2001 and 2007. Also, dengue is one of several diseases that may expand its range as a result of global climate change (Chapter 6).

Dengue symptoms include sudden onset of fever with severe joint and muscle pain. Vomiting and diarrhea may occur, and a rash of small red spots often appears on the legs and chest. The dengue virus (a flavivirus similar to the agent of hepatitis C) has four distinct serological types: DENV-1, DENV-2, DENV-3, and DENV-4. A person who survives infection with one type of dengue becomes immune to that type only; if he or she later contracts another type, the immune system sometimes overreacts, causing DHF. Urbanization and travel have helped expand the range of dengue and have increased the probability of contracting more than one type. The international trade in used tires may also spread the disease, since tires trap water in which mosquito vectors can breed.

DHF symptoms include bleeding or bruising that results from an increase in vascular permeability. Complications such as dengue encephalitis and dengue hepatitis may also occur. DHF is classified according to the following grades:

- Grade I: Fever and minor hemorrhagic symptoms (positive tourniquet test or easy bruising).
- Grade II: Grade I symptoms plus spontaneous bleeding from skin or mucous membranes.
- Grade III: General circulatory failure; rapid, weak pulse, low blood pressure.
- Grade IV: Profound shock, undetectable blood pressure or pulse.

Dengue does not spread directly from one person to another. The vectors are mosquitoes in the genus *Aedes,* usually *Aedes aegypti* or *A. albopictus* (Figure 3.4). These insects are active in the daytime, often prefer urban areas, and have the peculiar habit of biting people on the ankles. To transmit dengue, a mosquito must survive for 10 to 14 days after taking blood from an infected person. Meanwhile, the mosquito continues to take blood meals while the virus multiplies inside its body. Once the virus reaches high enough levels, it finds its way into the bloodstream of the mosquito's next target. A 2008 study showed that wild rodents, bats, and marsupials near cities may serve as a reservoir. Nonhuman primates may also be hosts in some areas.

Who Is at Risk?

Worldwide, about 2.5 billion people are at risk for dengue. In the United States, recent autochthonous transmission—that is, transmission in the same place where the outbreak occurs—has been documented only in Texas and Hawaii. Many cases probably go unreported, since mild dengue resembles flu. *Aedes* mosquitoes capable of transmitting dengue are abundant year-round in many American cities, such as Tucson, Arizona, but not enough local residents are infected to start an outbreak.

As noted above, people who have already had dengue are at increased risk of DHF if they are exposed to a second dengue serotype. In 1977, for example, Cuba had an epidemic of DEN-1. When a DEN-2 epidemic visited the island in 1981, people who were immune to DEN-1 from the previous epidemic were nearly four times more likely to contract DHF than people who had no immunity at all.

A study of dengue risk factors in Laredo, Texas, and neighboring Nuevo Laredo, Mexico, showed that the higher risk in Mexico was largely attributable to the use of evaporative coolers

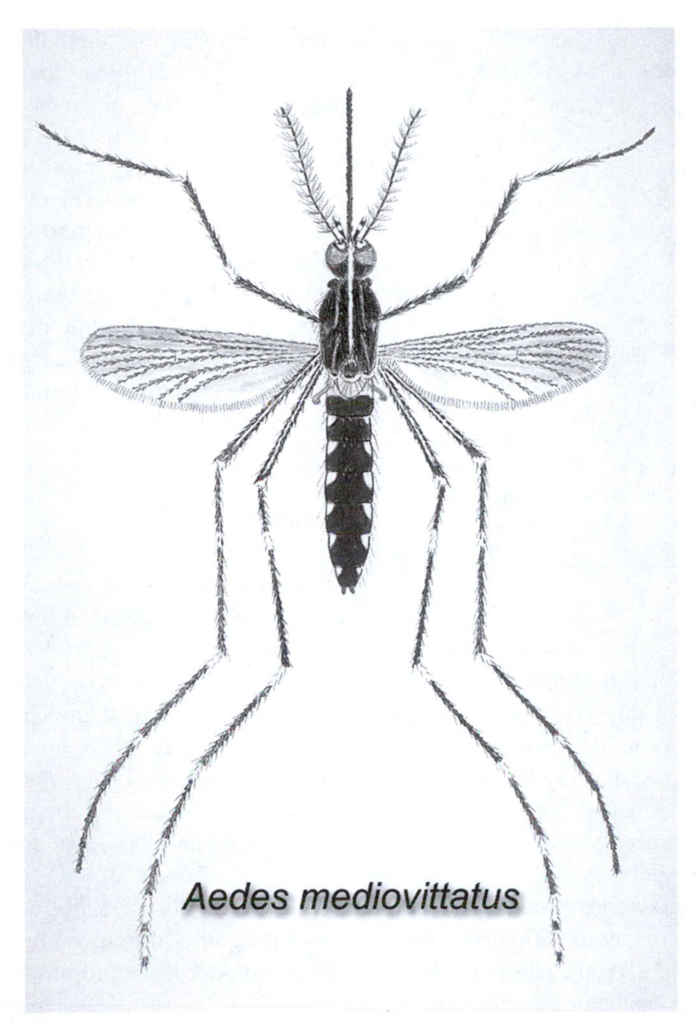

Figure 3.4 The mosquito *Aedes mediovittatus* is a known vector of dengue fever in humans.
Source: U.S. Centers for Disease Control and Prevention, Public Health Image Library.

instead of air conditioners. The dry, cool environment inside air-conditioned buildings is hostile to *Aedes* mosquitoes. The investigators found that 55 percent of dengue cases in Nuevo Laredo would not have occurred if all households had air conditioning. However, medical records showed that at least ten Laredo residents sought treatment for dengue (as later confirmed by blood tests), but received a diagnosis of "flu-like illness." In other words, part of the reason for the absence of dengue north of the border is that doctors are simply not familiar with the disease (Case Study 3-5).

The Numbers

Dengue infects 50 to 100 million people every year, but about 90 percent develop no symptoms or report mild flu-like symptoms. On average, about 500,000 people are hospitalized with

Case Study 3-5: NAFTA and Dengue

Among the many criticisms leveled at the North American Free Trade Agreement (NAFTA) of 1992 is the claim that immigrants and visitors will bring tropical diseases with scary names into the United States. Dengue, malaria, typhoid, cholera, yellow fever, and leprosy are the diseases most often cited as examples in anti-NAFTA literature, but dengue is the only one of these that appears to pose a significant threat north of the border. Studies suggest that even dengue is more amenable to vector control programs, home air conditioners, and window screens than to border security.

dengue every year. Although the death rate is low (some 20,000 worldwide in a typical year), dengue is numerically the most important vector-borne viral disease of humans (Table 3.3).

The case fatality rate in each outbreak depends on the serotypes involved, the availability of treatment, and the accuracy of reporting. In dengue shock syndrome (DSS), mortality can be as high as 40 percent. Cuba reported 354,515 dengue cases in 1981, with an estimated 24,000 cases of DHF, 10,000 cases of DSS, and 158 deaths. In 2000, Brazil reported 288,245 dengue cases with 91 deaths.

History

A Chinese medical encyclopedia published in A.D. 992 described a disease similar to dengue. Outbreaks reportedly took place in the French West Indies in 1635 and in Panama in 1699. In 1771, a Spanish doctor in Puerto Rico wrote about a fever called *quebranta huesos* ("it breaks bones"), probably a reference to dengue. Yet dengue was not well known until 1779, when major outbreaks occurred in northern Africa, Spain, and India. In 1780, the disease spread to North America on a ship, and Philadelphia had a large epidemic. By 1801, when Queen Luisa of Spain caught dengue and described it in a series of letters, the name and the disease apparently were widely recognized. After World War II, a dengue pandemic began in southeast Asia and spread around the globe.

Doctors became aware of dengue hemorrhagic fever (DHF) in the 1950s during dengue epidemics in the Philippines and Thailand. Vector control programs eliminated *Aedes aegypti* from many countries in the 1960s, but its range expanded again after these programs ended. The first known major DHF epidemic occurred in Cuba in 1981.

Table 3.3 Some Outbreaks of Dengue and Dengue Hemorrhagic Fever

Year	Location	Estimated Cases	Estimated Deaths
1981	Cuba	354,515	158
1989–1990	Venezuela	6,000+	78
1994	Nicaragua	20,469	6
1996	Delhi, India	8,900	374
1998	Argentina	818	0
1998	Vietnam	119,429	342
2000	Bangladesh	5,575+	90
2001–2002	Hawaii	122	0
2002	Malaysia	32,289	74+
2002	Easter Island, Chile	636	0
2006	India	12,317	184
2007	Jamaica	4,260	18
2008	Rio de Janeiro, Brazil	55,000	67
2009	Bolivia	56,000+	25+

Sources: World Health Organization (WHO), U.S. Centers for Disease Control and Prevention (CDC).

Prevention and Treatment

Dengue is a complicated disease to prevent or treat. *Aedes* mosquitoes are hard to avoid because they are active in the daytime, but people in endemic areas can use insect repellents and protective clothing. Mosquito control might sound like a no-brainer; however, a 2008 study showed that partially eliminating vectors might actually increase the incidence of DHF. Adult *Aedes* mosquitoes often live inside houses, but air conditioners (not evaporative coolers) greatly reduce risk. It is also important to eliminate containers of water where mosquitoes can breed. Mosquito fish (*Gambusia* and others) and small crustaceans called copepods can remove mosquito larvae from ponds, but even underground septic tanks can serve as habitat. In 2009, researchers reported preliminary success in controlling *Aedes* mosquitoes by infecting them with a bacterial symbiont that shortens the adult lifespan.

Efforts to develop a vaccine have been unsuccessful, mainly because of the existence of four different serotypes. One possible solution, proposed by Rice University researchers in 2008, would be to develop four vaccines and inject them simultaneously at different locations on the body. As of 2009, at least two major pharmaceutical companies and several government agencies (including the U.S. Army) were working to develop dengue vaccines.

Once a person contracts dengue, treatment is mainly supportive. Corticosteroids apparently are not helpful in preventing shock. Patients should not take aspirin or other anticoagulant drugs without consulting a physician because of the danger of increased bleeding.

Popular Culture

The Consul's File, a 1977 novel by Paul Theroux, describes the feverish visions of a young American teacher who contracts dengue in Malaysia and nearly dies. The same author also explored this theme in his 1975 short story "Dengue Fever," a dark tale of fever and the supernatural.

Another Theroux novel, *The Mosquito Coast,* inspired the 1986 motion picture with Harrison Ford as an obsessive inventor-turned-survivalist. After inventing a machine that uses fire to create ice, he moves his family to coastal Honduras, a region with a high incidence of dengue and malaria. He expects his gift of ice to transform this pristine tropical wilderness into paradise, by cooling homes and promoting health, but things don't work out that way. Instead, one necessity follows another, air conditioning and mosquito nets and patriarchy and violence, until at last he finds himself on the verge of recreating the very world he has rejected.

One traditional cure for dengue fever is raw pegaga leaf juice. Pegaga or ulam pegaga (*Centella asiatica*) grows wild in many parts of Asia and is a popular cure for many ailments, particularly those that tend to resolve in a few days without treatment. Other sources claim that the miracle dengue cure is papaya leaf juice (or tea).

A plant called boneset (*Eupatorium perfoliatum*) was once popular in the southern United States as a treatment for the pain and fever of dengue and other illnesses. People kept bouquets of dried boneset hanging from the house rafters to remind children to dress warmly and keep dry, in order to avoid getting sick and being required to drink bad-tasting boneset tea.

For a poetic advertisement involving dengue, see Case Study 3-6.

The Future

Some authorities predict that dengue will expand its range into the United States as a result of global warming, immigration, and international travel. Others point out that dengue was here

Case Study 3-6: Dengue Doggerel

The following poem, which appeared in the *Vicksburg* (Mississippi) *Herald* newspaper in 1873, illustrates public perceptions of dengue in the American South of that era, before the advent of dengue hemorrhagic fever and NAFTA. The poet was not named, but we can probably rule out Sidney Lanier:

Silent and cold on his lonely cot the lover bold did lay,
His limbs were stiff, and the dengue pains had wasted his form away.
While the fever raged he cried aloud, in incoherent strain
"Oh, that my drooping eyes might see that angel face again!"
He cried in the anguish of dengue pain for a sight of his love once more,
Till her heart was touched and she came to the side of his bed, which was spread on the floor.
Through dreary days and weary nights she sat by her lover's couch,
And proved that at nursing and tending him she wasn't any slouch.
She soothed his spirit and drove away all trace of dengue blues,
By reading the Herald every day aloud—all the local news.[1]

[1] *Northern Vindicator*, 6 December 1873.

before, and that our air conditioners chased it away. What the world really needs is an effective tetravalent vaccine that would make it possible to eradicate the disease. Also, since doctors outside endemic areas often fail to recognize dengue, a better diagnostic test is needed. Otherwise, undiagnosed dengue might be transferred through blood transfusions.

References and Recommended Reading

Barrera, R., et al. "Unusual Productivity of *Aedes aegypti* in Septic Tanks and Its Implications for Dengue Control." *Medical and Veterinary Entomology*, Vol. 22, 2008, pp. 62–69.

Barreto, M. L., and M. G. Teixeira. "Dengue Fever: A Call for Local, National, and International Action." *Lancet*, Vol. 372, 2008, p. 205.

daSilva, L., and R. Richtmann. "Vaccines under Development: Group B Streptococcus, Herpes-Zoster, HIV, Malaria and Dengue." *Jornal de Pediatria*, Vol. 82, 2006, pp. S115–124.

de la Sierra, B., et al. "Race: A Risk Factor for Dengue Hemorrhagic Fever." *Archives of Virology*, Vol. 152, 2007, pp. 533–542.

Domingo-Carrasco, C., and J. Gascon-Bustrenga. "Dengue and Other Hemorrhagic Viral Fevers." *Enfermedades Infecciosas y Microbiologia Clinica*, Vol. 23, 2005, pp. 615–626.

Fuller, T. "The War on Dengue Fever." *New York Times*, 3 November 2008.

Gibbons, R. V., and D. W. Vaughn. "Dengue: An Escalating Problem." *British Medical Journal*, Vol. 324, 2002, pp. 1563–1566.

Gould, E. A., and T. Solomon. "Pathogenic Flaviviruses." *Lancet*, Vol. 371, 2008, pp. 500–509.

Gubler, D. J. "Dengue/Dengue Hemorrhagic Fever: History and Current Status." *Novartis Foundation Symposium*, Vol. 277, 2006, pp. 3–16.

Gulati, S., and A. Maheshwari. "Atypical Manifestations of Dengue." *Tropical Medicine and International Health*, Vol. 12, 2007, pp. 1087–1095.

Halstead, S. B. "Dengue." *Lancet*, Vol. 370, 2007, pp. 1644–1652.

Janes, C. R. "Theorizing Global-Local Linkages in Global Health Studies." Paper presented to the 2003 Fulbright Visiting Scholars Conference, "International Cooperation in a Borderless World," Washington, DC, 2–5 April 2003.

Katz, T. M., et al. "Insect Repellents: Historical Perspectives and New Developments." *Journal of the American Academy of Dermatology*, Vol. 58, 2008, pp. 8865–8871.

Leong, A. S., et al. "The Pathology of Dengue Hemorrhagic Fever." *Seminars in Diagnostic Pathology*, Vol. 24, 2007, pp. 227–236.

Lo, R. V., and S. J. Gluckman. "Fever in the Returned Traveler." *American Family Physician*, Vol. 68, 2003, pp. 1343–1350.

Lum, L. C. "Dengue Encephalitis: A True Entity?" *American Journal of Tropical Medicine and Hygiene*, Vol. 54, 1996, pp. 256–259.

Maroun, S. L., et al. "Case Report: Vertical Dengue Transmission." *Jornal de Pediatria*, 23 October 2008.
McMeniman, C. J., et al. "Stable Introduction of a Life-Shortening *Wolbachia* Infection into the Mosquito *Aedes aegypti*." *Science*, Vol. 323, 2009, pp. 141–144.
Nobuchi, H. "The Symptoms of a Dengue-like Illness Recorded in a Chinese Medical Encyclopedia." *Kanpo Rinsho*, Vol. 26, 1979, pp. 422–425. [Japanese.]
Noqueira, R. M., et al. "Dengue Viruses in Brazil, 1986–2006." *Pan American Journal of Public Health*, Vol. 22, 2007, pp. 358–363.
Ooi, E. E., et al. "Dengue Prevention and 35 Years of Vector Control in Singapore." *Emerging Infectious Diseases*, Vol. 12, 2006, pp. 887–893.
Peña, G., et al. "Underdiagnosis of Dengue—Laredo, Texas, 1999." *Lancet*, Vol. 285, 2001, p. 877.
Rigau-Perez, J. G. "The Early Use of Break-Bone Fever (*Quebranta Huesos*, 1771) and Dengue (1801) in Spanish." *American Journal of Tropical Medicine and Hygiene*, Vol. 59, 1998, pp. 272–274.
Rush, A. B. 1789. *An Account of the Bilious Remitting Fever, as It Appeared in Philadelphia in the Summer and Autumn of the Year 1780*. Philadelphia, PA: Prichard and Hall.
"Severe Dengue Infections May Go Unrecognized in International Travelers." *Medical News Today*, 1 April 2007.
Stephenson, J. R. "Understanding Dengue Pathogenesis: Implications for Vaccine Design." *Bulletin of the World Health Organization*, Vol. 83, 2005, pp. 308–314.
Stevenson, M. "Spread of Dengue Fever Reaches Fever Pitch in Mexico." Associated Press, 1 April 2007.
"Study: Dengue Fever Is Underreported." United Press International, 16 October 2007.
Thaha, M., et al. "Acute Renal Failure in a Patient with Severe Malaria and Dengue Shock Syndrome." *Clinical Nephrology*, Vol. 70, 2008, pp. 427–430.
Thammapalo, S., et al. "Relationship between Transmission Intensity and Incidence of Dengue Hemorrhagic Fever in Thailand." *PLoS Neglected Tropical Diseases*, Vol. 2, 2008, p. e263.
Thoisy, B. D., et al. "Dengue Infection in Neotropical Forest Mammals." *Vectorborne and Zoonotic Diseases*, 22 October 2008.
Wilder-Smith, A., and J. L. Deen. "Dengue Vaccines for Travelers." *Expert Review of Vaccines*, Vol. 7, 2008, pp. 569–578.
World Health Organization. "Fact Sheet No. 117: Dengue and Dengue Haemorrhagic Fever." Revised May 2008.
Zhou, H., and M. W. Deem. "Sculpting the Immunological Response to Dengue Fever by Polytopic Vaccination." *Vaccine*, Vol. 24, 2006, pp. 2451–2459.

BAD BUGS AND MIRACLE DRUGS

Summary of Threat

Many disease-causing bacteria are becoming resistant to antibiotics. In addition, many viruses, fungi, parasites, disease vectors (such as mosquitoes), and reservoir hosts (such as rodents) are becoming resistant to the drugs and poisons used to control their numbers. Fighting the war on disease is far more difficult without effective weapons, and the problem is becoming critical.

Other Names

Pathogens in this category are popularly called "resistant bugs," "superbugs," or "megabugs." A few of their names are depressingly familiar to the general public, such as MRSA (methicillin-resistant *Staphylococcus aureus*). Many others have names that are familiar only to healthcare workers and undertakers. Table 3.4 lists some of the worst drug-resistant bacteria making the rounds as of 2009.

Table 3.4 Examples of Drug-Resistant Pathogenic Bacteria

Species	Name or Strain	Resistant To
Acinetobacter baumannii	MDR-AB	Carbapenem, others
Acinetobacter baumannii	XDR-AB	MDR + ampicillin/sulbactam
Bacillus anthracis	Anthrax	Penicillin
Clostridium difficile	C-Diff or CDF	Metronidazole
Enterobacter sakazakii	ENB_IM	Imipenem or meropenem
Enterobacter sakazakii	ENB_CF	Ceftazidime, cefotaxime, or ceftriaxone
Enterococcus faecium	VRE or ENC_VM	Vancomycin
Escherichia coli	EC_CF3	Ceftazidime, cefotaxime, or ceftriaxone
Escherichia coli	EC_FQ	Ciprofloxacin, ofloxacin, or levofloxacin
Klebsiella pneumoniae	ESBL-KP	Cephalosporins
Klebsiella pneumoniae	KP_IM	Imipenem, carbapenem
Klebsiella pneumoniae	KP_CF	Ceftazidime, cefotaxime, or ceftriaxone
Mycobacterium tuberculosis	MDR-TB	Isoniazid, rifampicin
Mycobacterium tuberculosis	XDR-TB	Isoniazid, rifampicin, fluoroquinolones + amikacin, kanamycin, or capreomycin
Neisseria gonorrhoeae	MDR-GC	Fluoroquinolones
Pseudomonas aeruginosa	PA_CIP	Ciprofloxacin
Pseudomonas aeruginosa	PA_IMI	Imipenem
Pseudomonas aeruginosa	PA_CF	Ceftazidime
Pseudomonas aeruginosa	PA_PIP	Piperacillin
Pseudomonas aeruginosa	PA_MER	Meropenem
Pseudomonas aeruginosa	PA_OFL	Ofloxacin
Pseudomonas aeruginosa	PA_LEV	Levofloxacin
Salmonella typhi	CT18	Fluoroquinolones
Staphylococcus aureus	MRSA or SA_ME	Methicillin
Stenotrophomonas maltophila	Sm or SM	Trimethoprim-sulfamethoxazole
Streptococcus pneumoniae	STP_CF	Cefotaxime or ceftriaxone
Streptococcus pneumoniae	STP_PN	Penicillin

Sources: World Health Organization (WHO), U.S. Centers for Disease Control and Prevention (CDC).

Description

Antibiotics, hailed as "miracle drugs" in the 1940s and 1950s, have become less effective in recent decades due to the evolution of resistant bacteria. Figure 3.5 shows an antibiogram—the result of a laboratory test in which small disks containing antibiotics are placed on a bacterial culture. The antibiotic diffuses from the disk onto the surface of the plate, and if the bacteria grow up to the edge of the disk, they are probably resistant to that antibiotic.

Certain resistant bacterial strains have become a major problem in U.S. hospitals, where these bacteria infect an estimated 2 million patients every year. As bacteria evolve, pharmaceutical companies respond by inventing ever more powerful and expensive antibiotics. But bacteria are not the only organisms that have become resistant to common drugs. Many pathogenic viruses, such as HIV, and protozoans, such as the malaria parasite, have shown the same disturbing trend. Governments have stockpiled Tamiflu® in preparation for influenza pandemics, only to discover that many flu viruses are already resistant to it (Chapter 2). Some fungi that cause pneumonia and systemic infections, such as *Aspergillus*, have also become resistant to antifungal drugs.

To make matters worse, the mosquitoes and ticks that vector many diseases are showing increased resistance to DDT and other pesticides. Even the rodents that serve as reservoir hosts

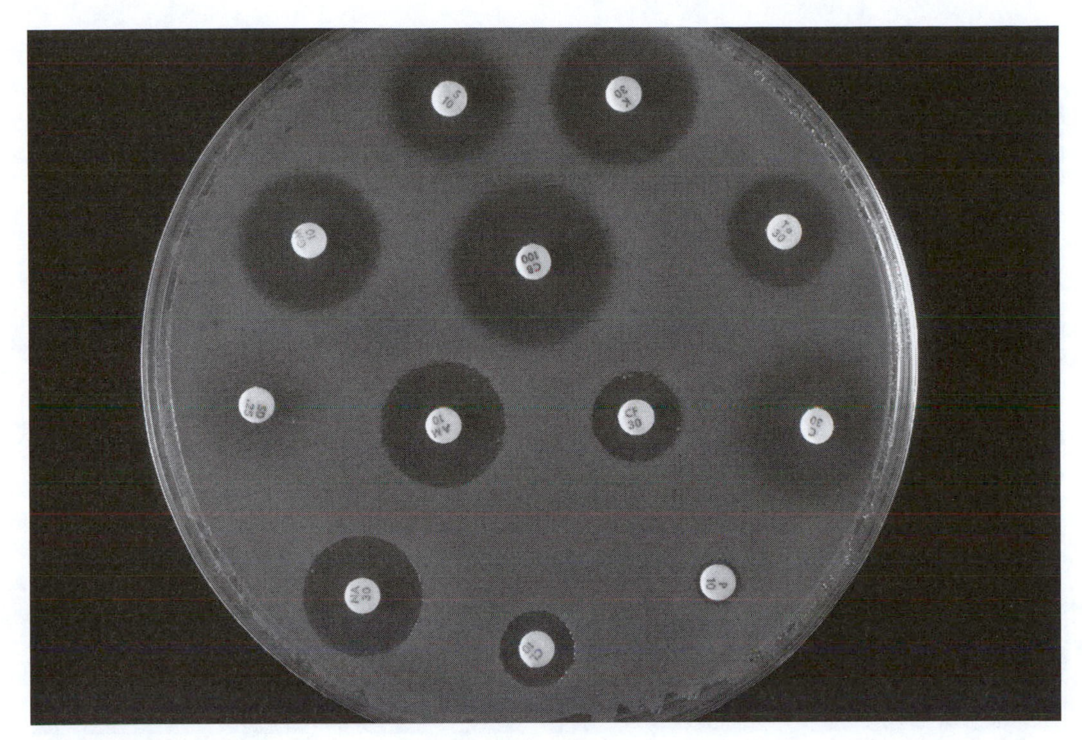

Figure 3.5 Antibiogram study (to measure antibiotic resistance) using a plate culture of the bacterium *Enterobacter sakazakii*.

Source: U.S. Centers for Disease Control and Prevention, Public Health Image Library.

are often resistant to anticoagulants and other common rodenticides. This section, however, will focus mainly on drug-resistant bacteria.

In many cases, doctors must resort to potentially dangerous antibiotics because the safer ones no longer work. In other cases, no drug is effective, and doctors find themselves in the same situation they faced a century ago, watching helplessly as patients die of untreatable infections. Resistant bacteria that infect wounds or cause pneumonia have emerged as perhaps the most deadly long-term biological threat we face today.

Who Is at Risk?

Antibiotic resistance represents a threat to everyone. As usual, the most vulnerable groups are the very young, the very old, the poor and malnourished, and those with suppressed immune systems. Since studies have shown that infection rates vary widely from one hospital to another, living near a bad hospital might be the most important risk factor of all. Unfortunately, community-acquired drug-resistant infections (those acquired outside hospitals) are also becoming common.

In one study, poultry workers were 32 times as likely as others to carry gentamicin-resistant *E. coli*. Pig farmers are reportedly at risk for MRSA (Figure 3.6). People who buy herbal supplements from health food stores represent another, less expected risk group. According to a 2008 study, potentially dangerous antibiotic-resistant bacteria (including the famous *Stenotrophomonas maltophila*) were present in samples of several herbal products. Football players are at risk for MRSA, apparently because of injuries and shared equipment.

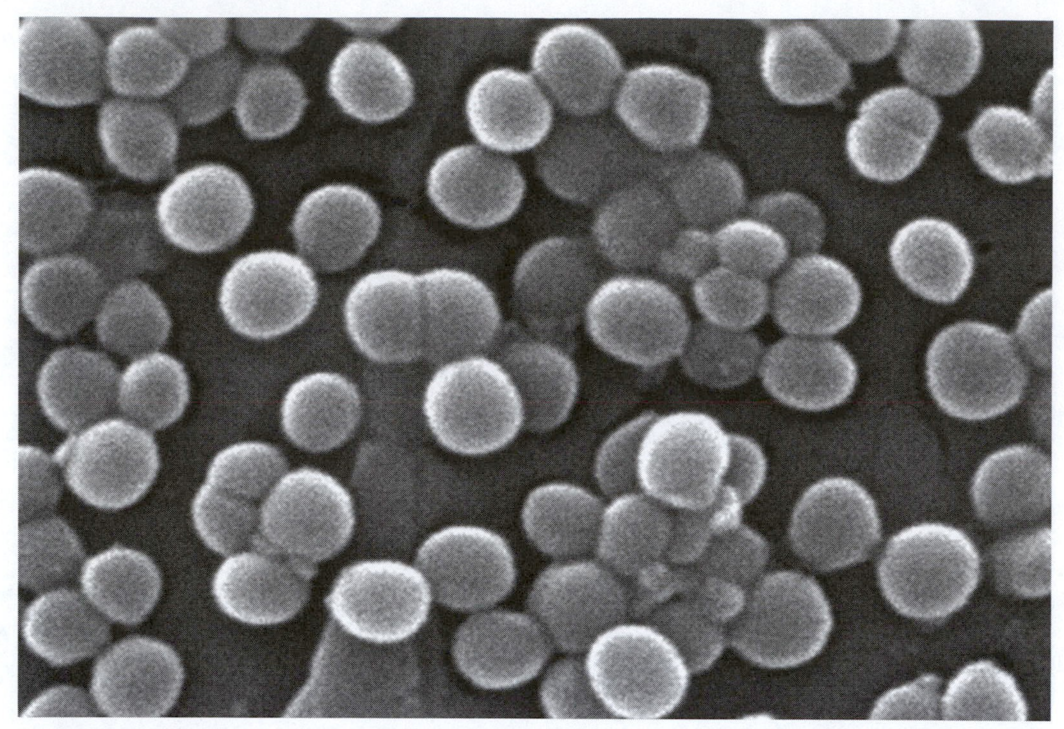

Figure 3.6 Scanning electron micrograph showing clumps of a methicillin-resistant strain (MRSA) of the bacterium *Staphylococcus aureus*.

Source: U.S. Centers for Disease Control and Prevention, Public Health Image Library.

The Numbers

Every year, an estimated 2 million Americans acquire bacterial infections while hospitalized, and about 100,000 die as a result. One of the worst offenders is *Clostridium difficile*, a bacterium that causes diarrhea and is resistant to most known antibiotics. It infects some 300,000 hospitalized patients in the United States annually, and an estimated 1 in every 1,000 patients in Europe—but the death rate was low until about 2006, when a highly virulent strain began to emerge. As a result, *C. difficile* did not even make the "top six" in a 2006 report by the Infectious Diseases Society of America:

1. *Acinetobacter baumanni* first gained public attention because of infections in soldiers returning from Iraq—between 2002 and 2004. It now accounts for about 7 percent of all hospital-acquired cases of bacterial pneumonia in the United States, and it is resistant to most drugs. Reported mortality rates range from 19 to 54 percent.
2. *Aspergillus* is a fungus that often infects people with compromised immune systems, but it was already a killer when AIDS was unknown. In 1978–1982, for example, it caused 1 in every 200 hospital deaths in Germany. Drug-resistant strains emerged in about 1990, and the death rate now exceeds 50 percent.
3. Methicillin-resistant *Staphylococcus aureus* (MRSA) infected about 880,000 hospital patients in 2007, and it is also common in community-acquired infections. It can survive for up to 12 days on contaminated surfaces. In 2005, about 18,650 Americans died from

MRSA (and 16,000 from AIDS). In 2008, Targanta reported that a new antibiotic called oritavancin was effective against MRSA, at least temporarily.

4. ESBL-producing bacteria (which produce the enzyme extended-spectrum beta lactamase, or ESBL) as a group are resistant to antibiotics called cephalosporins. The two most common examples, *E. coli* and *Klebsiella*, cause thousands of urinary tract and wound infections every year, with mortality as high as 64 percent.
5. Vancomycin-resistant *Enterococcus faecium* (VRE) is a major cause of bloodstream infections, meningitis, and endocarditis in hospital patients. In a 2006 survey, 10 percent of patients harbored this bacterium, which is resistant to most antibiotics. Some hospitals reportedly refuse to admit patients with this infection.
6. The bacterium *Pseudomonas aeruginosa* is particularly dangerous for patients with HIV or cystic fibrosis, because it can quickly become resistant to any antibiotic. In some cases, the only available treatment is a lung transplant.

History

Scottish scientist Alexander Fleming discovered penicillin in 1928, and German chemists created the first sulfa drugs in 1932. Other "miracle drugs" followed in rapid succession. When these drugs first appeared, doctors and the general public tended to use them indiscriminately. But whenever populations of living organisms are exposed to a poison, resistance is likely to evolve. Some bacteria, for example, have genes that enable them to produce enzymes (such as ESBL) that break down the antibiotic molecule. Bacteria with this capability are more likely than others to survive and reproduce in the presence of antibiotics. Some antibiotic-resistant bacteria can even transfer the resistance genes to other species of bacteria.

As a result, many bacterial infections that were once treatable are now resistant. But it would not be fair to blame medical science; the first doctors who used antibiotics were simply trying to save lives. Parents must also accept their share of responsibility, for demanding antibiotics every time their child has the sniffles. Some survivalist groups encourage their members to stockpile antibiotics that they have no idea how to use, while other people reportedly take antibiotics for their anti-inflammatory effect. All these are minor players compared with factory farms, which use antibiotics to prevent disease outbreaks and promote growth in livestock. In the early twenty-first century, such farms use about 70 percent of all antibiotics in the United States.

Prevention and Treatment

If the entire patient population converged on a few excellent hospitals (such as Johns Hopkins in Baltimore) to avoid infection with superbugs, the healthcare system would soon collapse. Alternatively, perhaps the state and federal agencies that regulate hospitals could simply do a better job of enforcing sanitation requirements. New methods of disinfection might help if consistently applied; for example, a recent study showed that a specific wavelength of blue light can kill MRSA on hospital surfaces.

Pharmaceutical companies are in the process of developing more effective antibiotics, but there are obstacles. Companies need to earn a profit, and the return on investment is reportedly low for anti-infective drugs in general. There is also the age-old dance of predator and prey; as soon as a new antibiotic is invented and bacteria encounter it, resistance is likely to evolve (Case Study 3-7).

Another approach would focus on convenient, affordable tests that would enable doctors to distinguish between bacterial and viral infections. Studies have shown that communities can reduce the prevalence of resistant bacteria by avoiding the unnecessary use of antibiotics. For example, in Finland in the 1990s, concerns about erythromycin-resistant *Streptococcus* A led to a campaign to reduce erythromycin use. The result was a reduction in resistance levels from 19 percent to 9 percent in only three years.

Yet another approach to defeating enteric superbugs is the fecal transplant. This procedure is exactly what it sounds like. The physician obtains a sample of fecal material from a healthy person and injects it into the colon of a person infected with *Clostridium difficile*. Despite aesthetic problems, the reported success rate is as high as 90 percent.

Case Study 3-7: Back to the Drawing Board

Villagers in the South American nation of Guyana (famous for the 1978 Jonestown incident) have used the drug chloroquine for many years to prevent and treat malaria. The unexpected result is that most of these people now harbor *E. coli* and other bacteria that are resistant to the antibiotic ciprofloxacin. Why? It turns out that cipro is chemically related to chloroquine, although the latter is not an antibiotic. Thus, bacteria exposed to chloroquine also become resistant to cipro. Doctors have always assumed that there was no interaction between treatments used for bacterial and parasitic infections, but that assumption clearly was wrong. Fortunately, bacterial resistance is often reversible. As soon as an effective malaria vaccine becomes widely available, the use of chloroquine will probably decline.

Popular Culture

In 2007, fans of the TV series *Project Runway* were shocked to learn that a participant was receiving hospital treatment for a life-threatening MRSA infection. This one incident probably generated more public controversy and concern than all the hundreds of thousands of similar infections that afflict U.S. residents every year.

Robin Cook's 2007 novel *Critical* explores the personal and economic impacts of a devastating MRSA epidemic at three New York hospitals. The characters include a conflicted physician turned CEO, dastardly Mafiosos, and a pathologist whose husband needs knee surgery. Giving away the ending would be just wrong.

In the 2004 motion picture *The Day after Tomorrow*, a young woman in flooded Manhattan cuts her leg on submerged debris and develops a severe bacterial infection. She recovers quickly after a single shot of penicillin, but in real life she might not be so lucky. Similar wound infections that occurred in the wake of Hurricane Katrina were fatal in about 50 percent of cases, and were mostly resistant to penicillin.

In the 1958 motion picture *Earth versus the Spider*, a really large spider goes on a killing rampage until the bad sheriff's minions squirt it with DDT. The spider is stunned at first, but unexpectedly recovers, and the hero ends up electrocuting it instead. Scientists already knew about DDT resistance in the 1950s, but people were so busy congratulating themselves on the chemical defeat of malaria that they tended not to listen. (Unfortunately, electric bug zappers have not lived up to expectations either.)

For the origin of "Bugzilla," see Case Study 3-8.

The Future

New drugs to prevent (rather than cure) resistant bacterial infection would help. In 2008, researchers were conducting human trials of a bacteriocidal gel that can be applied to the nostrils. New classes of antibiotics may also be on the horizon, but bacteria will probably become resistant to these drugs as well. Another approach is the use of designer probiotics—genetically engineered bacteria that attack pathogens. Again, the pathogens will undoubtedly fight back. Also, there is no scientific advance that cannot be turned to mischief. The same technologies that give us designer drugs and bugs may enable terrorists to create drug-resistant pathogens.

Case Study 3-8: Superbug Hype

Real superbugs are bad enough, but exaggerated ones are worse. In 2008, the press christened the opportunistic bacterium *Stenotrophomonas maltophila* "the megabug" or "Bugzilla" and hinted darkly at mass casualties. To put the matter in perspective, *S. maltophila* was responsible for less than 1 percent of all reported bloodstream infections in 2007. It does not appear to spread easily between patients, and in 2008 it remained treatable with at least one antibiotic. The misunderstanding apparently started when researchers published a description of the *S. maltophila* genome and referred to its "remarkable capacity for drug and heavy metal resistance." Scientists and science writers need to choose words carefully.

References and Recommended Reading

Adler, J., and J. Interlandi. "Caution: Killing Germs May Be Hazardous to Your Health." *Newsweek*, 29 October 2007.

Blakeslee, H. W. "Some Germs Sneer at Sulfa." *Daily Capital News*, 5 August 1943.

"Blue Light Destroys MRSA." United Press International, 30 January 2009.

Coates, A. R., and Y. Hu. "Novel Approaches to Developing New Antibiotics for Bacterial Infections." *British Journal of Pharmacology*, Vol. 152, 2007, pp. 1147–1154.

Croft, A., et al. "Update on the Antibacterial Resistance Crisis." *Medical Science Monitor*, Vol. 13, 2007, pp. RA103–RA118.

Datta, R., and S. S. Huang. "Risk of Infection and Death Due to Methicillin-Resistant *Staphylococcus aureus* in Long-Term Carriers." *Clinical Infectious Diseases*, Vol. 47, 2008, pp. 176–81.

Donnellan, E. "Dangers of Antibiotic Overuse Highlighted." *Irish Times*, 18 November 2008.

Engel, M. "Proposal Targets a Deadly Infection." *Los Angeles Times*, 1 January 2008.

"Experts Concerned about Flu, MRSA Combo." United Press International, 27 April 2008.

"Flies May Be Spreading MRSA from Fowl Feces." Reuters, 16 March 2009.

Goldburg, R., et al. "The Risks of Pigging Out on Antibiotics." *Science*, Vol. 321, 2008, p. 1294.

Gould, I. M. "The Epidemiology of Antibiotic Resistance." *International Journal of Antimicrobial Agents*, 29 August 2008.

Groopman, J. "Superbugs." *The New Yorker*, 11 August 2008.

Gulshan, K., and W. S. Moye-Rowley. "Multidrug Resistance in Fungi." *Eukaryotic Cell*, Vol. 6, 2007, pp. 1933–1942.

Hawkey, P.M. "The Growing Burden of Antimicrobial Resistance." *Journal of Antimicrobial Chemotherapy*, Vol. 62, Supplement 1, 2008, pp. 1–9.

Hellemans, R., et al. "Fecal Transplantation for Recurrent *Clostridium difficile* Colitis, an Underused Treatment Modality." *Acta Gastro-Enterologica Belgica*, Vol. 72, 2009, pp. 269–270.

"Influenza A Showing Tamiflu Resistance." United Press International, 19 December 2008.

Ishizuka, M., et al. "Pesticide Resistance in Wild Mammals—Mechanisms of Anticoagulant Resistance in Wild Mammals." *Journal of Toxicological Sciences*, Vol. 33, 2008, pp. 283–91.

Johnson, A. P., and G. J. Duckworth. "The Emergence of *Stenotrophomonas maltophila*." *British Medical Journal*, Vol. 336, 2008, p. 1322.

Juncosa, B. "Antibiotic Resistance: Blame It On Livesaving Malaria Drug?" *Scientific American*, 21 July 2008.

Kuchment, A. "Trapping the Superbugs." *Newsweek*, 13 December 2004.

Larson, E. "Community Factors in the Development of Antibiotic Resistance." *Annual Review of Public Health*, Vol. 28, 2007, pp. 435–447.

Linares, J. F., et al. "Antibiotics as Intermicrobial Signaling Agents Instead of Weapons." *Proceedings of the National Academy of Sciences* (U.S.), Vol. 103, 2006, pp. 19484–19489.

Merpol, S. B. "Valuing Reduced Antibiotic Use for Pediatric Acute Otitis Media." *Pediatrics*, Vol. 121, 2008, pp. 669–673.

"New Test Shows Promise at Reducing Antibiotic Use." Associated Press, 18 February 2004.
"Out of Patience with Hospital Infections." United Press International, 24 July 2008.
Ozolins, M., et al. "Comparison of Five Antimicrobial Regimens for Treatment of Mild to Moderate Inflammatory Facial Acne Vulgaris in the Community: Randomised Controlled Trial." *Lancet*, Vol. 364, 2004, pp. 2188–2195.
Park, A. 2007. "Fighting Drug-Resistant Bugs." *Time*, 7 June 2007.
Patrick, D. M. "Antibiotic Use and Population Ecology: How You Can Reduce Your 'Resistance Footprint.'" *Canadian Medical Association Journal*, Vol. 180, 2009, pp. 416–421.
"Possible MRSA Cure Undergoing Trials." United Press International, 18 May 2008.
"Powerful Antibiotic Battles MRSA." United Press International, 22 October 2008.
Roberts, D. R., and R. G. Andre. "Insecticide Resistance Issues in Vector-Borne Disease Control." *American Journal of Tropical Medicine and Hygiene*, Vol. 50 (6 Suppl.), 1994, pp. 21–34.
Roghmann, M. C., and L. McGrail. "Novel Ways of Preventing Antibiotic-Resistant Infections: What Might the Future Hold?" *American Journal of Infection Control*, Vol. 34, 2006, pp. 469–475.
Rubinstein, E., et al. "Pneumonia Caused by Methicillin-Resistant *Staphylococcus aureus*." *Clinical Infectious Diseases*, Vol. 46, Supplement 5, 2008, pp. S378–S385.
Sachs, J. "The Superbugs Are Here." *Prevention*, December 2006.
Santamour, B. "As a Superbug Spreads, So Does Misinformation." *Hospitals and Health Networks*, Vol. 82, 2008, pp. 36–40.
Seppälä, H., et al. "The Effect of Changes in the Consumption of Macrolide Antibiotics on Erythromycin Resistance in Group A Streptococci in Finland." *New England Journal of Medicine*, Vol. 337, 1997, pp. 441–446.
Sleator, R. D., and C. Hill. "Battle of the Bugs." *Science*, Vol. 321, 2008, pp. 1294–1295.
Stobbe, M. "Gut Superbug Causing More Illnesses, Deaths." Associated Press, 28 May 2008.
Talbot, G. H., et al. "Bad Bugs Need Drugs: An Update on the Development Pipeline from the Antimicrobial Availability Task Force of the Infectious Diseases Society of America." *Clinical Infectious Diseases*, Vol. 42, 2006, pp. 657–668.
Ward, T. "Spread of MRSA: Past Time for Action." *Medscape Journal of Medicine*, Vol. 10, 2008, p. 32.
"Yeast May Combat Antibiotic-Resistant Pneumonia and Malaria." News Release, Dartmouth-Hitchcock Medical Center, 21 January 2004.

EMERGING DISEASES

Summary of Threat

An *emerging disease* is one that has recently appeared in a population for the first time, or one that previously existed but is rapidly increasing in incidence or geographic range. These diseases tend to be media favorites, because they are unfamiliar to most people and therefore frightening. This section will examine seven examples: SARS, West Nile, Ebola, Lyme disease, "pig strep," the arenaviruses, and diphtheria.

Other Names

Diphtheria is a re-emerging disease that humans have known and dreaded for centuries. Its older names include Bretonneau's disease, angina diphtherica, and diphtheritic malignant angina. The agent was formerly called the Krebs-Loeffler bacillus.

The other six examples are too new or unfamiliar to have acquired many names. SARS (severe acute respiratory syndrome) has no other English name as of 2009. West Nile virus is commonly called WNV. Ebola has four isolates called Ebola Reston, Ebola Sudan, Ebola Zaire, and Ebola Tai; "green monkey disease" refers to the closely related Marburg virus. Lyme disease is often called Lyme borreliosis. Pig strep is human infection with the swine bacterium *Streptococcus suis*. Arenaviruses include the agents of lymphocytic choriomeningitis (LCM), Lassa fever, and other diseases.

Description

The first known cases of severe acute respiratory syndrome (SARS) in 2003 started with a high fever, headache, and cough that rapidly progressed to severe pneumonia with a high fatality rate (10 to 20 percent). Many patients also had diarrhea. The virus (Figure 3.7) spread by contact

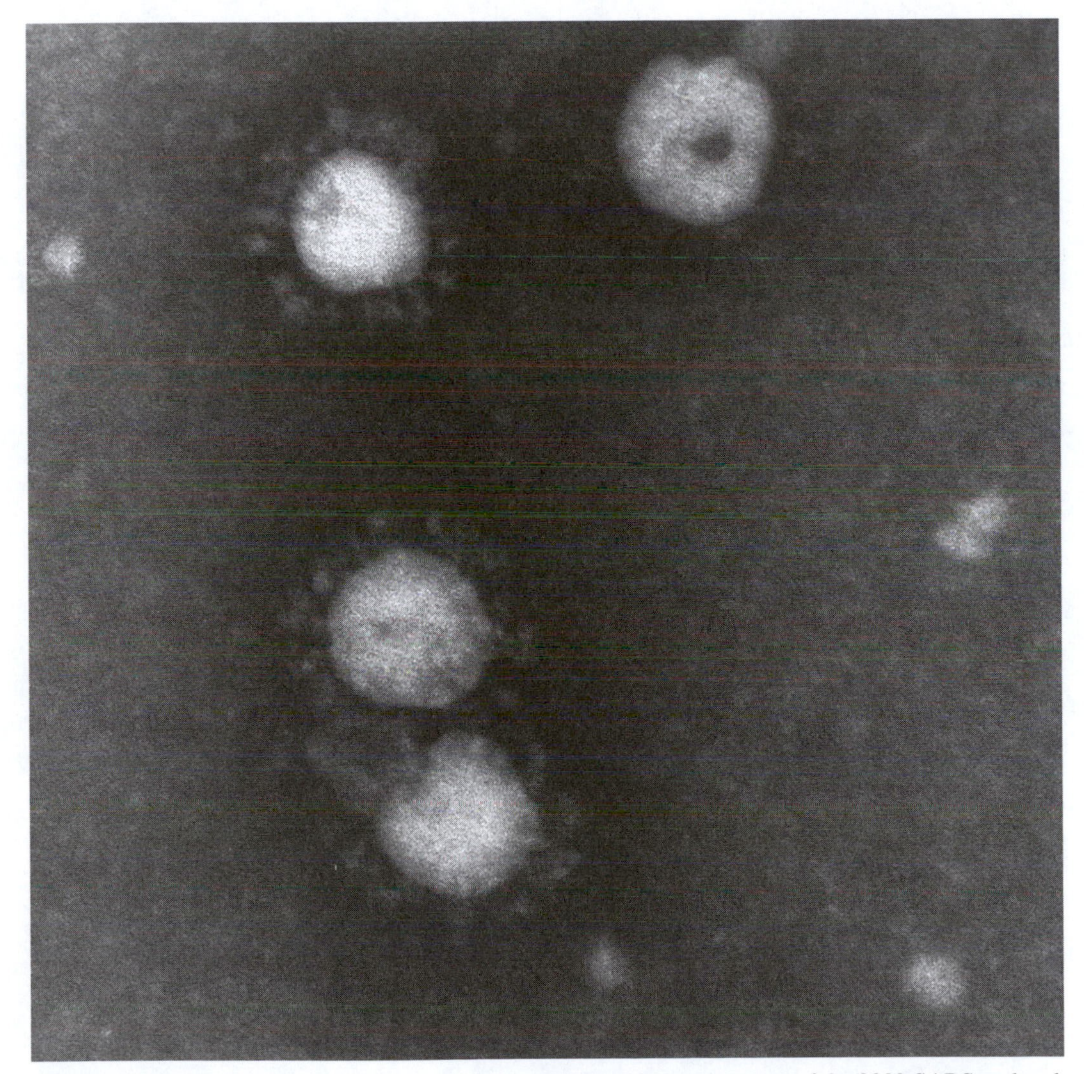

Figure 3.7 Electron micrograph showing a coronavirus identified as the cause of the 2003 SARS outbreak in Asia.

Source: Swedish Institute for Infectious Disease Control.

with contaminated surfaces and also by airborne transmission. A 2004 study determined that aerosol droplets from a flushed toilet transmitted the disease to residents of an apartment building in Hong Kong.

West Nile Virus (WNV) is just one of several forms of mosquito-borne viral encephalitis, and not the most deadly. It is famous because of its abrupt arrival in New York in 1999 and its rapid invasion of all 48 conterminous states. Most infected people have no symptoms or only a mild illness. Severe cases (less than 1 percent) may cause convulsions, paralysis, or coma, but the death rate for hospitalized patients is below 4 percent. Long-term effects, if any, may include polio-like limb weakness. WNV also infects horses, dogs, cats, and a number of wild mammals and birds.

Ebola hemorrhagic fever has been the darling of shock journalists since its 1976 discovery in Zaire (now called the Democratic Republic of the Congo). Symptoms include a high fever, headache, vomiting, and diarrhea, often followed by bleeding from every orifice, multiple organ failure, and death. The agent is a filovirus (Figure 3.8), similar to those that cause some other hemorrhagic fevers. Fruit bats apparently serve as reservoir hosts.

Lyme disease has achieved a high level of public awareness in the United States, although the death rate is near zero. The agent is a spirochete, and the clinical course is comparable to that of syphilis. But Lyme does not spread directly from one person to another; it is transmitted by ticks, specifically the immature stages of certain deer ticks. Early symptoms often include fever, joint pain, and the famous bulls-eye rash (Figure 3.9). Long-term effects of Lyme are controversial.

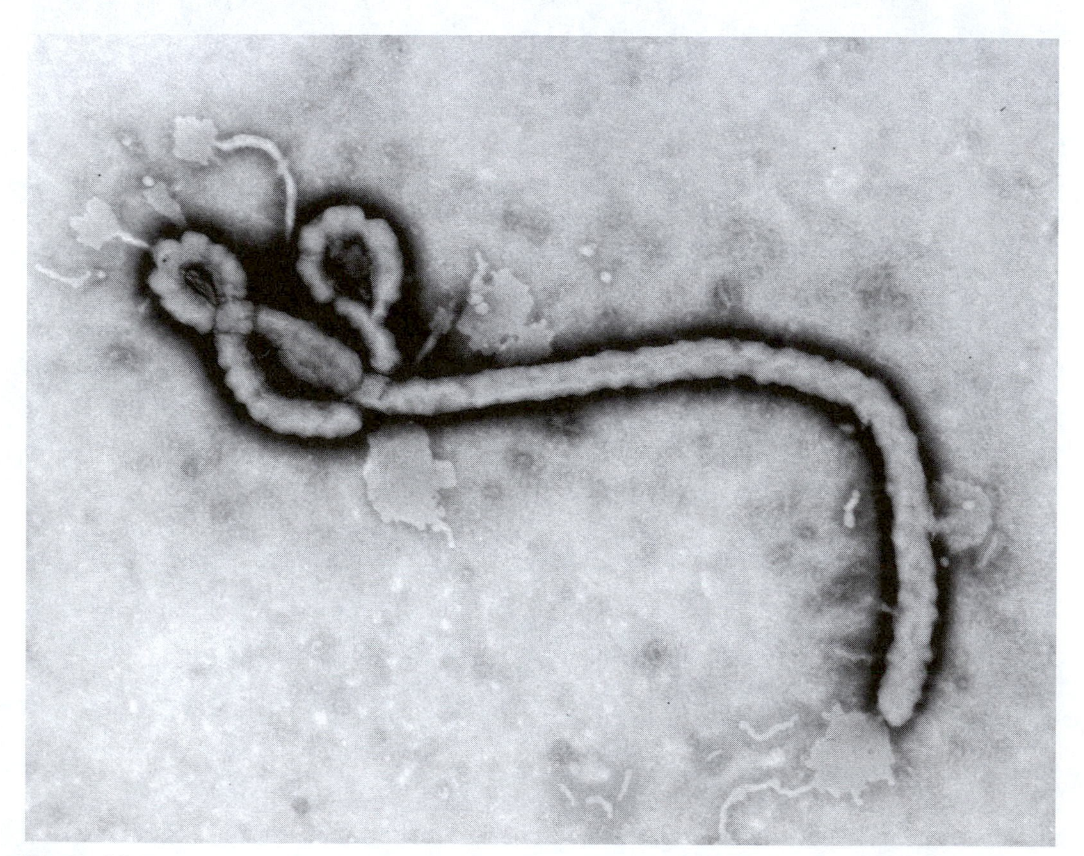

Figure 3.8 Transmission electron micrograph showing the virus that causes Ebola hemorrhagic fever. *Source:* U.S. Centers for Disease Control and Prevention, Public Health Image Library.

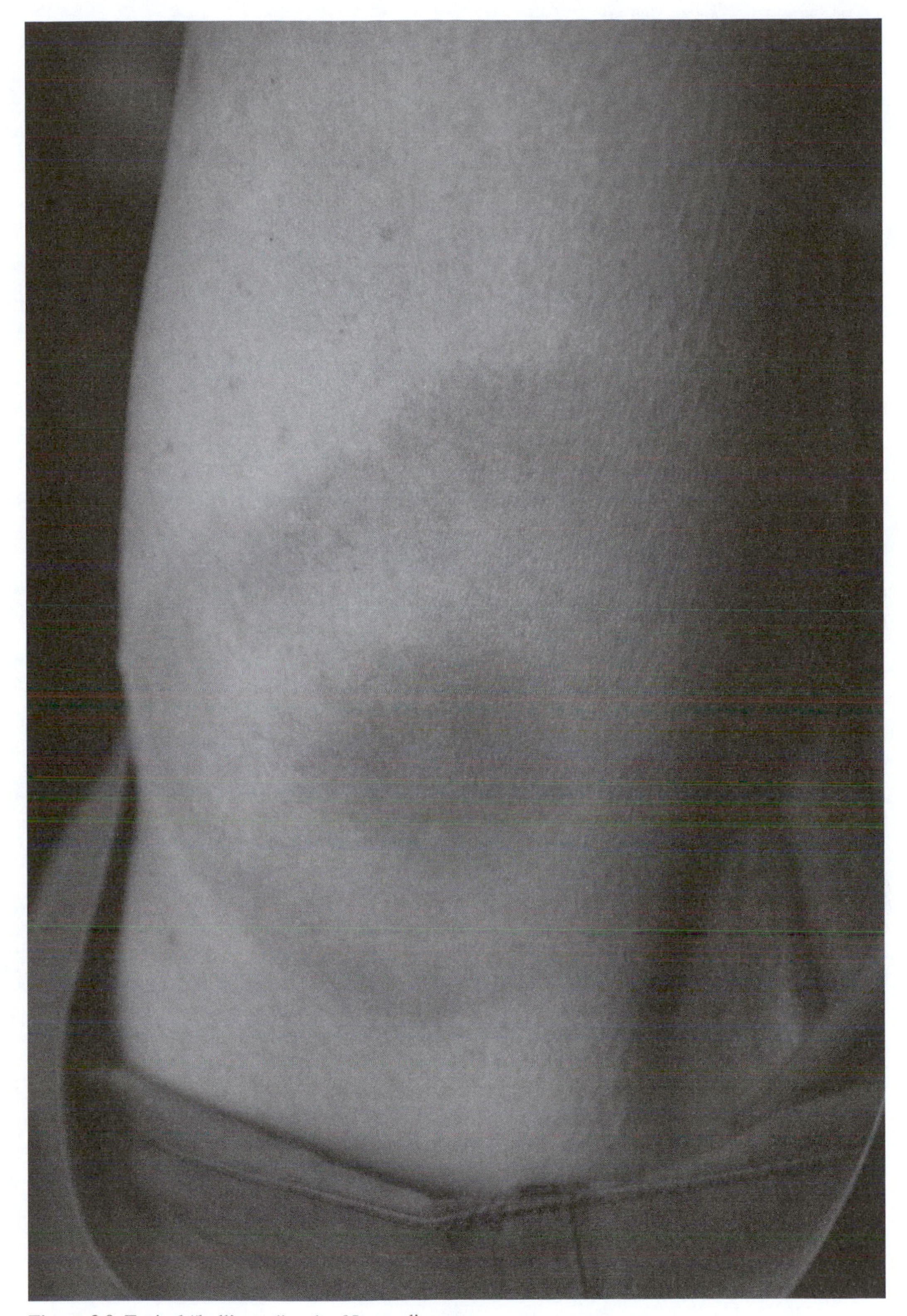

Figure 3.9 Typical "bull's-eye" rash of Lyme disease.

Source: U.S. Centers for Disease Control and Prevention, Public Health Image Library.

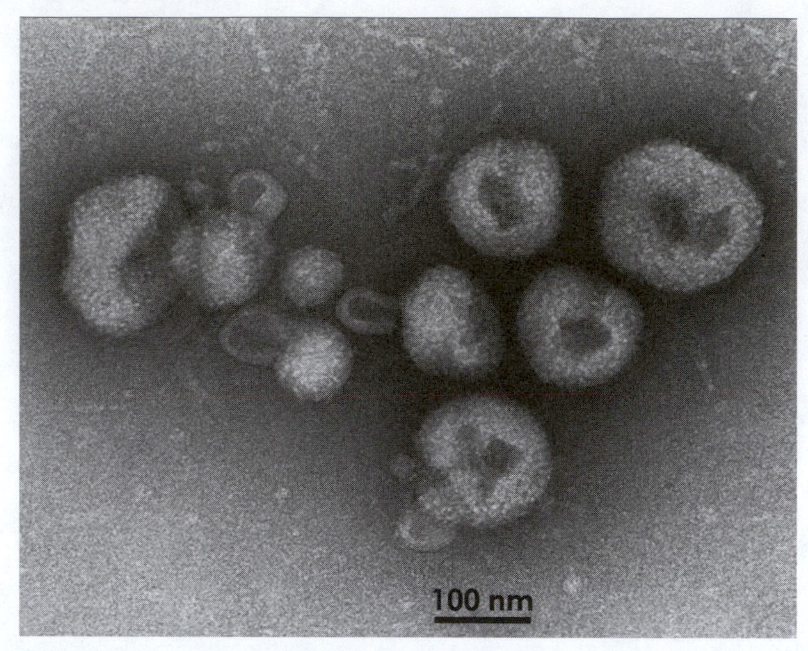

Figure 3.10 Transmission electron micrograph showing a recently discovered arenavirus that can cause fatal hemorrhagic fever in humans.

Source: U.S. Centers for Disease Control and Prevention, Public Health Image Library.

The bacterium *Streptococcus suis* mainly infects pigs and other domesticated animals. In recent years, however, pig strep has caused outbreaks among swine workers in China, southeast Asia, and Europe. Many people in North America are at risk, but only sporadic human cases have been reported. Pig strep can cause hemorrhagic fever, meningitis, septicemia, arthritis, or endocarditis.

Arenaviruses include the agents of Lassa fever, lymphocytic choriomeningitis, Argentine hemorrhagic fever, and several other diseases (Figure 3.10). Most cause the proverbial flu-like symptoms followed by neurological signs; some also cause bleeding, and at least one (LCM) can cause miscarriage or birth defects. Arenaviruses spread directly between humans by airborne droplet, from mother to fetus, or by organ transplantation. Some outbreaks appear to be associated with wild rodents.

There is no nonpolitical reason for anyone to have diphtheria nowadays, because the vaccine has been available for nearly a century, and it is highly effective and cheap. The agent, a bacterium called *Corynebacterium diphtheriae*, produces its deadly toxin only if a phage (a virus that reproduces in bacteria) is present. Diphtheria usually affects the tonsils and throat and can cause death by asphyxiation. It can also infect the skin (Figure 3.11).

Who Is at Risk?

The only known risk factor for SARS is exposure to people with SARS, or possibly exposure to infected fruit bats. As of 2009, no country has reported SARS for the past five years, but it would be a mistake to assume that the disease no longer exists.

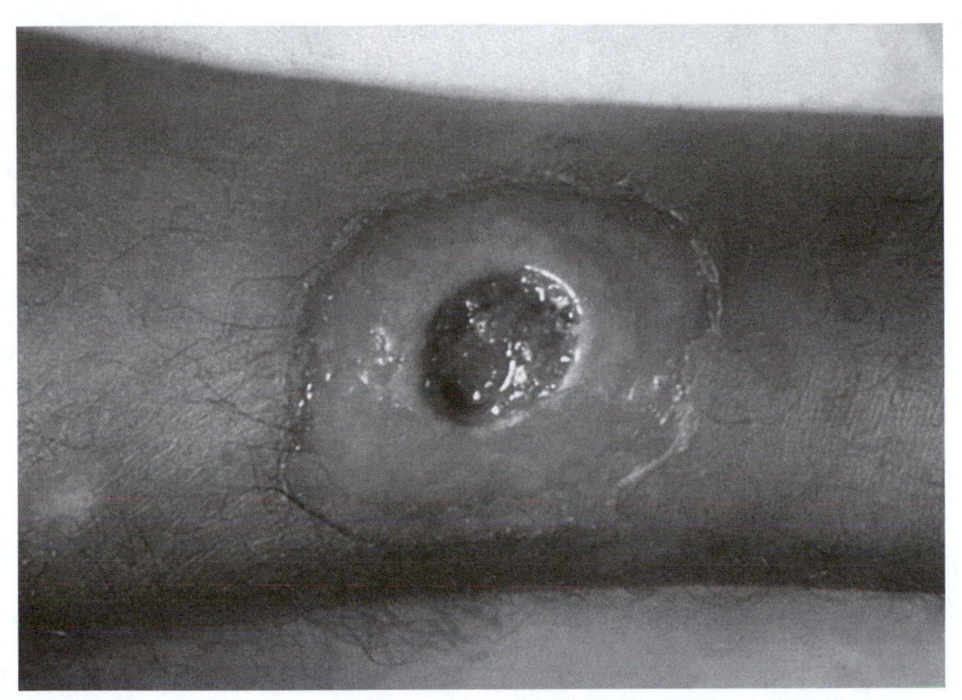

Figure 3.11 A diphtheria skin lesion on a human leg. Once nearly eradicated in most of the world, diphtheria is now a re-emerging disease

Source: U.S. Centers for Disease Control and Prevention, Public Health Image Library.

Mosquitoes transmit West Nile virus, and people who work or live outdoors are at increased risk of being bitten (Case Study 3-9). Risk factors for severe disease include old age, high blood pressure, diabetes, and immune suppression.

Thanks to an overzealous press, many people think they are at risk for Ebola. In fact, it seems to be a difficult disease to catch, even in endemic areas. Visitors to Africa should avoid obvious hazards such as handling dead animals. In 2009, an American visiting Uganda contracted the closely related Marburg virus after touring a cave occupied by fruit bats. (He recovered, and none of his contacts became ill.)

People who do not check themselves carefully for ticks after walking in the woods are most likely to be at risk for Lyme disease. An immature deer tick is extremely small and easy to miss (see Figure 6.6, page 239). Parents should teach children how to check for ticks and how to remove them safely.

Case Study 3-9: West Nile and the Homeless

The arrival of West Nile in North America added one more item to the list of biohazards facing the homeless community. Without window screens (or windows), the homeless cannot avoid exposure to mosquitoes. A 2007 study of nearly 400 homeless people in Houston, Texas, showed that about 7 percent had been infected with West Nile. Those who spent most of their time outdoors due to less stable housing arrangements were at greatest risk, with seroprevalence of about 13 percent. Other studies have confirmed this finding. Unexpectedly, subjects in the 2007 study who used mosquito precautions were *more* likely than others to be infected, possibly because the precautions available to them (such as candles or swatting) were ineffective.

Case Study 3-10: An Alaskan Hero

In the winter of 1925 a diphtheria epidemic was raging in Nome, Alaska. To save the town, a relay team of ten dogsled drivers transported a supply of diphtheria antitoxin 674 miles through blizzards that made air travel impossible. Edgar Nollner drove the last leg of the dog relay and delivered the serum in record time. Mr. Nollner died in 1999 at age 94, survived by 20 children and an estimated 200 grandchildren. A widely held belief—recently repeated on the History Channel—is that the annual Iditarod race commemorates the 1925 serum run. Other sources claim that the purpose of the race (established in 1967) was to commemorate the 1909 Alaskan gold rush or the 1908 All Alaskan Sweepstakes Race. The purpose of today's Iditarod is a matter of conjecture, but 1967 newspaper articles about the first race do not mention the serum run.

The only known risk factor for pig strep is exposure to sick pigs, goats, or other livestock. Breaks in the skin may increase risk. A 2008 study showed that nearly 10 percent of Iowa swine workers were seropositive for pig strep, but none of them had developed clinical illness. As of 2009, there are no reports of direct transmission from one human to another.

As of 2009, the risk factors for arenavirus diseases are poorly understood, but may include contact with the feces or urine of wild or domesticated rodents. Some cases of lymphocytic choriomeningitis have been traced to pet hamsters, house mice, or other rodents. Lassa fever is unknown outside Africa, where the incidence is highest during the dry season. Other arenaviruses occur in North and South America.

Unvaccinated persons are at risk during diphtheria outbreaks, which usually occur either in isolated populations (Case Study 3-10), in groups that oppose vaccination, or in countries where political instability has disrupted the healthcare system. Older people may need booster shots.

The Numbers

According to WHO, the 2003–2004 SARS outbreak caused 8,450 reported cases (from 29 countries) and 810 deaths. Thus, the death rate was nearly 10 percent.

In 2007, the United States had about 35,000 reported cases of West Nile (1,227 of them serious) and 117 reported deaths. The statistics for other years have been similar, but the mortality rate is hard to determine because many mild cases are undetected.

As of 2009, the largest Ebola outbreak to date was in Uganda in 2000–2001, with 425 reported cases and 224 deaths (53 percent). Other outbreaks have ranged in size from a few cases to a few hundred, with mortality rates of 25 to 90 percent.

In 2006, the CDC reported nearly 20,000 new cases of Lyme disease in the United States. The incidence doubled between 1991 (when Lyme became a notifiable disease) and 2007. Some sources attribute the increase to rising deer populations (Chapter 6); others think the numbers are inflated, due to media coverage and inaccurate testing; others claim that Lyme is underreported due to a lack of public awareness.

During the 2005 pig strep outbreak in China, there were 215 human cases and 38 deaths (about 18 percent). The death rate was highest among those who developed toxic shock syndrome.

Case fatality rates for arenavirus infections range from about 5 percent to over 30 percent, depending on the specific virus and the form of the infection.

The former Soviet Union had a major resurgence of diphtheria in 1990–1996, apparently because of the breakdown of healthcare services, with some 150,000 cases and 5,000 deaths. If these numbers are accurate, the death rate was about 3 percent. In 1921, by comparison, the United States had about 200,000 cases of diphtheria, with death rates of 5 percent to 10 percent depending on the age group.

History

A Chinese cook who prepared wild animal dishes may have started the SARS epidemic in 2002, but one hopes that his name will not join Mrs. O'Leary's cow and AIDS "Patient Zero" on history's scapegoat roster. SARS spread from China to other Asian countries and eventually to Canada, causing global panic and various measures that later proved ineffective, such as the wholesale slaughter of civet cats (a suspected reservoir host). But no new cases appeared after April 2004, and as of 2009 the disease has not returned. The rapid identification of the SARS coronavirus, by laboratories in three countries working in cooperation, ranks among the greatest of public health achievements.

First discovered in Uganda in 1937, West Nile made its debut in New York in 1999. Some people blamed its arrival on global warming, others on terrorism; it was more likely the usual story of risks and benefits from international travel. Within a few years, the range of WNV included most of North America and part of South America. Human cases in Mexico have been rare, although Mexican horses have contracted the disease.

The first official Ebola outbreaks occurred in Zaire and Sudan in 1976–1977. A small outbreak in Sudan followed in 1979, but except for a series of highly publicized laboratory accidents (in the Philippines and the United States), Ebola did not resurface until the 1990s. The high mortality rate and dramatic symptoms have made Ebola a greatly feared disease.

European doctors described a disease similar to Lyme before 1900 and associated it with tick bites in 1909. The disease did not receive a name until 1975, after an outbreak of juvenile rheumatoid arthritis in the town of Lyme, Connecticut.

A 2005 disease outbreak in China inspired international rumors of a virulent new strain of bird flu, dengue hemorrhagic fever, or even Ebola. The world had not forgotten SARS. But the hysteria subsided when the agent was identified as the well-known bacterium *Streptococcus suis*. Sporadic human cases had previously occurred in Asia and Europe; the first human cases in North America appeared in 2006.

Researchers studying St. Louis encephalitis in 1933 discovered the first known arenavirus (LCM) by accident. Since then, new arenaviruses have turned up every few years. These viruses tend to cause small outbreaks, often associated with rodents. For example, the Whitewater Arroyo virus caused fatal hemorrhagic fever in three California women in 1999–2000. Woodrats (*Neotoma*) apparently serve as a reservoir host for this arenavirus, but the specific circumstances of transmission are unclear.

During the nineteenth century, diphtheria claimed the lives of several thousand American children every year. The history of diphtheria should have ended when the vaccine became available in the 1920s, but this disease tends to return whenever public health programs falter. The largest recent epidemic was in the former Soviet Union, as previously discussed.

Prevention and Treatment

Doctors in 2003–2004 found no effective treatment for SARS. Steroids and antiviral drugs proved ineffective. The best preventive measures were among the oldest: quarantine and protective masks. According to a 2003 report, healthcare personnel wearing surgical masks or N95 particle respirator masks were 13 times less likely to contract SARS than those without masks.

As of 2009, there is no West Nile vaccine approved for human use in the United States, but there is one for horses. The best preventive measure is to avoid mosquito bites, by wearing mosquito repellent, and by making sure all windows have intact screens.

Experimental Ebola vaccines have been effective in laboratory animals, but as of 2009, treatment of infected human patients focuses on supportive care (see also Case Study 7-15, page 282). Males who recover are advised to avoid sexual relations until semen is free of virus.

GlaxoSmithKline introduced a Lyme disease vaccine in 1999 and pulled it off the market in 2002, citing low demand and poor sales. According to a 2006 *Nature* editorial, the withdrawal of this vaccine was a prime example of "unfounded public fears" blocking vaccine development. Oral antibiotics usually cure Lyme disease, but only if given within a few days after a tick bite, and only if the person really has Lyme.

Chinese officials reportedly stopped the 2005 *Streptococcus suis* outbreak by prohibiting "backyard slaughtering." Workers in commercial slaughterhouses wore protective clothing that was not available to most infected farmers.

Some arenavirus infections are treatable with ribavirin. Pet rodents should not be allowed to hang out with wild rodents. Pregnant women should wear gloves and masks when handling rodents or, better yet, avoid handling them at all.

Doctors usually administer diphtheria vaccine to children in a trivalent vaccine called DTaP (diphtheria and tetanus toxoids and acellular pertussis). An older, less expensive version called DPT is also in use. An unvaccinated person who is exposed to diphtheria should receive diphtheria antitoxin as soon as possible, followed by antibiotics if necessary.

Popular Culture

In the 1995 movie *Outbreak*, an Ebola-like viral disease becomes airborne and infects a town. Such an outbreak might happen, but it's highly unlikely that two Army doctors could whip up an effective antiserum in a few hours and quickly restore people to full health from the end stage of multiple organ failure.

Modern parents who oppose vaccination might wish to read Stewart O'Nan's 1999 novel *A Prayer for the Dying*, which deals with the horrors of a diphtheria epidemic in Wisconsin after the Civil War.

In 2007, a television program aired the theory that the Jersey Devil—a legendary creature of the Mid-Atlantic states—might be a hammer-headed fruit bat that escaped from the Plum Island Animal Disease Center. This large bat is a host for the Ebola virus in Africa, and it would certainly scare the daylights out of any inebriated duck hunter who came upon it in the New Jersey woods. But—(The author was about to divulge the true identity of the Jersey Devil, when three visitors in black suits warned her to keep quiet.)

In his 1842 short story "The Masque of the Red Death," Edgar Allen Poe described a fictitious but seemingly Ebola-like disease:

> The 'Red Death' had long devastated the country. No pestilence had ever been so fatal, or so hideous. Blood was its Avatar and its seal—the redness and the horror of blood. There were sharp pains, and sudden dizziness, and then profuse bleeding at the pores, with dissolution. The scarlet stains upon the body and especially upon the face of the victim, were the pest ban which shut him out from the aid and from the sympathy of his fellow-men. And the whole seizure, progress and termination of the disease, were the incidents of half an hour.[1]

Did Poe foresee Ebola? Or has a similar disease surfaced in the past, leaving a vivid collective memory? Did Ebola, like pizza and the Internet, come along and fill a vacant niche in the human soul? To paraphrase a familiar quotation: if Ebola did not exist, would mankind have felt obliged to invent it? Edgar Allen Poe apparently did.

---

1. E. A. Poe, *Short Stories* (NY: Editions for the Armed Services, 1945).

The Future

According to several infectious disease experts quoted in a 2009 report, the global recession could hasten the spread of exotic diseases by interfering with vector control and surveillance programs. By the time this book is published, perhaps we will know whether they were right or not. The study of global climate change (Chapter 6) is another perennial source of emerging disease warnings.

References and Recommended Reading

Amman, B. R., et al. "Pet Rodents and Fatal Lymphocytic Choriomeningitis in Transplant Patients." *Emerging Infectious Diseases*, Vol. 13, 2007, pp. 719–725.

Badiaga, S., et al. "Preventing and Controlling Emerging and Reemerging Transmissible Diseases in the Homeless." *Emerging Infectious Diseases*, Vol. 14, 2008, pp. 1353–1359.

Bausch, D. G., et al. "Treatment of Marburg and Ebola Hemorrhagic Fevers: A Strategy for Testing New Drugs and Vaccines under Outbreak Conditions." *Antiviral Research*, Vol. 78, 2008, pp. 150–161.

Chastel, C. "Global Threats from Emerging Viral Diseases." *Bulletin de l'Académie Nationale de Médecine*, Vol. 191, 2007, pp. 1563–1577. [French]

Choi, C.Q. "Going to Bat: Natural Reservoir for Emerging Viruses May Be Bats." *Scientific American*, March 2006, pp. 24–24B.

Dong, J., et al. "Emerging Pathogens: Challenges and Successes of Molecular Diagnostics." *Journal of Molecular Diagnostics*, Vol. 10, 2008, pp. 185–197.

Enserink, M. "New Arenavirus Blamed for Recent Deaths in California." *Science*, Vol. 289, 2000, pp. 842–843.

European Centre of Disease Prevention and Control. "Arenaviruses—Factsheet." Updated 14 October 2008.

"First U.S. Case of Marburg Fever Confirmed." Associated Press, 8 February 2009.

Gonzalez, J. P., et al. "Arenaviruses." *Current Topics in Microbiology and Immunology*, Vol. 315, 2007, pp. 253–288.

Gonzalez, J. P., et al. "Ebolavirus and Other Filoviruses." *Current Topics in Microbiology and Immunology*, Vol. 315, 2007, pp. 363–387.

Jamieson, D. J., et al. 2006. "Lymphocytic Choriomeningitis Virus: an Emerging Obstetric Pathogen?" *American Journal of Obstetrics and Gynecology*, Vol. 194, 2006, pp. 1532–1536.

Knobler, S., et al. (Eds.) 2004. *Learning from SARS: Preparing for the Next Disease Outbreak*. Washington, D.C.: National Academies Press.

Le Guenno, B. "Haemorrhagic Fevers and Ecological Perturbations." *Archives of Virology Supplement* 13, 1997, pp. 191–199.

Lee, G. T., et al. "*Streptococcus suis* Meningitis, United States." *Emerging Infectious Diseases*, Vol. 14, 2008, pp. 183–185.

Lemonick, M. D. "A Deadly Mystery." *Time*, 26 April 2007.

Leroy, E. M., et al. "Fruit Bats as Reservoirs of Ebola Virus." *Nature*, Vol. 438, 2005, pp. 575–576.

Lindsey, N. P., et al. "West Nile Virus Activity—United States, 2007." *Morbidity and Mortality Weekly Report*, Vol. 57, 2008, pp. 720–723.

Loeb, M., et al. "Prognosis After West Nile Virus Infection." *Annals of Internal Medicine*, Vol. 149, 2008, pp. 232–241.

Meyer, T. E., et al. "West Nile Virus Infection Among the Homeless, Houston, Texas." *Emerging Infectious Diseases*, Vol. 13, 2007, pp. 1500–1503.

Motavelli, J. "Connecting the Dots: the Emerging Science of Conservation Medicine Links Human and Animal Health with the Environment." *E: The Environmental Magazine*, November–December 2004.

Nakazibwe, C. "Marburg Fever Outbreak Leads Scientists to Suspected Disease Reservoir." *Bulletin of the World Health Organization*, Vol. 85, 2007, pp. 654–656.

Oglesby, C. "West Nile Virus a North American Fixture." CNN, 6 June 2005.

Oldstone, M. B. "A Suspenseful Game of "Hide and Seek" between Virus and Host." *Nature Immunology*, Vol. 8, 2007, pp. 325–327.

Omi, S. 2006. *SARS: How a Global Epidemic Was Stopped*. Geneva: World Health Organization.

Osterhaus, A. D. M. E., et al. "The Aetiology of SARS: Koch's Postulates Fulfilled." *Philosophical Transactions of the Royal Society of London, Series B, Biological Sciences*, Vol. 359, 2004, pp. 1081–1082.

Peters, C. J. "Emerging Infections: Lessons from the Viral Hemorrhagic Fevers." *Transactions of the American Clinical and Climatological Association*, Vol. 117, 2006, pp. 189–197.

Qiu, W.-G., et al. "Wide Distribution of a High-Virulence *Borrelia burgdorferi* Clone in Europe and North America." *Emerging Infectious Diseases*, Vol. 14, 2008, pp. 1097–1104.

"Scientists Develop SARS Vaccine." United Press International, 19 July 2006.

Skovgaard, N. "New Trends in Emerging Pathogens." *International Journal of Food Microbiology*, Vol. 120, 2007, pp. 217–224.

Snelson, H. "PRRS and Ebola Virus Reported in Philippine Pigs." News Release, American Association of Swine Veterinarians, 15 December 2008.

"Study: Bats, Not Civets, Source of SARS." United Press International, 20 February 2008.

Reilley, B., et al. "SARS and Carol Urbani." *New England Journal of Medicine*, Vol. 348, 2003, pp. 1951–1952.

"Russian Scientist Dies of Ebola after Lab Accident." *CIDRAP News*, 25 May 2004.

Schnirring, L. 2008. "Tests Indicate an Arenavirus in South African Deaths." *CIDRAP News*, 13 October 2008.

Shapiro, E. D. 2008. "Lyme Disease." *Advances in Experimental Medicine and Biology*, Vol. 609, 2008, pp. 185–195.

Thomas, R. M. "Edgar Nollner, 94, Dies; Hero in Epidemic." *New York Times*, 24 January 1999.

U.S. Centers for Disease Control and Prevention. "Fact Sheet: Basic Information About SARS," 13 January 2004.

U.S. Centers for Disease Control and Prevention. "Brief Report: Lymphocytic Choriomeningitis Virus Transmitted through Solid Organ Transplantation—Massachusetts, 2008." *Morbidity and Mortality Weekly Report*, Vol. 57, 2008, pp. 799–801.

U.S. Centers for Disease Control and Prevention. "Fact Sheet: Lymphocytic Choriomeningitis" (undated).

University of California, Berkeley. "Arenavirus Infection Linked to Deaths in California." Media release, 4 August 2000.

Wang, L. F., and B. T. Eaton. "Bats, Civets and the Emergence of SARS." *Current Topics in Microbiology and Immunology*, Vol. 315, 2007, pp. 325–344.

Weinstein, R. A. "Planning for Epidemics—the Lessons of SARS." *New England Journal of Medicine*, Vol. 350, 2004, pp. 2332–2334.

Weiss, R. A., and A. J. McMichael. "Social and Environmental Risk Factors in the Emergence of Infectious Diseases." *Nature Medicine*, Vol. 10 (Suppl. 12), 2004, pp. S70–S76.

WHAT ABOUT PNEUMONIA?

Summary of Threat

Pneumonia is a major cause of death worldwide, but it is not a single disease. Many infectious diseases (including most of those described in this book) can result in pneumonia, defined as inflammation and consolidation in one or both lungs. In fatal cases of measles and influenza, the cause of death is often secondary pneumonia. Other diseases and some chemical exposures can cause primary pneumonia.

Other Names

Pneumonia with inflammation of the bronchi is called bronchopneumonia. Lobar pneumonia is bacterial pneumonia that involves only one lobe of a lung. So-called walking pneumonia is a just a mild case of pneumonia, often caused by *Mycoplasma*; but the similarly named wandering pneumonia is an infection that invades successive parts of the lung. Aspiration pneumonia results from inhalation of food, vomit, or other material.

The terms organizing pneumonia and organized pneumonia refer to the presence of fibrous tissue in the alveoli. When the bronchioles are also affected, the condition is bronchiolitis obliterans organizing pneumonia (BOOP). Specific forms of pneumonia bear the names of their agents, such as pneumococcal, chlamydial, or PC (*Pneumocystis carinii*) pneumonia. Older names for pneumonia include winter fever, lung fever, and old man's friend. Most names in other languages are cognates: *niwmonia* in Welsh, *niumoan* in Manx, *pneumonie* in French, *lunginflammation* in Swedish, and so forth.

Description

Many diseases and chemical exposures can cause pneumonia (Table 3.5), which means inflammation of the lungs with consolidation—a condition in which some of the alveolar lung

Table 3.5 Forms of Pneumonia

Name	Agent
Bacterial	
Acinetobacter pneumonia	*Acinetobacter baumannii*
Actinomycosis	*Actinomyces israelii*, others
Branhamellosis	*Moraxella (Branhamella) catarrhalis*
Chlamydial pneumonia	*Chlamydia trachomatis* or *C. pneumoniae*
Chronic pneumonia	*Pseudomonas aeruginosa*, others
E. coli pneumonia	*Escherichia coli*
Friedlander's pneumonia	*Klebsiella pneumoniae*
Haemophilus (Hib) pneumonia	*Haemophilus influenzae*
Inhalation anthrax	*Bacillus anthracis*
Legionellosis	*Legionella pneumophila*, others
Melioidosis	*Pseudomonas pseudomallei*
Meningococcal pneumonia	*Neisseria meningitides*, others
Mycoplasmal pneumonia	*Mycoplasma pneumoniae*
Pasteurellosis	*Pasteurella multocida*
Pertussis pneumonia	*Bordetella pertussis*
Pneumococcal pneumonia	*Streptococcus pneumoniae*
Pneumonic plague	*Yersinia pestis*
Psittacosis	*Chlamydia psittaci*
Pulmonary nocardiosis	*Nocardia asteroides*, others
Q fever (may lead to BOOP)	*Coxiella burnetii*
Serratia pneumonia	*Serratia marcescens*
Staphylococcal pneumonia	*Staphylococcus aureus* (including MRSA)
Tularemic pneumonia	*Francisella tularensis*
Typhoid pneumonia	*Salmonella typhi*
Walking pneumonia	Any mild pneumonia, often *Mycoplasma*
Viral	
Adenovirus pneumonia	Adenovirus 7, others
Cytomegalovirus pneumonia	Human herpesvirus 5
Epstein-Barr virus pneumonia	Epstein-Barr virus

(*Continued*)

Table 3.5 (*Continued*)

Name	Agent
Giant cell pneumonia	Measles (rubeola) morbillivirus
Hantavirus pulmonary syndrome	Sin Nombre virus, others
Herpes pneumonia	Human herpesvirus 1 or others
hMPV pneumonia	Human metapneumovirus
Influenza pneumonia	Influenza A, B viruses
Measles pneumonia	Measles (rubeola) morbillivirus
Parainfluenza pneumonia	Parainfluenza viruses
RSV pneumonia	Respiratory syncytial virus
Severe acute respiratory syndrome	SARS coronavirus
Varicella pneumonia	Human herpesvirus 3
Fungal	
Aspergillosis	*Aspergillus flavus*, *A. fumigatus*
Blastomycosis	*Blastomyces dermatitidis*
Coccidioidomycosis	*Coccidioides immitis*
Cryptococcosis	*Cryptococcus neoformans*
Histoplasmosis	*Histoplasma capsulatum*
Moniliasis (candidiasis)	*Candida albicans*, others
Mucormycosis	Fungi of family Mucoraceae
Paracoccidiomycosis	*Paracoccidioides brasiliensis*
Penicilliosis	*Penicillium marneffei*
Pneumocystosis (may lead to BOOP)	*Pneumocystis carinii*, others
Sporotrichosis	*Sporothrix schenckii*
Parasitic	
Ancylostomiasis	*Ancylostoma duodenale, Necator americanus*
Ascariasis	*Ascaris lumbricoides*
Paragonimiasis	*Paragonimus westermani*
Pulmonary echinococcosis	*Echinococcus granulosis, E. multilocularis*
Pulmonary schistosomiasis	*Schistosoma*, several species
Strongyloidiasis	*Strongyloides stercoralis*
Toxoplasmosis	*Toxoplasma gondii*
Noninfectious	
Acute eosinophilic pneumonia	Cause unknown; drugs or allergens?
Aspiration pneumonia	Inhalation of food, drink, or vomit
Chemical pneumonia	Inhalation of irritants or toxins
Desquamative pneumonia	Cause unknown
Epler's pneumonia	Cause unknown; autoimmune?
Exogenous lipoid pneumonia	Aspiration of oil
Extrinsic pneumonia	Allergic reaction

spaces are filled with blood cells and fibrin. Typical symptoms include a cough, fever, chest pain, and breathing difficulty.

About 40 percent of all human cases of pneumonia result from infection with the bacterium *Streptococcus pneumoniae* (Figure 3.12). Many other cases follow viral diseases such as influenza. Several highly publicized disease outbreaks in recent decades were specific forms of pneumonia, although the media did not call them by that name. Examples include

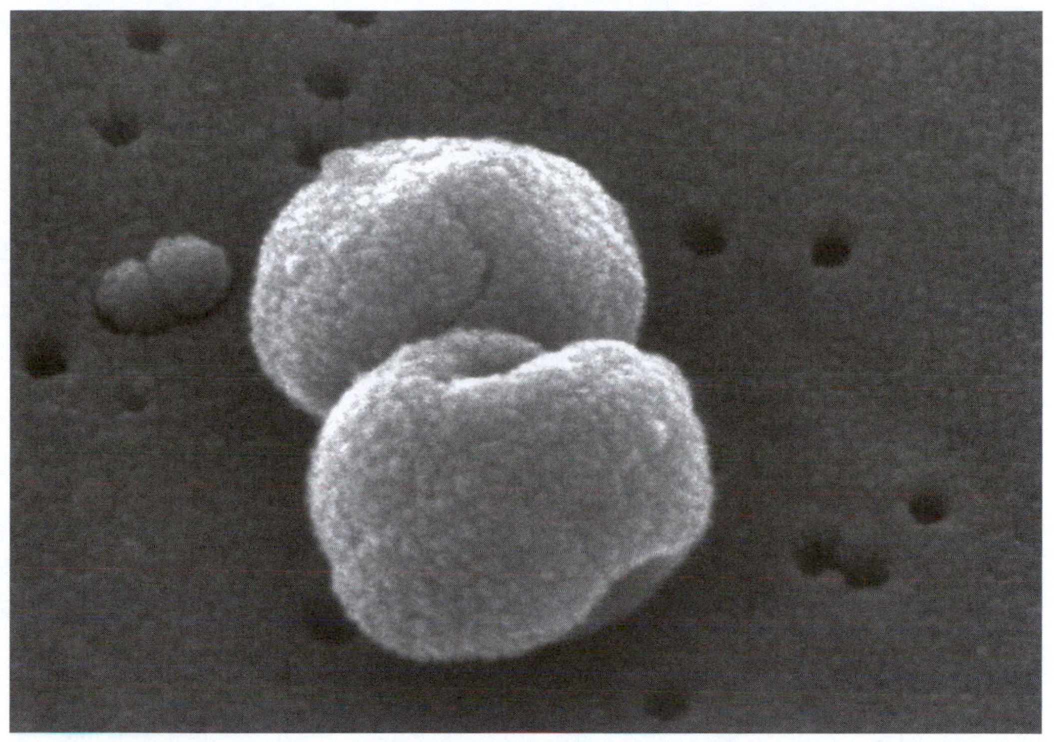

Figure 3.12 Transmission electron micrograph of *Streptococcus pneumoniae*, the most common cause of pneumonia in humans.

Source: U.S. Centers for Disease Control and Prevention, Public Health Image Library.

Legionnaires' disease or legionellosis in 1976, hantavirus pulmonary syndrome in 1993, and SARS in 2003. The 1918 influenza pandemic may have owed its high death toll to secondary bacterial pneumonia.

Who Is at Risk?

Children, elderly people, and those with compromised immune systems are at highest risk for severe pneumonia. Unusual forms of pneumonia often turn up in people with AIDS and organ transplant recipients taking immunosuppressant drugs. An example is *Pneumocystis carinii* (PC) pneumonia, one of the infections that doctors recognize as a "red flag" for possible AIDS or other conditions affecting the immune system (Case Study 3-11). People who work with livestock may also be at risk for some forms of pneumonia (Case Study 3-12).

The Numbers

An estimated 4.3 million people die from pneumonia every year, including more than 60,000 in the United States. In 2008, WHO reported that pneumonia kills more children every year than AIDS, malaria, and measles combined. Even "rare" forms of pneumonia are not really rare; for example, some 25,000 U.S. residents contract legionellosis every year, and about 1,000 of them die.

Case Study 3-11: Déjà Vu

When AIDS made its debut in 1981 and its victims began to die from *Pneumocystis* pneumonia, few people realized that something similar had happened at least once before—an epidemic of this same atypical pneumonia, possibly associated with a retroviral disease and acquired T-cell deficiency. According to a 2005 article, Europe had such an epidemic between 1920 and 1963. It resembled AIDS, except that most victims were premature infants. The authors of the study hypothesized that the infectious agent might be a retrovirus that originated in West Africa and arrived in Germany in the 1920s with returning colonists. Likely routes of spread were similar to those later observed for HIV: sexual contact between adult carriers (who did not become ill), blood transfusions and other hospital procedures in wartime Europe, human donor milk, and placental transmission. For unknown reasons, this tragic epidemic peaked in the 1950s and then ended. Carleton Gajdusek proposed a similar explanation in 1976, the same year he received the Nobel Prize for his work on prions.

Case Study 3-12: Splat!

In 2003 a pregnant ewe gave birth to twin lambs at an outdoor farmer's market in Germany. After the ewe ate the placenta, the market staff covered the parturient fluids with straw but did not otherwise clean the area. Several hundred visitors stood near the pen and apparently inhaled contaminated aerosols. As it turned out, many of the sheep at the exhibit were infected with a bacterium called *Coxiella burnetii*, and nearly 300 people contracted Q fever—an atypical pneumonia also known as Balkan grippe, abattoir fever, coxiellosis, or Nine Mile agent. This was not the largest such outbreak; in Germany in 1954, more than 500 human cases were traced to an infected cow that aborted at a farmer's market.

History

By the time Hippocrates (460–375 B.C.) became the father of modern medicine, pneumonia was an old story. The ancient Egyptians did a lot of digging and rock quarry work when they built the pyramids, and many developed silicosis (lung disease from dust inhalation), which is often associated with pneumonia. Egyptian papyri also refer to "remedies to drive out cough" and "putrefaction of mucus." But even walking upright, living in caves, and building fires can adversely affect the lungs, and a full history of pneumonia would be a history of humankind.

Prevention and Treatment

Proper nutrition, including adequate vitamin D, may help prevent pneumonia and other diseases. To avoid hantavirus pulmonary syndrome, wear a dust mask when sweeping areas that may contain rodent droppings. Water heaters should be set at a high enough temperature to inhibit the bacteria that cause legionellosis (without scalding people or wasting energy). The film of crud that builds up on shower curtains may contain bacteria that cause pneumonia, so it's a good idea to clean or replace them.

The pneumococcus (*Streptococcus pneumoniae*) vaccine protects babies against seven strains of this bacterium. Since the introduction of this vaccine in 2000, rates of infection with these strains have declined by 85 percent, but the other strains have become more common. Another vaccine protects against *Haemophilus influenzae* B (Hib), which also causes pneumonia. Influenza may lead to pneumonia, so annual flu shots are a good idea.

Popular Culture

Everyone has heard that we can "catch pneumonia" by getting chilled, but the truth is more complicated. When cold weather drives people indoors, they are more likely to share respiratory diseases. Also, people feel chills when ambient temperatures are low, but also when a fever is starting. Cold, dry air can cause a runny nose that might be mistaken for a respiratory infection.

According to an old joke, the best way to cure a cold is to get wet and stand in front of an open window until the cold becomes pneumonia, which doctors know how to treat. But becoming chilled doesn't cause pneumonia, and doctors can't always cure pneumonia anyway.

A 2007 study concluded that overweight people are less likely than others to die from pneumonia. Popular discourse translated this finding to mean that obesity confers resistance. In fact, the same study found that obese people are more likely than others to die from diabetes or kidney disease, so they might simply die of something else before pneumonia gets them.

American and European folk remedies for pneumonia include snakeroot (*Aristolochia serpentaria* and others) goldenseal (*Hydrastis canadensis*), peach, elderflower tea, catnip, deer blood mixed with wine, the feces of a white dog, or a poultice containing any of the following: mustard and flaxseed, potato, cabbage, hops, lobelia, onion, cornmeal mixed with onion, or mashed garlic. Other dubious cures include inhaling nettle fumes in a sweat lodge or placing a boiled onion in each armpit.

Traditional Chinese herbal remedies for bacterial pneumonia include concoctions of reed (*Phragmites communis*), Job's-tears (*Coix lacryma-jobi*), honeysuckle (*Lonicera* sp), dandelion (*Taraxacum officinale*), almond (*Prunus dulcis*), and ma huang (*Ephedra sinica*), plus inorganic ingredients such as gypsum powder.

The Future

Air pollution is increasing in many parts of the world, and irritated or damaged lungs are susceptible to infection. The economic recession of 2009 may be another risk factor; unemployed people sometimes turn off their air conditioners during summer smog alerts, burn wood to keep warm in winter, avoid seeing doctors, or accept "dirty" jobs with employers who try to cut corners by not providing adequate personal protective equipment. Malnourished children are also susceptible to pneumonia.

References and Recommended Reading

Bartram, J., et al. (Eds.) "*Legionella* and the Prevention of Legionellosis." World Health Organization, 2007.

Brooks, W. A., et al. "Human Metapneumovirus Infection among Children, Bangladesh." *Emerging Infectious Diseases*, Vol. 13, 2007, pp. 1611–1613.

Cunha, B. A. "The Atypical Pneumonias: Clinical Diagnosis and Importance." *Clinical Microbiology and Infection,* Vol. 12 (Suppl 3), 2006, pp. 12–24.

Fulhorst C. F., et al. "Hantavirus and Arenavirus Antibodies in Persons with Occupational Rodent Exposure." *Emerging Infectious Diseases*, Vol. 13, 2007, pp. 532–538.

Goldman, A. S., et al. "What Caused the Epidemic of *Pneumocystis* Pneumonia in European Premature Infants in the Mid-20th Century?" *Pediatrics*, 2 May 2005.

Greenwood, B. "A Global Action Plan for the Prevention and Control of Pneumonia." *Bulletin of the World Health Organization*, Vol. 86, 2008, pp. 322–323.

Hatfield, G. 2004. *Encyclopedia of Folk Medicine*. Santa Barbara, CA: ABC-CLIO.

Hay, D. "Beware of *Legionella* Bacteria." *Seattle Times*, 3 May 2008.

Heikkinen, T., et al. "Human Metapneumovirus Infections in Children." *Emerging Infectious Diseases*, Vol. 14, 2008, pp. 101–106.

Hsieh, Y. C., et al. "The Transforming *Streptococcus pneumoniae* in the 21st Century." *Chang Gung Medical Journal*, Vol. 31, 2008, pp. 117–124.

Huss, A., et al. "Efficacy of Pneumococcal Vaccination in Adults: a Meta-Analysis." *Canadian Medical Association Journal*, Vol. 180, 2009, pp. 48–58.
Jonsson, C. B., et al. "Treatment of Hantavirus Pulmonary Syndrome." *Antiviral Research*, Vol. 78, 2008, pp. 162–169.
Kelt, D. A., et al. "Threat of Hantavirus Pulmonary Syndrome to Field Biologists Working with Small Mammals." *Emerging Infectious Diseases*, Vol. 13, 2007, pp. 1285–1287.
Klugman, K. P. "Time from Illness Onset to Death, 1918 Influenza and Pneumococcal Pneumonia." *Emerging Infectious Diseases*, Vol. 15, 2009, pp. 346–347.
Levine, J. R., et al. "Occupational Risk of Exposure to Rodent-Borne Hantavirus at US Forest Service Facilities in California." *American Journal of Tropical Medicine and Hygiene*, Vol. 78, 2008, pp. 352–357.
Mangiarotti, P., and E. Pozzi. "Emergence of New Pneumonia: Besides Severe Acute Respiratory Syndrome." *Minerva Medica*, Vol. 97, 2006, pp. 395–409. [Italian]
McQuiston, J. H., et al. "Q Fever." *Journal of the American Veterinary Medical Association*, 15 September 2002.
"Pneumonia Kills More Children than AIDS, Malaria and Measles." Press Trust of India Ltd., 3 May 2008.
Porten, K., et al. "A Super-Spreading Ewe Infects Hundreds with Q Fever at a Farmer's Market in Germany." *BMC Infectious Diseases*, 6 October 2006.
Roth, D. E., et al. "Acute Lower Respiratory Infections in Childhood: Opportunities for Reducing the Global Burden through Nutritional Interventions." *Bulletin of the World Health Organization*, Vol. 86, 2008, pp. 356–364.
Rubinstein, E., et al. "Pneumonia Caused by Methicillin-Resistant *Staphylococcus aureus*." *Clinical Infectious Diseases*, Vol. 46 (Suppl 5), 2008, pp. S378–S385.
Rudan, I., et al. "Epidemiology and Etiology of Childhood Pneumonia." *Bulletin of the World Health Organization*, Vol. 86, 2008, pp. 408–416.
Scott, J. A. G., et al. "Pneumonia Research to Reduce Childhood Mortality in the Developing World." *Journal of Clinical Investigation*, Vol. 118, 2008, pp. 1291–1300.
Straus, W. L., et al. "Risk Factors for Domestic Acquisition of Legionnaires Disease." *Archives of Internal Medicine*, Vol. 156, 1996, pp. 1685–1692.
Zeier, M., et al. "New Ecological Aspects of Hantavirus Infection: A Change of a Paradigm and a Challenge of Prevention—a Review." *Virus Genes*, Vol. 30, 2005, pp. 157–180.
Zuger, A. "'You'll Catch Your Death!' An Old Wives' Tale? Well . . ." *New York Times*, 4 March 2003.

WHAT ABOUT MENINGITIS AND ENCEPHALITIS?

Summary of Threat

Meningitis and encephalitis are dangerous, but neither one is a single disease. Meningitis is inflammation of the membranes that cover the brain and spinal cord, and encephalitis is inflammation of the brain itself. Many diseases, including most of the examples in Chapters 2 and 3, can lead to meningitis or encephalitis as a complication. Many other diseases can cause primary meningitis or encephalitis.

Other Names

Believe it or not, spinal meningitis was once called Simple Smiling Jesus—either because the two phrases have a vaguely similar cadence, or because meningitis victims supposedly grimace and bear their burdens gracefully (particularly if they are in a coma).

Names such as spinal meningitis or cerebral meningitis refer to the part of the central nervous system affected. Aseptic meningitis is caused by a virus rather than a bacterium. Bacterial meningitis was once called spotted fever or purples because of the rash. Names in other languages are mostly cognates, such as *méningite* (French), *menenjit* (Turkish), and *meningjit* (Albanian). The Czech term for meningitis, *zánět mozkových blan,* means literally "cerebral membrane inflammation."

Names for encephalitis include cephalitis, phrenitis, brain fever, swelling of the brain, dropsy of the brain, sleeping sickness, and swamp sickness. Historians describe the form that caused the great epidemic of 1915–1926 (and others) as encephalitis lethargica, von Economo's syndrome, Economo-Cruchet disease, or Redlich's syndrome. Names for encephalitis in other languages include *encefalitis* (Spanish), *ansefalit* (Turkish), and *enkefaliitti* (Finnish). The Indonesian term, *penyakit otak,* means "brain disease."

Description

Like pneumonia, these two terms—meningitis and encephalitis—refer to processes and symptoms rather than specific diseases. As stated earlier, meningitis is inflammation of the membranes that cover the brain and spinal cord, and encephalitis is inflammation of the brain itself. As a general (but not absolute) rule, most serious cases of meningitis are bacterial, and most serious cases of encephalitis are viral. In the United States, the best-known examples are meningococcal meningitis and West Nile encephalitis. Toxic chemicals, injuries, and autoimmune disease can also cause meningitis or encephalitis (Tables 3.6 and 3.7).

Bacterial meningitis usually starts with a high fever, stiff neck, headache, confusion, nausea, and a rash. Even with treatment, many cases result in deafness, blindness, mental retardation, seizures, or even shock and death within as little as 24 hours. Political activist Helen Keller (1880–1968) became blind and deaf as a result of childhood meningitis. Yet most people who carry the meningococcus in their nasal passages do not become sick.

Encephalitis often begins with similar symptoms—fever, neck pain, headache, drowsiness, and nausea. Severe cases may cause convulsions, coma, or paralysis. Some forms of encephalitis,

Table 3.6 Forms of Meningitis

Name	Agent
Bacterial	
Actinomycosis meningitis	*Actinomyces israeli* and related species
Anthrax meningitis	*Bacillus anthracis*
Bacteroides meningitis	*Bacteroides* species
Bartonellosis meningitis	*Bartonella henselae*
Campylobacter meningitis	*Campylobacter jejuni* and related species
Dog bite fever	*Capnocytophaga canimorsus*
Enterobacter meningitis	*Enterobacter sakazakii*
GBS Meningitis	Group B *Streptococcus*
Gonococcal meningitis	*Neisseria gonorrhoeae*
Hib meningitis	*Haemophilus influenzae* serotype B
Klebsiella meningitis	*Klebsiella pneumoniae*
Legionella meningitis	*Legionella pneumophila*, others
Listerial meningitis	*Listeria monocytogenes*
Lyme neuroborreliosis	*Borrelia burgdorferi*

(*Continued*)

Table 3.6 (*Continued*)

Name	Agent
Melioidosis	*Pseudomonas pseudomallei*
Meningococcal meningitis	*Neisseria meningitides*
Mycoplasma meningitis	*Mycoplasma pneumoniae*
Neonatal meningoencephalitis	*Bacillus cereus*, others
Pseudomonas meningitis	*Pseudomonas aeruginosa*
Pneumococcal meningitis	*Streptococcus pneumoniae*
Rhodococcus meningitis	*Rhodococcus equi*
Salmonella meningitis	*Salmonella typhi* or related species
Serratia meningitis	*Serratia marcescens*
Staphylococcal meningitis	*Staphylococcus epidermidis*
Streptococcal meningitis	*Streptococcus pneumoniae, S. suis*, others
Syphilitic meningitis	*Treponema pallidum*
Tuberculous meningitis	*Mycobacterium tuberculosis*
Typhoid meningitis	*Salmonella typhi*
Yersiniosis	*Yersinia enterocolitica*
Viral	
Adenovirus meningitis	Adenovirus
California meningitis	California virus (same as La Crosse)
Colorado tick fever meningitis	Colorado tick fever virus
Congenital rubella meningitis	Rubella virus
Coxsackievirus meningitis	Coxsackievirus B, others
Eastern equine meningitis	Eastern equine encephalitis virus
Echovirus meningitis	Echovirus 13, others
Enterovirus meningitis	Enterovirus 71, others
Hendra virus meningitis	Hendra virus
Herpes meningitis	HHV-1, -2, -3, -4, -5, -6
La Crosse meningitis	La Crosse virus
Lymphocytic choriomeningitis	Lymphocytic choriomeningitis virus
Measles meningitis	Measles (rubeola) morbillivirus
Mumps meningitis	Mumps virus
Poliovirus meningitis	Poliovirus
Rabies meningoencephalitis	Rabies virus
St. Louis meningitis	St. Louis encephalitis virus
TOSV meningitis	Toscana virus
Western equine meningitis	Western equine encephalitis virus
Fungal	
Aspergillosis	*Aspergillus fumigatus* and related species
Blastomycotic meningitis	*Blastomyces dermatitidis*
Cryptococcal meningitis	*Cryptococcus neoformans*
Histoplasmosis meningitis	*Histoplasma capsulatum*
Scedosporiosis	*Scedosporium apiospermum*
Systemic candidiasis	*Candida albicans* and related species
Torulopsis meningitis	*Torulopsis glabrata*
Parasitic	
African trypanosomiasis	*Trypanosoma brucei*
Chagas disease	*Trypanosoma cruzi*
Human myiasis	Fly larvae, various species
Primary amebic meningoencephalitis	*Naegleria fowleri, Balamuthia mandrillaris*

Noninfectious	
Biotinidase deficiency	Genetic
Complement receptor deficiency	Genetic
Hemophagocytic reticulosis	Genetic
Traumatic meningitis	Head injury
Vogt-Koyanagi-Harada syndrome	Cause unknown, possibly infectious

Note: Many pathogens can cause both meningitis and encephalitis (see Table 3.7).

Table 3.7 Forms of Encephalitis

Name	Agent
Bacterial	
Actinomycosis encephalitis	*Actinomyces israeli* and related species
Bacteroides encephalitis	*Bacteroides* species
Bartonellosis encephalitis	*Bartonella henselae*
Campylobacter encephalitis	*Campylobacter jejuni* and related species
Dog bite fever	*Capnocytophaga canimorsus*
E. coli encephalitis	*Escherichia coli*
Enterobacter encephalitis	*Enterobacter sakazakii*
GBS encephalitis	Group B *Streptococcus*
Hib encephalitis	*Haemophilus influenzae* serotype B
Human monocytic ehrlichiosis	*Ehrlichia chaffeensis*
Klebsiella encephalitis	*Klebsiella pneumoniae*
Listeriosis encephalitis	*Listeria monocytogenes*
Lyme encephalitis	*Borrelia burgdorferi*
Meningococcal encephalitis	*Neisseria meningitides*
Mycobacterial encephalitis	*Mycobacterium avium* or *M. intracellulare*
Neonatal meningoencephalitis	*Bacillus cereus*, others
Pneumococcal encephalitis	*Streptococcus pneumoniae*
Pontiac fever encephalitis	*Legionella pneumophila*
Pseudomonas encephalitis	*Pseudomonas aeruginosa*
Staphylococcal encephalitis	*Staphylococcus aureus*
Streptococcal encephalitis	*Streptococcus pneumoniae, S. suis*, others
Syphilis (secondary or congenital)	*Treponema pallidum*
Typhoid encephalitis	*Salmonella typhi*
Yersiniosis	*Yersinia enterocolitica*
Viral	
Adenovirus encephalitis	Adenovirus
Bosin's disease	Measles (rubeola) morbillivirus
California encephalitis	California virus (same as La Crosse)
Central European tick-borne encephalitis	TBE virus
Chandipura virus encephalitis	Chandipura virus
Colorado tick fever encephalitis	Colorado tick fever virus
Cytomegalovirus encephalitis	Cytomegalovirus (HHV-5)
Dawson's encephalitis	Measles (rubeola) morbillivirus
Dengue encephalitis	Dengue 2 and 3 viruses
Eastern equine encephalitis	Eastern equine encephalitis virus
Echovirus encephalitis	Echovirus Type 9 and others
Enterovirus encephalitis	Enterovirus 71, others

(*Continued*)

Table 3.7 *(Continued)*

Name	Agent
Hendra virus encephalitis	Hendra virus
Hepatitis C encephalomyelitis	Hepatitis C virus
Herpes encephalitis	HHV-1, -2, -3, -4, -5, or -6
HIV encephalitis	HIV-1
Human Vilyuisk encephalitis	Vilyuisk human encephalitis virus
Influenza encephalitis	Influenza A and B viruses
Japanese encephalitis	Japanese encephalitis
Kumlinge virus encephalitis	Kumlinge virus
La Crosse encephalitis	La Crosse virus
Louping Ill encephalitis	Louping Ill virus
Lymphocytic choriomeningitis	Lymphocytic choriomeningitis virus
Measles encephalitis	Measles (rubeola) morbillivirus
Mumps encephalitis	Mumps virus
Murray Valley encephalitis	Murray Valley encephalitis virus
Nipah virus encephalitis	Nipah virus
Poliovirus encephalitis	Poliovirus
Powassan encephalitis	Powassan virus
Progressive rubella panencephalitis	Rubella virus
Rabies encephalitis	Rabies virus
St. Louis encephalitis	St. Louis encephalitis virus
Van Bogaert encephalitis	Measles (rubeola) morbillivirus
Varicella-zoster encephalitis	Varicella-zoster virus (HHV-3)
Venezuelan equine encephalitis	Venezuelan equine encephalitis virus
West Nile encephalitis	West Nile virus
Western Equine encephalitis	Western Equine encephalitis virus
Fungal	
Aspergillosis	*Aspergillus fumigatus* and related species
Blastomycotic encephalitis	*Blastomyces dermatitidis*
Chaetomium encephalitis	*Chaetomium atrobrunneum*
Cryptococcal encephalitis	*Cryptococcus neoformans*
Histoplasmosis	*Histoplasma capsulatum*
Systemic candidiasis	*Candida albicans* and related species
Valley fever	*Coccidioides immitis*
Parasitic	
Amebic encephalitis	*Balamuthia mandrillaris*
Amebic encephalitis	*Sappinia diploidea*
Angiostrongyliasis	*Angiostrongylus cantonensis*
Granulomatous amebic encephalitis	*Acanthamoeba keratitis*
Malarial encephalitis	*Plasmodium falciparum* and related species
Schistosomiasis	*Schistosoma japonicum*
Toxoplasmosis	*Toxoplasma gondii*
Trypanosomiasis	Trypanosoma brucei, T. cruzi
Noninfectious	
Acute disseminated encephalitis	Autoimmune disease
Behçet's disease	Autoimmune disease
Cholesteatoma	Genetic or after injury
Familial histiocytic reticulosis	Genetic
Hashimoto's encephalitis	Autoimmune disease?

Kawasaki disease	Cause unknown, possibly infectious
Rasmussen encephalitis	Autoimmune disease?
Systemic lupus erythematosus	Autoimmune disease
Traumatic encephalitis	After head injury

Note: Many pathogens can cause both encephalitis and meningitis (see Table 3.6).

such as rabies and Nipah (Figure 3.13), have a very high death rate. In the United States, the most dangerous viral mosquito-borne encephalitis is probably eastern equine, but there are only about five reported cases each year. In 2007, a visiting Scotsman caught eastern equine encephalitis and fell into a coma, but eventually recovered.

Who Is at Risk?

Everyone is susceptible, but young people in crowded situations are at higher risk for bacterial meningitis, whereas older adults and those exposed to mosquito bites are prime candidates for viral encephalitis. AIDS is also a risk factor, both because the immune system is suppressed and because HIV itself can cause encephalitis. Since measles, mumps, and other diseases can cause encephalitis or meningitis, unvaccinated people are also at risk. For unknown reasons, African Americans are more likely than Caucasians to contract meningitis, but less likely to become deaf as a result.

Having epidural anesthesia, handling live bats, drinking unpasteurized milk, eating inadequately cooked garden slugs, sharing tattoo needles, playing beer pong, swimming in a warm lake without a nose clip, and aspirating polluted water are risk factors for specific (and mostly rare) forms of meningitis or encephalitis.

The Numbers

Meningococcal meningitis often affects college students, military recruits, or other healthy young adults (see Case Study 3-13). There are about 3,000 reported cases every year in the United States (including 100 to 125 on college campuses) and 500 cases per year in New Zealand. The largest recorded outbreak was in western Africa in 1996, with 250,000 reported cases and 25,000 deaths. In most outbreaks, the mortality rate ranges from 4 percent to 40 percent, and 11 to 19 percent of survivors have long-term deficits.

The estimated annual incidence of acute encephalitis is 6 per 100,000 population worldwide. In 2005, Japanese encephalitis killed more than 1,100 people in northern India and Nepal, most of them children. (There is an effective vaccine, but it was not available to most people at the time.)

Case Study 3-13: Going Forth

At least once a year, the press terrifies American families by highlighting a college student's tragic death from meningococcal meningitis. Within 24 hours after contracting this disease, often in a crowded setting such as a college dormitory, a previously healthy young adult is dead. And the debate starts over: Maybe I shouldn't live in a dorm? Look what happened to that other guy. But I can't afford my own apartment. Maybe I should get a shot? But they say it doesn't work half the time. Besides, the shot might make me sick, and midterms are next week. And Mom lost her insurance, and the shot costs $120. Would I rather have a chemistry textbook or a meningitis shot? Let's wait and see if anybody else on campus gets sick.

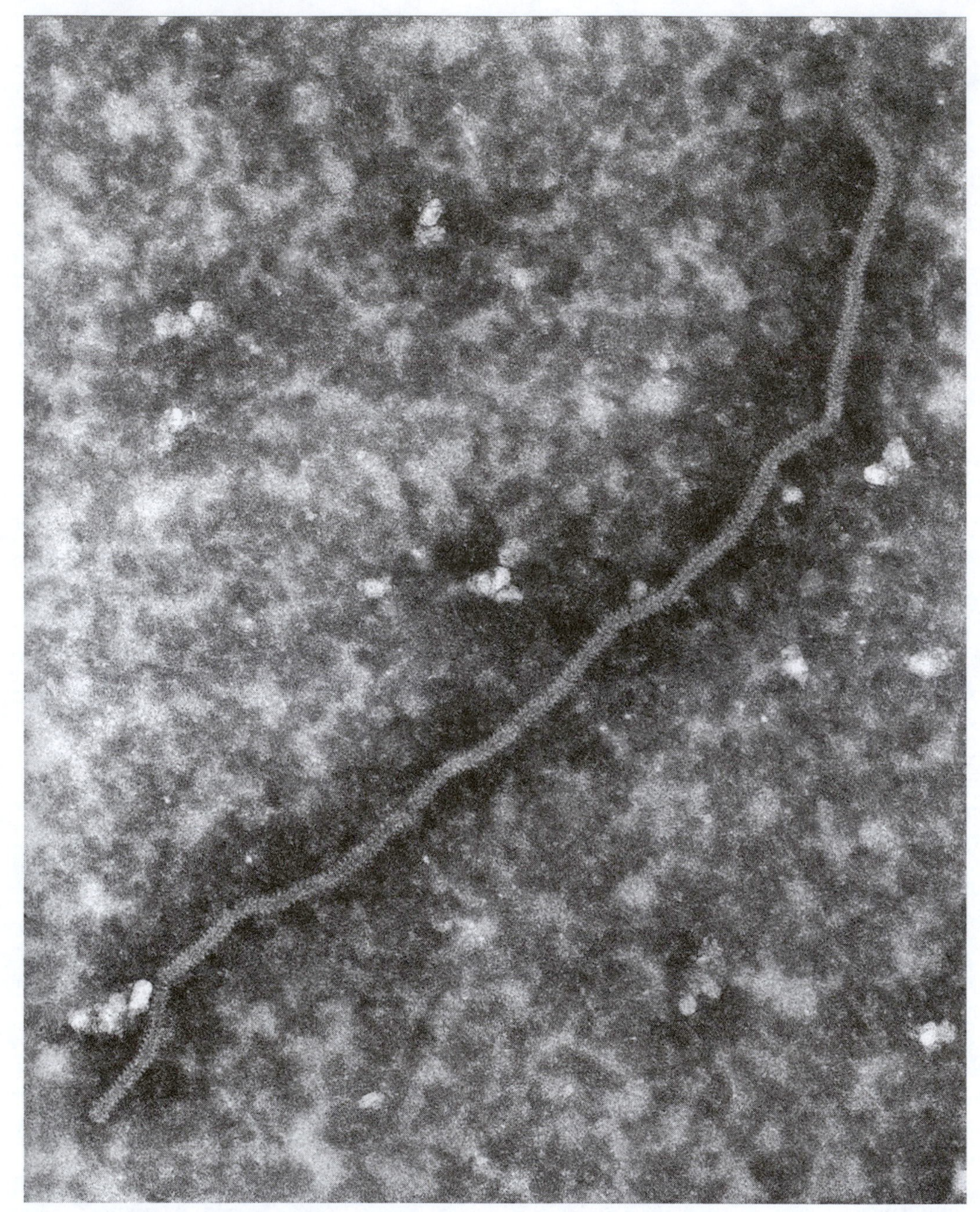

Figure 3.13 Transmission electron micrograph showing Nipah virus, the agent of an emerging disease in humans.

Source: U.S. Centers for Disease Control and Prevention, Public Health Image Library.

In 1998–1999, Nipah encephalitis infected 265 people in Malaysia and killed 105. The death rate was 40 percent, and many survivors had permanent brain damage. Nipah also infected horses, cats, dogs, goats, and pigs. Public health authorities reportedly ended the outbreak by slaughtering nearly 1 million pigs. In 2005, a food-borne Nipah outbreak killed 11 of 12 infected people (92 percent).

History

The Austrian scientist Anton Weichselbaum (1845–1920) discovered the agent of meningococcal meningitis in 1887, and the American physician Simon Flexner (1863–1946) was the first to treat it successfully using an antiserum obtained from horses. The prognosis improved with the discovery of sulfa drugs in the 1930s and penicillin in the 1940s, but mortality remains at about 10 percent.

In the 1990s, Disney World in Florida had to close earlier in the evening than usual due to repeated outbreaks of St. Louis encephalitis (which is spread by mosquitoes that are active after dark). In 1990, Florida had 184 cases and seven deaths. The problem might have resulted from the reduction of vector control programs.

Prevention and Treatment

The most common causes of encephalitis are probably the herpes simplex virus, which is hard to avoid, and the varicella-zoster virus, which lurks in most people who have had chickenpox. Bacterial meningitis is also hard to avoid, because people usually catch it by contact with one another. An element of luck is also involved; for every young adult who contracts fatal meningitis, thousands of others are infected without ever having symptoms.

As of 2009, the most widely used meningococcus vaccine protects against only four of the seven most common *Neisseria meningitidis* serogroups (A, C, Y, and W-135). This vaccine does not prevent infection with serogroup B, which causes nearly half of all cases in the United States and more than half of all fatal cases in Sweden. The vaccine is still better than nothing. In 2009, human vaccines are also available for Japanese encephalitis and Venezuelan equine encephalitis, but not for St. Louis, La Crosse, eastern equine, western equine, or West Nile encephalitis.

Popular Culture

Oliver Sacks' 1973 book *Awakenings*, and the 1990 motion picture, explore the lives of real people who woke from a semi-comatose state some fifty years after contracting encephalitis lethargica in the 1915–1930 epidemic (Case Study 3-14).

Case Study 3-14: Waking Up

Between 1915 and 1930, a mysterious pandemic known popularly as "sleepy sickness" caused thousands of people to lapse into a semiconscious state with a masklike facial appearance. About 20 to 40 percent of these patients died; some recovered, but later developed neurological problems; others simply remained semiconscious, in some cases for decades, until doctors awakened them (temporarily) with drugs used to treat Parkinson's disease. Similar outbreaks occurred in Italy in 1889–1890 and in Iceland in 1948–1949. The agent was never identified, and some sources claim it no longer exists, but sporadic cases continue to turn up, sometimes after a disease such as herpes zoster, bartonellosis, or a *Streptococcus* infection.

North American and European folk remedies for "brain fever" (encephalitis) include ginger and boneset tea. Chinese herbal remedies for meningitis include shepherd's-purse (*Capsella bursa-pastoris*), Japanese holly fern (*Cyrtomium fortunei*), mulberry leaves (*Morus* sp.), weeping forsythia (*Forsythia suspensa*), mint, licorice, almonds, honeysuckle flowers (*Lonicera* sp.), or bamboo shoots; or, if high fever is present, a concoction of rhinoceros or buffalo horn with figwort, sage, Chinese peony, and other herbs. For mosquito-borne encephalitis, treatments include purple giant hyssop (*Agastache rugosa*), Persian shield (*Strobilanthes* sp.), gardenia (*Gardenia florida*), aster (*Aster trinervius*), or border grass (*Liriope* sp.). Rhinoceros or buffalo horn was added to the mix in severe cases.

The Future

In 2007, pharmaceutical companies in the United Kingdom and India announced progress in developing a vaccine that would protect against all five serogroups of meningococcal meningitis. As of 2009, no such vaccine is yet available. The world also needs a meningococcal vaccine for infants and toddlers and a human vaccine for West Nile.

References and Recommended Reading

Bulakbasi, N., and M. Kocaoglu. "Central Nervous System Infections of Herpesvirus Family." *Neuroimaging Clinics of North America*, Vol. 18, 2008, pp. 53–84.

Byrd, T. F., and L. E. Davis. "Multidrug-Resistant Tuberculous Meningitis." *Current Neurology and Neuroscience Reports*, Vol. 7, 2007, pp. 470–475.

Chua, K. B., et al. "Nipah Virus: A Recently Emergent Deadly Paramyxovirus." *Science*, Vol. 288, 2000, pp. 1432–1435.

Czermak, M., and T. Jean. "Von Economo-Cruchet Lethargic Encephalitis and its Relation to HIV Infection." *L'Encéphale*, Vol. 16, 1990, pp. 375–82. [French.]

Dourmashkin, R. R. "What Caused the 1918–30 Epidemic of Encephalitis Lethargica?" *Journal of the Royal Society of Medicine*, Vol. 90, 1997, pp. 515–520.

Ewald, A. J., and D. B. McKeag. "Meningitis in the Athlete." *Current Sports Medicine Reports*, Vol. 7, 2008, pp. 22–27.

Fitch, M. T., et al. "Emergency Department Management of Meningitis and Encephalitis." *Infectious Disease Clinics of North America*, Vol. 22, 2008, pp. 33–52.

Garrett, L. A. "Complacency Boosts West Nile Peril." *Los Angeles Times*, 6 September 2004.

Gould, E. A., and T. Solomon. "Pathogenic Flaviviruses." *Lancet*, Vol. 371, 2008, pp. 500–509.

Gurley, E. S., et al. "Person-to-Person Transmission of Nipah Virus in a Bangladeshi Community." *Emerging Infectious Diseases*, Vol. 13, 2007, pp. 1031–1037.

Halperin, J. J. "Nervous System Lyme Disease." *Infectious Disease Clinics of North America*, Vol. 22, 2008, pp. 261–274.

Hviid, A., et al. "Mumps." *Lancet*, Vol. 371, 2008, pp. 932–944.

Jmor, F., et al. "The Incidence of Acute Encephalitis Syndrome in Western Industrialized and Tropical Countries." *Virology Journal*, October 2008.

Katragkou, A., et al. "*Scedosporium apiospermum* Infection after Near-Drowning." *Mycoses*, Vol. 50, 2007, pp. 412–421.

Keynan, Y., and E. Rubinstein. "The Changing Face of *Klebsiella pneumoniae* Infections in the Community." *International Journal of Antimicrobial Agents*, Vol. 30, 2007, pp. 385–389.

Lo, M. K., and P. A. Rota. "The Emergence of Nipah Virus, a Highly Pathogenic Paramyxovirus." *Journal of Clinical Virology*, Vol. 43, 2008, pp. 396–400.

Luby, S. P., et al. "Foodborne Transmission of Nipah Virus, Bangladesh." *Emerging Infectious Diseases*, Vol. 12, 2006, pp. 1888–1894.

McCall, S., et al. "The Relationship Between Encephalitis Lethargica and Influenza: A Critical Analysis." *Journal of Neurovirology*, Vol. 14, 2008, pp. 177–185.

Pace, D., and A. J. Pollard. "Meningococcal A, C, Y and W-135 Polysaccharide-Protein Conjugate Vaccines." *Archives of Disease in Childhood*, Vol. 92, 2007, pp. 909–915.

Salleh, A. "Man's Brain Infected by Eating Slugs." ABC Science Online, 20 October 2003.

Schneider, J. I. "Rapid Infectious Killers." *Emergency Medicine Clinics of North America*, Vol. 22, 2004, pp. 1099–1115.

Schut, E. S., et al. "Community-Acquired Bacterial Meningitis in Adults." *Practical Neurology*, Vol. 8, 2008, pp. 8–23.

Seijvar, J. J. "The Long-Term Outcomes of Human West Nile Virus Infection." *Clinical Infectious Diseases*, Vol. 44, 2007, pp. 1617–1624.

Smith, T. C., et al. "Exposure to *Streptococcus suis* Among U.S. Swine Workers." *Emerging Infectious Diseases*, Vol. 14, 2008, pp. 1925–1927.

Stone, M. J., and C. P. Hawkins. "A Medical Overview of Encephalitis." *Neuropsychological Rehabilitation*, Vol. 17, 2007, pp. 429–449.

Teyssou, R., and E. Muros-Le Rouzic. "Meningitis Epidemics in Africa: A Brief Overview." *Vaccine*, Vol. 25 (Suppl. 1), pp. A3–A7.

Weingartl, H. M., et al. "Recombinant Nipah Virus Vaccines Protect Pigs Against Challenge." *Journal of Virology*, Vol. 80, 2006, pp. 7929–7938.

Woodard, J. L., and D. M. Berman. "Prevention of Meningococcal Disease." *Fetal and Pediatric Pathology*, Vol. 25, 2006, pp. 311–319.

Yu, H., et al. "Human *Streptococcus suis* Outbreak, Sichuan, China." *Emerging Infectious Diseases*, Vol. 12, 2006, pp. 914–920.

CONCLUSION

We aren't quite finished yet. These ten direct biological threats, and thousands more, occupy an ever-changing world that largely determines the level of risk. So what sort of world do these threats share with us? What can humans do about indirect biological threats that destroy food and other key resources? And are there any global trends in progress that might increase or decrease the associated risk?

4

Food Insecurity

> I am aware of some of the tragic repercussions of the chemical fight against insects taking place in France and elsewhere and I deplore them. Modern man no longer knows how to foresee and forestall. He will end by destroying the earth from which he and other living creatures draw their food. Poor bees, poor birds, poor men.
>
> —Albert Schweitzer, 1956 letter to a French beekeeper

"Food insecurity" is a modern euphemism for hunger or the reasonable expectation of hunger. As of 2009, it affects billions of people.

Chapters 2 and 3 describe infectious diseases that directly threaten humans, but the story doesn't end there. Hundreds of diseases and pests team up with weather and mismanagement every year to attack the human food supply (Table 4.1), more successfully in some regions than others. Starving people become highly vulnerable to infectious diseases, such as tuberculosis, malaria, and pneumonia.

This chapter describes some biological threats to livestock species that represent an important component of the human food supply, including cattle, sheep, pigs, chickens—and bees.

WHAT ABOUT BEES?

In recent years, the Internet has buzzed with a quotation attributed to Albert Einstein: "If the bee disappeared then man would have only four years to live. No more bees, no more pollination, no more plants, no more animals, no more men." The statement is nonsense, and there is no evidence that Einstein said it; but the warning persists, often in conjunction with true reports of honeybee disease epidemics.

The most likely source of this urban legend is a vaguely similar statement by Albert *Schweitzer*, not Einstein (quoted at the top of this page). Although written in 1956, this warning first reached a large audience at a Schweitzer symposium in 1992. By 1994, the quotation attributed to Einstein appeared in a pamphlet distributed by the National Union of French Apiculture. In other words, it appears that somebody got the names mixed up.

Table 4.1 Some Outbreaks of Animal Diseases

Disease	Year	Location	Loss or Cost
Anthrax	1979–1980	Zimbabwe	Many cattle, 182 humans
Anthrax	2004	Zimbabwe	Thousands of wild mammals
Anthrax	2006	Saskatchewan	804 farm animals
Rinderpest	1600s	Europe	200 million cattle
Rinderpest	1890s	South Africa	90% of all cattle
Rinderpest	1983	Nigeria	1 million cattle
Mad Cow/BSE	1986–1996	England	179,000 cattle*
Foot-and-mouth	2001	England	4 million cattle
Heartwater	1998	Eastern Cape	214 million Rand
Classical swine fever	1997–1998	Netherlands	11 million pigs
Blue-ear pig disease	2007	China	1 million pigs
Newcastle disease	1971–1974	Southern California	12 million chickens
Newcastle disease	1973	Northern Ireland	260,000 chickens
Avian influenza	1983–1984	Northeastern United States	17 million chickens
Varroa bee mite	1985–1995	North America	95% of wild honeybees

*Plus an estimated 4.4 million cattle slaughtered as a control measure, and 140+ human deaths from vCJD.

Neither Schweitzer nor Einstein mentioned a four-year time frame, but many sources state that the loss of all bees could reduce food production by 25 to 33 percent. Might someone have interpreted the lower number to mean that one-fourth of humanity (including one-fourth of the farmers) would starve every year, and in four years we would all be gone? This is the type of goofy reasoning that underlies many pseudoscientific claims. Besides ignoring the rules of exponential decay, the scenario overlooks obvious alternatives, such as planting more crops that are not dependent on bees, using artificial pollination, or simply sharing food. Most adults in developed nations could survive a 25 percent reduction in daily caloric intake. (In 2003, according to the United Nations Food and Agriculture Organization, the average American adult consumed 3,770 calories per day and needed about 2,200.)

Even if mankind is not facing imminent extinction, however, many crops do need bees, and we will return to this topic. But since bees are seldom uppermost in the public consciousness, we will start with a scary disease that everyone has read about.

MAD COW DISEASE

Summary of Threat

Bovine spongiform encephalopathy (BSE), or mad cow disease, is a poorly understood, fatal illness that apparently results from a change in the shape of certain proteins in the brain. Similar diseases occur in other mammals, including humans, and the infectious agents may cross species boundaries in food or blood products. BSE is classified as a biosecurity threat under the Bioterrorism Protection Act of 2002.

Other Names

Similar diseases in other mammals include scrapie in sheep and goats; chronic wasting disease (CWD) in deer, elk, and moose; transmissible mink encephalopathy (TME) in farmed mink;

feline spongiform encephalopathy (FSE) in cats; an unnamed prion disease in squirrels; and Creutzfeldt-Jakob disease (CJD) and kuru in humans. As a group, these diseases are called transmissible spongiform encephalopathies (TSE). When people contract a TSE from contaminated meat, the result is called new variant Creutzfeldt-Jakob disease (vCJD or nvCJD). Mad cow disease is *la vaca loca* ("crazy cow") in Spanish, *gekke-koeienziekte* ("cow madness") in Dutch, and *maladie des vaches folles* ("disease of crazy cows") in French.

Description

BSE usually begins with an unsteady, trembling cow that loses its appetite and generally seems out of sorts. The cow may lick its nose, grind its teeth, or stand around with its head down (Figure 4.1). Eventually, the animal must be euthanized.

BSE and related diseases are associated with abnormal membrane proteins called prions, which appear to transmit disease by inducing normal proteins to fold incorrectly. As of 2009, this is the most widely accepted explanation, but some researchers have proposed that prions are not the whole story. One theory is that small bacteria called spiroplasma are the underlying cause. Others claim that pesticides, trace metals, or toxins in animal feed are contributing factors.

Figure 4.1 A "mad cow," later found to have bovine spongiform encephalopathy. The animal showed abnormal posture, weight loss, and other symptoms of this disease.

Source: USDA Animal and Plant Health Inspection Service (APHIS).

Case Study 4-1: Mad Squirrels

Despite their popular image, wild squirrels and chipmunks often eat carrion or even kill live prey. As a result, they are exposed to the same biological threats as any other predator, including prion diseases. A 1997 paper in the journal *Lancet* described five patients who developed the neurological disease CJD after a history of eating squirrel brains, a culinary tradition in rural Kentucky. The authors of the paper reported that the brains are often scrambled with eggs or else cooked in a meat and vegetable dish known as burgoo.

Which Animals Are at Risk?

Animals that eat infected meat or bone meal are the main risk group. Infected cows may also transmit the disease to their calves through the placenta. Studies suggest that a genetic mutation can increase the risk of BSE.

If prions cause TSE, these diseases should appear only in species that eat meat or its byproducts (and those that undergo organ transplants or blood transfusions). Mink, cats, and humans are meat eaters, and until recently, mass-produced cattle and sheep were often fed recycled carrion. Wild squirrels also eat meat on a fairly regular basis (Case Study 4-1), and even wild deer have been observed munching on the carcasses of dead deer.

The Numbers

As of 2009, about 300 people had died of variant Creutzfeldt-Jakob disease (Figure 4.2), including three in the United States. Most probably ate BSE-contaminated meat, but at least four contracted vCJD from transfusions. These numbers represent about 1 percent of the people who die of "normal" CJD during the same time period.

At the height of the BSE epidemic in England, an estimated 179,000 cattle were infected, and 4.4 million were slaughtered to stop the epidemic. Since about 470,000 infected cattle had already entered the human food chain, no one can predict how many more human cases will appear.

History

In 1997, Dr. Stanley B. Prusiner won a Nobel Prize for his discovery of prions. A generation earlier, in 1976, Dr. Daniel Carleton Gajdusek won a Nobel Prize for determining that an infectious agent similar to a slow virus caused both kuru and scrapie, but the exact nature of that agent was Prusiner's contribution.

On 19 September 1985, doctors in England examined Cow Number 133 (the first "mad cow") and found the brain lesions now known as spongiform encephalopathy. The British government acknowledged the outbreak in 1986, and it peaked in 1992–1993. Yet there is evidence that a similar disease has existed since ancient times. The author's 2002 book cites a nineteenth-century account of abnormal behavior in Irish cows. In 2007, the journal *New Scientist* reported that the Roman writer Publius Flavius Vegetius Renatus described a similar cattle disease in the fifth century A.D.

About ten years after the 1986 BSE outbreak, England suffered an outbreak of a similar neurological disease (CJD) in humans. First described in 1920, CJD was usually sporadic. Thus, the new disease was called variant or new variant CJD.

The original source of the 1986 outbreak is unknown, but it might have been an infected antelope that died at a safari park and was made into meat and bone meal. Since about 1926, farmers had recycled animal remains into livestock food, but this practice was officially discontinued after scientists suspected a link to BSE. Feeding dead animals to cows might seem strange, but the rendering process usually kills bacteria or other pathogens. Prions are harder to kill, because they are not alive.

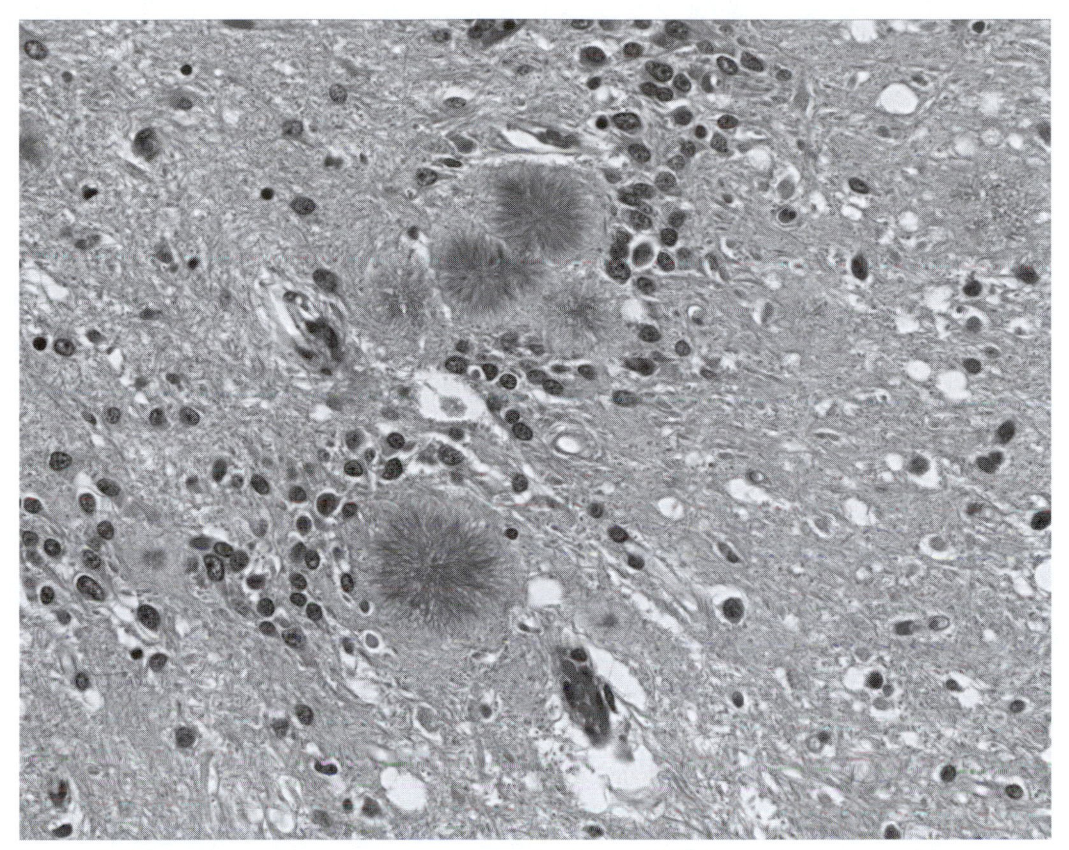

Figure 4.2 Light micrograph of human brain tissue showing amyloid plaques found in variant Creutzfeldt-Jakob disease, attributed to the prion that causes bovine spongiform encephalopathy (BSE) in cattle.

Source: U.S. Centers for Disease Control and Prevention, Public Health Image Library.

Another theory is that cows caught the disease from humans. During the first half of the twentieth century, British cows allegedly ingested human remains that were imported from India as a component of animal feed.

Prevention and Treatment

There is no known treatment for BSE in cattle or vCJD in humans. If future studies determine that TSE diseases require spiroplasma infection and do not result solely from prions, treatment with bactericidal agents may be an option. Meanwhile, the only known preventive measure is to avoid the use of mammalian proteins in farm animal food. Special decontamination procedures are necessary to ensure that surgical instruments are free of prions. In 2008, Canadian researchers reported a new urine test that may identify BSE biomarkers in living cattle.

In 2007, a Japanese company announced the creation of genetically engineered cows with no prion proteins, normal or otherwise. This development is promising, but it is too recent to evaluate. Since prion proteins apparently occur in all unmodified mammals, one wonders if their absence might cause unexpected problems. Recent studies suggest that normal prion proteins may be involved in olfaction, memory, and other brain functions.

Popular Culture

Prion diseases are perfect candidates for rumor and misconception. They are associated with hamburgers; they can make people go crazy and die; and they are as poorly understood today as bacterial and viral infections were a century ago.

In the 2006 motion picture *Mad Cowgirl*, the chain-smoking, hard-drinking female protagonist—a slaughterhouse health inspector whose brother operates a meatpacking business—is so worried about contracting vCJD that she eats large quantities of rare beef, jumps in the sack with a series of disturbingly creepy men, fantasizes about extreme violence, and ends up losing her mind anyway. It turns out that she has a brain tumor, or perhaps a pseudotumor or a fantasy about a pseudotumor. One message is clear: people can destroy themselves without any help from exotic diseases.

A 2004 book claims that Alzheimer's disease is the result of a secret government study of kuru and BSE, as evidenced by decades of widely publicized cattle mutilations (which others have attributed to everything from space aliens to the legendary predator known as Chupacabra). The book claims that these mutilations are proof of an illegal sampling program. In fact, Alzheimer's disease could be related to prions, but biologists who have examined the remains of allegedly mutilated cattle have come away singularly unimpressed.

In the spring of 2008, South Korea decided to resume its imports of U.S. beef. In an effort to dispel public health fears of BSE after months of street protests, a group of South Korean doctors and business executives ate American sirloin steak at a highly publicized banquet. To our knowledge, none of them got sick.

The Future

Every time a mad cow turns up, the host nation's beef is suspect, and the beef industry suffers. The beef byproducts industry is another victim, because meat-and-bone meal is no longer acceptable as animal feed. Since not even the biodiesel reaction can destroy its infectivity, this material is largely wasted. Worse, the inability of farmers to sell animal carcasses may encourage illegal disposal. Improved test methods may resolve this problem, and one of the alternative theories may even yield a cure.

References and Recommended Reading

Altman, L.K. "U.S. Scientist Wins Nobel for Controversial Work." *New York Times*, 7 October 1997.

Belay, E. D., and L. B. Schonberger. "The Public Health Impact of Prion Diseases." *Annual Review of Public Health*, Vol. 26, 2005, pp. 191–212.

Berger, J. R., et al. "Creutzfeldt-Jakob Disease and Eating Squirrel Brains." *Lancet*, Vol. 350, 1997, p. 642.

Brown, D. "The 'Recipe for Disaster' that Killed 80 and Left a £5bn Bill." *The Telegraph*, 19 June 2001.

Broxmeyer, L. "Thinking the Unthinkable: Alzheimer's, Creutzfeldt-Jakob and Mad Cow Disease: The Age-Related Reemergence of Virulent, Foodborne, Bovine Tuberculosis or Losing Your Mind for the Sake of a Shake or Burger." *Medical Hypotheses*, Vol. 64, 2005, pp. 699–705.

Bruederle, C. E., et al. "Prion Infected Meat-and-Bone Meal Is Still Infectious after Biodiesel Production." *PLoS ONE*, 13 August 2008.

Callahan, J. R. "Squirrels as Predators." *Great Basin Naturalist*, Vol. 53, 1993, pp. 137–144.

Callahan, J. R. 2002. *Biological Hazards: An Oryx Sourcebook*. Westport, CT: Oryx Press (imprint of Greenwood Publishing Group).

Colchester, A. C., and N. T. Colchester. "The Origin of Bovine Spongiform Encephalopathy: The Human Prion Disease Hypothesis." *Lancet*, Vol. 366, 2005, pp. 856–861.

Cosseddu, G. M., et al. "Advances in Scrapie Research." *Revue Scientifique et Technique*, Vol. 26, 2007, pp. 657–668.

"Fourth Case of Transfusion-Associated vCJD Infection in the United Kingdom." Eurosurveillance, 18 January 2007.
Heaton, M. P., et al. "Prevalence of the Prion Protein Gene E211K Variant in US Cattle." *BMC Veterinary Research*, Vol. 4, 2008, p. 25.
Imrie, C. E., et al. "Spatial Correlation Between the Prevalence of Transmissible Spongiform Diseases and British Soil Geochemistry." *Environmental Geochemistry and Health*, 22 April 2008.
Lasmeras, C. J. "The Transmissible Spongiform Encephalopathies." *Revue Scientifique et Technique*, Vol. 22, 2003, pp. 23–36.
Lemmer, K., et al. "Decontamination of Surgical Instruments from Prions. II. In Vivo Findings with a Model System for Testing the Removal of Scrapie Infectivity from Steel Surfaces." *Journal of General Virology*, Vol. 89, 2008, pp. 348–358.
MacKenzie, D. "New Twist in Tale of BSE's Beginnings." *New Scientist*, 17 March 2007, p. 11.
Marks, K. "Imported Antelope May Have Caused BSE Epidemic." *The Independent*, 19 April 2001.
Meikle, J. "Sudden Rise in BSE Alarms Scientists." *The Guardian*, 24 November 2003.
"New Mad-Cow Rule Poses Its Own Health Dangers." Associated Press, 7 December 2008.
"New Version of Mad Cow Suspected." United Press International, 18 December 2008.
Pennington, H. "Origin of Bovine Spongiform Encephalopathy." *Lancet*, Vol. 367, 2006, pp. 297–298.
Purdey, M. "The UK Epidemic of BSE: Slow Virus or Chronic Pesticide-Initiated Modification of the Prion Protein?" *Medical Hypotheses*, Vol. 46, 1996, pp. 445–454.
Quaid, L. "U.S. Mad Cow Cases are Mysterious Strain." Associated Press, 11 June 2006.
Race, B. L., et al. "Levels of Abnormal Prion Protein in Deer and Elk with Chronic Wasting Disease." *Emerging Infectious Diseases*, Vol. 13, 2007, pp. 824–830.
Richt, J. A., et al. "Production of Cattle Lacking Prion Protein." *Nature Biotechnology*, Vol. 25, 2007, pp. 132–138.
Richt, J. A., and S. M. Hall. "BSE Case Associated with Prion Protein Gene Mutation." *PLoS Pathogens*, 12 September 2008.
Simon, S. L., et al. "The Identification of Disease-Induced Biomarkers in the Urine of BSE Infected Cattle." *Proteome Science*, Vol. 6, 2008, p. 23.
Smith, P. G. "The Epidemics of Bovine Spongiform Encephalopathy and Variant Creutzfeldt-Jakob Disease: Current Status and Future Prospects." *Bulletin of the World Health Organization*, Vol. 81, 2003, pp. 123–130.
Yokoyama, T., and S. Mohri. "Prion Diseases and Emerging Prion Diseases." *Current Medicinal Chemistry*, Vol. 15, 2008, pp. 912–916.

FOOT-AND-MOUTH DISEASE

Summary of Threat

Foot-and-mouth disease (FMD) is a highly contagious viral disease of animals with cloven hoofs, including cattle, pigs, sheep, goats, and deer. It causes sores that interfere with feeding and movement. Although seldom fatal, it can spread rapidly through livestock populations and cause great economic losses. FMD is classified as a biosecurity threat under the Bioterrorism Protection Act of 2002.

Other Names

Foot-and-mouth disease is also called hoof-and-mouth disease, aphthous fever, aphtha disease, or aphthae epizooticae. Names in other languages include *fiebre aftosa* in Spanish, *Maul- und Klauenseuche* in German, *penyakit mulut dan kuku* in Indonesian, and *száj és körömfájás* ("mouth and hoof ache") in Hungarian. FMD is not related to a human disease called hand, foot,

Case Study 4-2: Phoenix the Calf

At the height of England's 2001 foot-and-mouth disease crisis, Prime Minister Tony Blair made a gesture fraught with ancient symbolism and PR genius. Yielding to public demand or to a higher authority, he officially spared the life of Phoenix—a newborn white calf that somehow survived for five days under a heap of dead cattle, which farmers had recently slaughtered as part of the campaign to stop the epidemic. In the midst of disaster, there is nothing like a lone survivor to buoy the spirit (see also Case Study 4-7, Strong Pig).

and mouth disease or to an all-too-human behavioral tendency called "foot in mouth" (a lapse of diplomacy).

Description

FMD causes fever and vesicles on the mouth, nipples, and feet (Figure 4.3), plus excessive salivation, lameness, and reluctance to move. The agent is one of the picornaviruses, a family that also includes the agents of polio and hepatitis A in humans. There are seven serotypes, each with many strains.

As of 2009, FMD may be the most economically devastating livestock disease. At least 95 percent of infected cattle recover without treatment, but the disease spreads rapidly and reduces meat and milk production. Also, the World Organisation for Animal Health (OIE) does not allow a country to export animals unless it has been free of FMD without vaccination for at least a year. Thus, the usual way to stop an epidemic is to slaughter millions of animals (Case Study 4-2). Although many question this policy, it may be defensible on economic grounds.

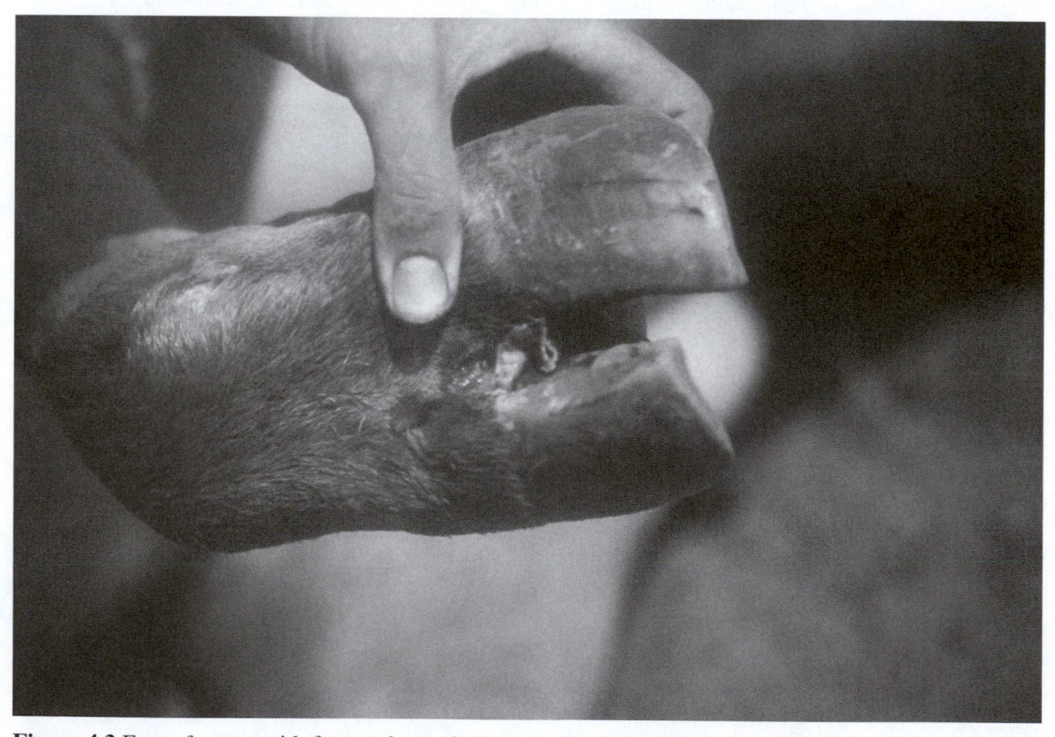

Figure 4.3 Foot of a cow with foot-and-mouth disease, showing a ruptured vesicle in the cleft.
Source: CSIRO Australian Animal Health Laboratory.

Which Animals Are at Risk?

FMD spreads by contact, aerosol transmission, or ingestion of contaminated material. In a susceptible population, all animals show symptoms during an outbreak. The most highly publicized outbreaks have involved cattle, but the disease also affects pigs, sheep, goats, and many wildlife species, including deer and giraffes. Camels and hippopotami appear to be resistant.

FMD is considered a zoonosis, and at least 50 human cases have occurred over the years. There are no reports of severe illness or transmission between humans.

The Numbers

The 2001 epidemic in England resulted in the slaughter of 10 million animals at an estimated cost of $20 billion pounds sterling, but that was the tip of the iceberg. Associated losses per week during that epidemic included $12 million pounds to agriculture and $140 million pounds to the tourism industry.

History

When German scientist Friedrich Loeffler discovered the FMD virus in 1897, farmers and veterinarians had already dealt with this disease for centuries. In 1834, three veterinarians in Germany each drank a quart of milk from an infected cow to test the hypothesis that humans could contract foot-and-mouth disease. It worked; all three developed symptoms. As of 2009, the most recent FMD outbreak in the United States was in 1914. Others occurred in England in 1967, 2001, and 2007, in Taiwan in 1997, and in China in 2005.

In 1946, an FMD outbreak in Mexico threatened to spread north into Texas, and the U.S. and Mexican governments cooperated in a plan to destroy a large proportion of animals in affected areas. But rural Mexican farmers and ranchers responded with such violent protest that, within a year, both governments switched to the slower and more expensive strategy of vaccinating healthy livestock.

Prevention and Treatment

The first FMD vaccines were ineffective and sometimes caused disease. A better vaccine—the world's first genetically engineered vaccine—has been available since 1981, but it works only for a few months and only against similar strains. The cost of vaccinating all animals would be prohibitive, and vaccinated animals can be carriers. Countries with vaccination programs lose OIE "disease-free" status and can no longer export animals. Thus, until a better method is found, farmers rely on captive bolt guns and backhoes to control FMD epidemics. The key to controlling FMD is to keep it out.

Popular Culture

In the 1963 motion picture *Hud*, the title character is the ruthless son of a wealthy Texas cattleman. When FMD infects the herd, Hud urges his father to sell the cattle quickly, before the health inspectors find out. Hud's question reflects an ongoing debate that continues to the present day: "You gonna let them shoot your cows out from under you on account of a schoolbook disease?" The law-abiding father prevails, and the cattle are slaughtered, in a scene that shocked 1963 audiences and critics.

One conspiracy theory holds that FMD does not really exist, but was invented by evil government scientists as an excuse for slaughtering millions of cows and ruining farmers. It is not clear why any government would want to do this.

The Future

Until recently, most FMD research in the United States took place at the Plum Island laboratory. In 2009, the Department of Homeland Security announced that a new facility in Manhattan, Kansas, will replace Plum Island starting in about 2014. A more effective and less expensive FMD vaccine, one that might make global eradication possible, is surely on every cattle rancher's wish list.

References and Recommended Reading

Alexandersen, S., and N. Mowat. "Foot-and-Mouth Disease: Host Range and Pathogenesis." *Current Topics in Microbiology and Immunology*, Vol. 288, 2005, pp. 9–42.

Bauer, K. "Foot-and-Mouth Disease as Zoonosis." *Archives of Virology* Supplement, Vol. 13, 1997, pp. 95–97.

Berríos, E. P. "Foot and Mouth Disease in Human Beings: a Human Case in Chile." *Revista Chilena de Infectología*, Vol. 24, 2007, pp. 160–163.

Brown, D., and A. McSmith. "Brown Scorns 'Urban Legend of Cover-up.'" *The Telegraph*, 28 June 2001.

"Contagious Cattle Disease Found in England." United Press International, 4 August 2007.

DeClerc, K., and N. Goris. "Extending the Foot-and-Mouth Disease Module to the Control of Other Diseases." *Developments in Biologicals*, Vol. 119, 2004, pp. 333–340.

"Foot and Mouth Disease Found in Kyrgyzstan." United Press International, 27 February 2004.

Grubman, M. J., and B. Baxt. "Foot-and-Mouth Disease." *Clinical Microbiology Reviews*, Vol. 17, 2004, pp. 465–493.

"Humans Test Negative for FMD in Britain, Animal Cases Continue to Decline." *Journal of the American Veterinary Medical Association*, 1 June 2001.

Kitching, P., et al. "Global FMD Control—Is It an Option?" *Vaccine*, Vol. 25, 2007, pp. 5660–5664.

Kitching, R. P., et al. "Use and Abuse of Mathematical Models: An Illustration from the 2001 Foot and Mouth Disease Epidemic in the United Kingdom." *Revue Scientifique et Technique*, Vol. 25, 2006, pp. 293–311.

Lombard, M., et al. "A Brief History of Vaccines and Vaccination." *Revue Scientifique et Technique*, Vol. 26, 2007, pp. 29–48.

Mahy, B. W. "Introduction and History of Foot-and-Mouth Disease Virus." Current *Topics in Microbiology and Immunology*, Vol. 288, 2005, pp. 1–8.

Musser, J. M. 2004. "A Practitioner's Primer on Foot-and-Mouth Disease." *Journal of the American Veterinary Medical Association*, Vol. 224, 2004, pp. 1261–1268.

Prempeh, H., et al. "Foot and Mouth Disease: The Human Consequences." *British Medical Journal*, Vol. 322, 2001, pp. 565–566.

Richardson, Z. "UC Davis Developing Model to Tackle Foot-and-Mouth Disease." *Food Chemical News,* 15 January 2007.

Rweyemamu, M., et al. "Planning for the Progressive Control of Foot-and-Mouth Disease Worldwide." *Transboundary and Emerging Diseases*, Vol. 55, 2008, pp. 73–87.

Sayler, C. "Border Makes New Mexico Vulnerable." *New Mexican*, 25 March 2001.

Schat, K. A., and E. Baranowski. "Animal Vaccination and the Evolution of Viral Pathogens." *Revue Scientifique et Technique*, Vol. 26, 2007, pp. 327–338.
"Second Outbreak of Foot-and-Mouth Disease Confirmed in Britain." CNN, 7 August 2007.
Sobrino, F., et al. "Foot-and-Mouth Disease: A Long Known Virus, But a Current Threat." *Veterinary Research*, Vol. 32, 2001, pp. 1–30.
"Soil from Government Lab Linked to Disease." United Press International, 14 December 2007.
Ugarte, R. "Strategy for the Control of Foot-and-Mouth Disease in Uruguay." *Developments in Biologicals*, Vol. 119, 2004, pp. 415–421.

ANTHRAX

Summary of Threat

Just the name says it all—but remember that anthrax is primarily a disease of cattle and other livestock. The agent is a bacterium that causes symptoms in cattle ranging from fever and convulsions to sudden death. In humans, anthrax more often presents as either a black scab on the skin or severe lung disease. Anthrax is classified as a biosecurity threat under the Bioterrorism Protection Act of 2002.

Other Names

Anthrax was once called woolsorter's disease, because it was an occupational hazard in the wool and hide industries. Other names are malignant pustule, malignant carbuncle, malignant edema, rag-picker's disease, splenic fever, black bane, black blood, Siberian pest, and Siberian ulcer. Anthrax is called anthrax in many languages. It is also known as *Milzbrand* ("spleen fire") in German, *pernarutto* ("spleen plague") in Finnish, and *charbon* ("coal") in French. The Greek word *anthrax*, which also means "coal," refers to the black lesions.

After the 2001 anthrax mailings, the media called the disease "thrax" for short. Soon "thrax" became a verb, meaning to infect with anthrax. This ghastly neologism evolved into the slang term "thraxed," which, according to the *Dictionary of American Slang*, refers to an unfortunate state of affairs, often followed by "yo."

Description

The agent of anthrax is a spore-forming bacterium called *Bacillus anthracis* (Figure 4.4) that can infect most mammals. The symptoms and course of the disease depend on the mode of transmission (usually by contact, fomites, or ingestion) and the species. Infected cattle may have fever, convulsions, and breathing difficulty, or they may appear normal until shortly before death. Bleeding from orifices may also occur. The incubation period is usually 3 to 7 days for herbivores and 1 to 2 weeks for pigs. Most infections in ruminants and horses are fatal, but pigs and carnivores often recover.

Most cattle are vaccinated in developed nations, but as of 2009, anthrax remains endemic in the Middle East, Africa, Central America, and South America. Sporadic cases and small outbreaks still occur in the United States, particularly in the southern Mississippi River Valley. Anthrax spores can survive in soil for many years.

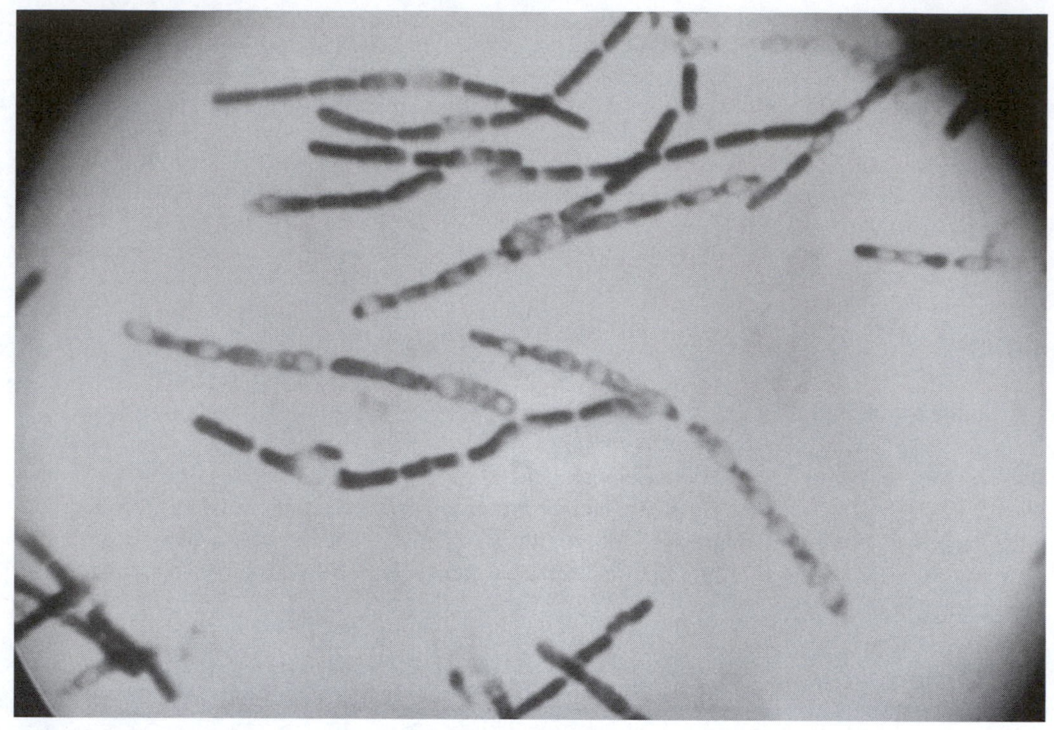

Figure 4.4 *Bacillus anthracis*, the bacterium that causes anthrax.
Source: U.S. Centers for Disease Control and Prevention, Public Health Image Library.

Which Animals Are at Risk?

Most domesticated and wild herbivorous mammals are at risk for anthrax. In 2004, it killed several hundred hippopotami in Uganda. Unvaccinated cattle, sheep, goats, camels, horses, and pigs are all susceptible; carnivores, great apes, and humans are incidental hosts. Birds are resistant (Case Study 4-3).

> **Case Study 4-3: Cold Chicken**
>
> Birds apparently are not susceptible to anthrax because their body temperature is too high for the bacteria to survive. We owe this discovery (among others) to Louis Pasteur, who did a famous experiment in 1878. When he exposed a chicken to anthrax and then chilled it in a basin of cold water, the chicken developed anthrax and died. (Thraxed, yo.) But if he retrieved the chicken from the basin and warmed it up before it became too sick, it recovered.

Risk may be highest in areas of neutral to mildly alkaline soil with periods of flooding followed by drought. Flooding brings the spores to the surface of the ground, and drought exposes them to grazing animals. Anthrax usually infects humans only in specific situations, such as wool processing plants and secret government labs. Sporadic cases of cutaneous anthrax, and a few fatal pulmonary infections, have resulted from exposure to imported wool or hides.

Some sources claim that vultures reduce anthrax risk by eating cows that die of anthrax before they rot and release spores. Others claim that vultures increase risk by transporting infected material. Yet others report that vultures are becoming extinct anyway, because cow carcasses often contain high levels of toxic veterinary drugs.

The Numbers

In 1945, more than 1 million sheep reportedly died from anthrax in Iran. In 1963, an anthrax outbreak killed an estimated 200 to 300 deer on an island in the Mississippi River in the United States, representing 60 to 90 percent of the herd. Global numbers are more readily available for humans; there are about 7,000 reported cases of human anthrax each year worldwide. The largest confirmed human outbreak was in Zimbabwe in 1979–1980, with 10,000+ cutaneous cases and at least 182 deaths. The largest known outbreak of human gastrointestinal anthrax was in Haiti in 1770, with 15,000 deaths.

In humans, the death rate for untreated inhalation anthrax is close to 100 percent, but more than half of infected persons survived with prompt treatment in the 2001 postal outbreak. About 95 percent of all human anthrax cases are the cutaneous form, which is less dangerous; 20 percent of untreated cases may be fatal.

History

Several ancient authors, including Virgil (70–19 B.C.), described livestock diseases that sound like anthrax. One of the plagues of Egypt in the Old Testament may be a reference to anthrax. But the actual "discovery" of anthrax came much later, probably in the 1700s. Until then, farmers often regarded anthrax, rinderpest, and FMD as aspects of one great plague.

In 1727, French physician Nicolas Fournier (1700–1781) described the forms of anthrax and its modes of transmission. In 1876, anthrax became the first disease linked to a microbial agent, thanks to the work of German physician Robert Koch (1843–1910). Louis Pasteur (1822–1895) developed the first anthrax vaccine for livestock in 1881.

Prevention and Treatment

Preventing anthrax requires annual vaccination of all grazing animals and postexposure prophylaxis of exposed animals. Adherence to reporting and quarantine measures, burning or burial of carcasses, and appropriate sanitary measures are essential. Antibiotics should not be given within the first week after the live vaccine.

Human anthrax vaccines require multiple doses and annual boosters, and their routine use is controversial. In 2007, Scripps Research Institute announced a combined anthrax vaccine-antitoxin that provides rapid treatment and long-term immunity with a single injection. Antibiotics often are effective if given within two days after exposure, either alone or in combination with a procedure to drain fluid from the chest.

Popular Culture

In the 1947 motion picture *Stallion Road*, a veterinarian named Larry (portrayed by Ronald Reagan) becomes infected with anthrax after treating diseased horses. A female rancher saves Larry by injecting him with the same antiserum he used on her horses, and they get married. Thus the repentant old doctor who had given him up for dead gets the line that will ultimately shape Larry's career: "All great progress has been made by unscientific people, mostly." Larry later becomes Governor of California and president of the United States.

In the 1939 motion picture *Home on the Prairie*, Texas Ranger Gene Autry tries to control an anthrax epidemic after the bad guys knowingly transport diseased cattle across the Mexican border into the United States. Some television westerns in the mid-1960s, including *The Big Valley*, also referred to anthrax as a disease of cattle. The same theme has appeared in several western novels, such as *Dixon's Edge*, by Dennis O'Keefe (2001), in which a rancher is told that his cattle have anthrax and must be slaughtered.

In 1981, two guitarists formed a heavy metal band that they called Anthrax, because it was the most evil-sounding name they could find in a biology book.

The Future

As more countries adopt livestock management standards that prevent outbreaks of anthrax, it may join smallpox on the short list of deadly diseases that are of interest only to biodefense researchers and bioterrorists (Chapter 6).

References and Recommended Reading

"Anthrax Epidemic Kills Man and Cows." *New York Times*, 17 July 1909.

Bakalar, N. "Discovering What Works on Anthrax." *New York Times*, 21 February 2006.

Banks, D. J., et al. "New Insights into the Functions of Anthrax Toxin." *Expert Reviews in Molecular Medicine*, Vol. 8, 2006, pp. 1–18.

Bullock, D. S. "Vultures as Disseminators of Anthrax." *Auk*, Vol. 73, 1956, pp. 283–284.

Clegg, S. B., et al. "Massive Outbreak of Anthrax in Wildlife in the Malilangwe Wildlife Reserve, Zimbabwe." *Veterinary Record*, Vol. 160, 2007, pp. 113–118.

Fasanella, A., et al. "Anthrax in Red Deer (*Cervus elaphus*), Italy." *Emerging Infectious Diseases*, Vol. 13, 2007, pp. 1118–1119.

Himsworth, C. G. "Anthrax in Saskatchewan 2006: An Outbreak Overview." *Canadian Veterinary Journal*, Vol. 49, 2008, pp. 235–237.

Hugh-Jones, M. E., and V. de Vos. "Anthrax and Wildlife." *Revue Scientifique et Technique*, Vol. 21, 2002, pp. 359–383.

Kellogg, F. E., and A. K. Prestwood. "Anthrax Epizootic in White-Tailed Deer." *Journal of Wildlife Diseases*, Vol. 6, 1970, pp. 226–228.

Leendertz, F. H., et al. "Anthrax in Western and Central African Great Apes." *American Journal of Primatology*, Vol. 68, 2006, pp. 928–933.

Meroney, J. "Ronald Reagan's Anthrax Encounter." *Washington Post*, 11 November 2001.

Morens, D. M. "Characterizing a 'New' Disease: Epizootic and Epidemic Anthrax, 1769–1780." *American Journal of Public Health*, Vol. 93, 2003, pp. 886–893.

Nishi, J. S., et al. "An Outbreak of Anthrax (*Bacillus anthracis*) in Free-Roaming Bison in the Northwest Territories, June–July 2006." *Canadian Veterinary Journal*, Vol. 48, 2007, pp. 37–38.

Odontsetseg, N., et al. "Anthrax in Animals and Humans in Mongolia." *Revue Scientifique et Technique*, Vol. 26, 2007, pp. 701–710.

Oncü, S., et al. "Anthrax—An Overview." *Medical Science Monitor*, Vol. 9, 2003, pp. RA276–283.

Onion, A. "Vultures on the Brink of Extinction." ABC News, 18 January 2006.

Selva, M. "Hippo Deaths Raise Fears of Anthrax Epidemic." *The Independent*, 12 November 2004.

Shadomy, S. V., and T. L. Smith. "Zoonosis Update: Anthrax." *Journal of the American Veterinary Medical Association*, Vol. 233, 2008, pp. 63–72.

Stratidis, J., et al. "Cutaneous Anthrax Associated with Drum Making Using Goat Hides from West Africa—Connecticut, 2007." *Morbidity and Mortality Weekly Report*, Vol. 57, 2008, pp. 628–631.

RINDERPEST

Summary of Threat

Rinderpest is a highly contagious cattle disease that has caused large epidemics. The virus is closely related to the human measles virus. WHO may soon declare rinderpest the second disease (after smallpox) that man has eradicated from the Earth. In 2009, it remains on the list of biosecurity threats designated under the Bioterrorism Protection Act of 2002.

Other Names

Rinderpest means "cattle plague" in German. English names include cattle plague, Russian cattle plague, steppe murrain, cattle murrain, contagious bovine typhus, and RPV (rinderpest virus). Names in most other languages translate as "cattle plague." An 1871 book lists 121 names for rinderpest in India alone; the name used in Bombay, for example, meant "the great disease" or "the worst disease." Wherever cattle represented wealth or security, rinderpest was greatly feared. It has sometimes been confused with a viral disease of sheep called ovine rinderpest or *peste des petits ruminants.*

Description

Rinderpest is (or was) among the worst of all cattle diseases. The death rate is high, and the disease spreads easily by contact or close-range airborne transmission. Cattle are the main hosts, but rinderpest can also infect sheep, goats, pigs, and many wild mammals. It was once prevalent in Europe and Africa, but it never became established in the New World, Australia, or New Zealand.

Symptoms include fever, discharge from the eyes and nose (Figure 4.5), lesions in the mouth, breathing difficulty, and diarrhea. Loss of appetite, decreased milk yield, and abortion of calves may also occur. There is only one known serotype, and at least two of its three lineages have been eradicated.

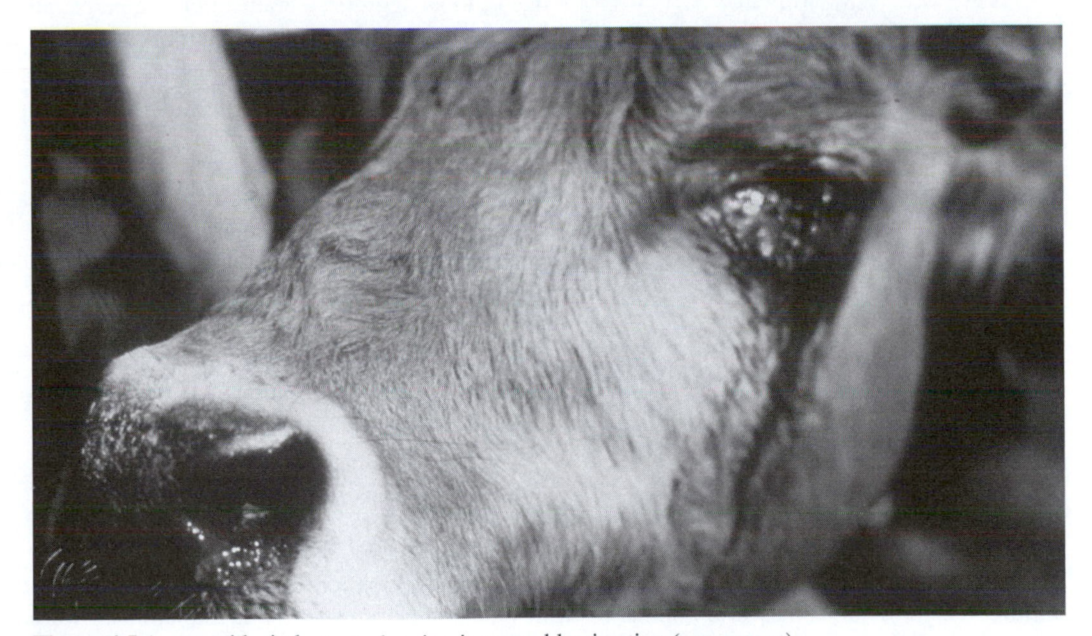

Figure 4.5 A cow with rinderpest, showing increased lacrimation (runny eyes).

Source: CSIRO Australian Animal Health Laboratory.

Which Animals Are at Risk?

Cattle that have never been vaccinated or exposed to the rinderpest virus are at highest risk. The zebu, an African cattle breed, appears to have partial resistance. Sheep, goats, pigs, and hippopotami are only mildly susceptible, and the disease is rare in camels. There are no credible reports of humans contracting rinderpest.

The Numbers

The number of cases of classical wild-type rinderpest reported in 2003–2008 appears to be zero, although pockets of infection may remain in Somalia and adjacent countries. In past epidemics, rinderpest infected hundreds of millions of cattle, and the case fatality rate ranged from about 20 percent to 90 percent.

History

In 1889–1896, rinderpest killed about 90 percent of cattle and many wild mammals in Africa, causing great hardship. In 2000, a conservationist wrote: "This animal pandemic could easily be the greatest catastrophe ever to strike the continent, rivaling that of the AIDS virus that is today decimating the population of this hapless land."[1] The pandemic (technically a panzootic) might have resulted from contact with infected Russian or Italian cattle, exacerbated by hot weather that forced animals to crowd together at water holes. Early vaccination methods were largely unsuccessful (Case Study 4-4).

Case Study 4-4: A Long Time Coming

In 1754 an anonymous correspondent to *Gentleman's Magazine* reported that he had successfully inoculated nine out of ten cattle against rinderpest by dipping a cloth in body fluids from an infected animal and inserting the cloth into an incision in the dewlap of a healthy animal. A series of similar experiments continued in several European countries during the eighteenth and nineteenth centuries, but a fully effective vaccine did not become available until after 1956, when British veterinarian Walter Plowright (1923–) used cell cultures to develop a live attenuated rinderpest virus. In 1999, Dr. Plowright won the World Food Prize for this achievement.

From 1969 to 1997, rinderpest swept through the Middle East, killing a large percentage of cattle in Iran, Iraq, Turkey, Lebanon, Syria, Israel, and the Arabian Peninsula. But thanks to an aggressive eradication campaign, all these countries were free of rinderpest by the end of 1999. Several countries initiated such campaigns in 1987, and in 1993 the United Nations coordinated these efforts under its aegis.

Unlike smallpox, rinderpest infects many species, including wild hosts that must also be tested. Civil unrest in Africa has further complicated the problem, by disrupting governments and increasing demand for meat. The vaccine is hard to deliver to remote areas in Africa because it breaks down in hot weather. Despite these problems, it appears that mass vaccination and improved public relations have finally done the trick.

Prevention and Treatment

Methods used to control past outbreaks include ring vaccination (using the attenuated cell culture vaccine), sometimes in combination with a slaughter program. Valuable animals were

1. Smithsonian Institution Digital Repository: Letter from the Desk of David Challinor, October 2000.

saved with supportive care and antibiotics to control secondary bacterial infections. Animals that recovered had long-lasting immunity.

Popular Culture

An engraving by François-Hippolyte Lalaisse (1812–1884), entitled "The Plague of the Cattle Murrain," shows dead cattle and distraught farmers in the aftermath of a European rinderpest outbreak. An 1866 British diarist wrote: "The cattle plague is spreading through the county like a roaring lion seeking whom it may devour. Some say it is a Russian disease, they call it the Renderpest."[2]

The result of Africa's 1889 rinderpest tragedy is visible in every Hollywood movie that depicts big game hunters hiring African porters in that era. The otherwise unaccountable supply of cheap, compliant labor represented cattle farmers who had recently lost their herds to rinderpest. According to Masai tradition, a medicine man named Mbatian predicted the cattle plague:

> He told the people to move their grazing grounds, "for," he said, "all the cattle will die. You will first of all see flies which make hives like bees, then the wild beasts will die, and afterwards the cattle." Both of these prophesies have come true: the Europeans have arrived, and the cattle died. Mbatian himself died while the cattle plague was raging (circa 1890).[3]

This unpleasantly Borg-like image of swarms of hive-building flies apparently refers to European colonists. (Another Masai tradition described them as green-skinned aquatic creatures with fat in their veins instead of blood.)

The famous Tsavo Man-Eaters—two lions that killed and ate numerous railway workers in Kenya in 1898—quite possibly resorted to snacking on humans because rinderpest had killed most of their usual prey, such as cattle and buffalo. Others have proposed that the lions had broken teeth that impaired their hunting ability. Nearly a century later, director Stephen Hopkins and screenwriter William Goldman immortalized this event in the 1996 motion picture *The Ghost and the Darkness*.

The Future

We are happy to report that the rinderpest virus may have no future, except in a few secure government freezers, alongside smallpox and other horrors of the past that may soon join them. But celebration may be premature. Rinderpest was nearly eradicated once before, in the mid-1970s, only to return after some nations relaxed their surveillance and vaccination programs.

References and Recommended Reading

Barrett, T., and P. B. Rossiter. "Rinderpest: the Disease and Its Impact on Humans and Animals." *Advances in Virus Research*, Vol. 53, 1999, pp. 89–110.

Center for Food Security and Public Health. "Rinderpest." Iowa State University College of Veterinary Medicine, updated August 2008.

---

2. Journal of John Ostle, January 1866.
3. Hollis, A.C. 1905. *The Masai: Their Language and Folklore*. Oxford: Clarendon Press.

DeClerc, K., and N. Goris. "Extending the Foot-and-Mouth Disease Module to the Control of Other Diseases." *Developments in Biologicals*, Vol. 119, 2004, pp. 333–40.

Diop, B. A., and P. Bastiaensen. "Achieving Full Eradication of Rinderpest in Africa." *Veterinary Record*, Vol. 157, 2005, pp. 239–240.

Matin, M. A., and M. A. Rafi. "Present Status of Rinderpest Diseases in Pakistan." *Journal of Veterinary Medicine B, Infectious Diseases and Veterinary Public Health*, Vol. 53, Supplement 1, 2006, pp. 26–28.

Mills, C. "The Wild, Wild Pest." *The Sciences*, March/April 1999.

Mukhopadhyay, A. K., et al. "Rinderpest: a Case Study of Animal Health Emergency Management." *Revue Scientifique et Technique*, Vol. 18, 1999, pp. 164–178.

Normile, D. "Rinderpest: Driven to Extinction." *Science*, Vol. 319, 2008, pp. 1606–1609.

Roeder, P. L., et al. "Experience with Eradicating Rinderpest by Vaccination." *Developments in Biologicals* (Basel), Vol. 119, 2004, pp. 73–791.

Roeder, P. L., and W. P. Taylor. "Rinderpest." *Veterinary Clinics of North America Food Animal Practice*, Vol. 18, 2002, pp. 515–547.

Rossiter, P., et al. "Rinderpest Seroprevalence in Wildlife in Kenya and Tanzania, 1982–1993." *Preventive Veterinary Medicine*, Vol. 75, 2006, pp. 1–7.

Spinage, C. A. 2003. *Cattle Plague: A History*. Philadelphia: J. B. Lippincott.

Tambi, E. N., et al. "Economic Impact Assessment of Rinderpest Control in Africa." *Revue Scientifique et Technique*, Vol. 18, 1999, pp. 458–477.

Taylor, W. P., et al. "The Principles and Practice of Rinderpest Eradication." *Veterinary Microbiology*, Vol. 44, 1995, pp. 359–367.

HEARTWATER

Summary of Threat

Heartwater is a deadly tick-borne disease of cattle, sheep, goats, and other animals, possibly including humans. Symptoms vary by species. The agent is a rickettsia (similar to a bacterium) that can live only as a parasite inside cells. As of 2009, heartwater has not reached North America, but it is listed as a biosecurity threat under the Bioterrorism Protection Act of 2002.

Other Names

Heartwater is also called cowdriosis, ehrlichiosis (one of several forms), or *Ehrlichia ruminantium* infection. Names in other languages include *malkopsiekte* or *bossiekte* (Afrikaans), *hidropericardio* (Spanish), *péricardite exsudative infectieuse* (French), *hidrocarditis infecciosa* (Portuguese), and *idropericardite dei ruminanti* (Italian). The Xhosa people of South Africa reportedly call this disease *inyongo*.

Description

As of 2009, heartwater is endemic in sub-Saharan Africa and offshore islands and has also reached several Caribbean islands. The agent (*Ehrlichia ruminantium)* belongs to a group of small bacteria-like organisms called rickettsiae that can live and reproduce only as parasites inside living cells. Most rickettsial diseases depend on ticks or other arthropod vectors, not only to transmit them between hosts but also to serve as hosts during part of the life cycle (Case Study 4-5). Heartwater vectors include the tropical bont tick (*Amblyomma variegatum*) and related species (Figure 4.6). The female bont tick, about the size of a grape when fully engorged, can also inflict severe bite wounds.

Symptoms are variable but often include fever, loss of appetite, convulsions, and prostration. Heartwater causes the blood vessels to become more permeable, and fluid or blood may collect in the lungs or in the pericardial sac, which surrounds the heart. Animals that recover may become carriers.

Which Animals Are at Risk?

Domestic cattle, sheep, and goats infected with heartwater become severely ill, but some other mammals, birds, and reptiles can carry the disease without showing signs. Heartwater has turned up in many species including Cape buffalo, giraffe, antelope, deer, guinea fowl, ostrich, and leopard tortoise. Studies suggest that heartwater can also infect dogs and possibly humans.

Case Study 4-5: The Life of a Tick

The tropical bont tick (TBT) is an example of a three-host tick. The female lays up to 20,000 eggs on the ground, where they hatch into groups of larvae that ascend blades of grass and wait. When these larvae sense the presence of a suitable host, they attach themselves to its muzzle or legs and feed on blood. The TBT at this stage prefers hairy parts of mammals, their principal hosts, but in a pinch they will accept a reptile or bird instead. Once engorged, the larvae drop to the ground and molt into nymphs, which then attach to a second host. The engorged nymphs again drop to the ground and molt into brightly colored adult ticks. The male TBT ascends the third host and often attaches near its anus, where he secretes a chemical that attracts the female tick. She climbs up and joins him, they mate, she becomes engorged with blood and drops to the ground, and the cycle starts over.

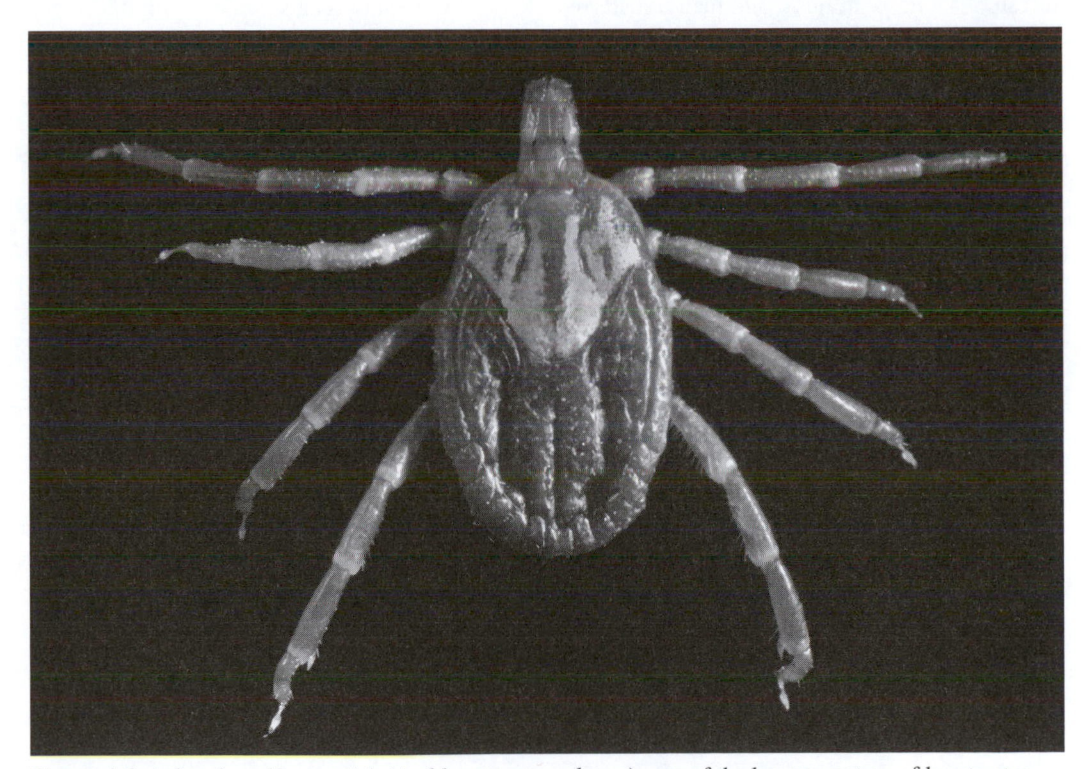

Figure 4.6 A female gulf coast tick (*Amblyomma maculatum*), one of the known vectors of heartwater.
Source: U.S. Centers for Disease Control and Prevention, Public Health Image Library.

The Numbers

According to the U.S. Department of Agriculture, if heartwater reaches the United States, it could cost the livestock industry at least $762 million per year. Since livestock in North America has no acquired resistance to this disease, and no vaccine is yet available, the mortality rate could approach 100 percent. In countries such as Zimbabwe, where heartwater is endemic, the largest component of economic loss is the cost of acaricides (tick-killing agents) and the dipping process used to kill ticks. Other costs include the loss of animals, their milk and other products and the cost of treatment.

History

British veterinarians who investigated heartwater in South Africa in the 1800s found that local settlers had beaten them to it. The ranchers already knew that bont ticks had arrived in the 1830s and somehow caused heartwater; they also knew heartwater was related to vegetation density, elevation, and overstocking. But the vets rejected these explanations in favor of a soil-borne anthrax-like agent, although the blood of animals with heartwater contained no bacteria that were visible using the instruments of the day. Thus, the researchers imported a batch of Pasteur's new anthrax vaccine and administered it to sheep, several of which promptly contracted anthrax and died. The study ended in frustration, and the team went home.

The role of ticks did not become clear until American scientists published a study of babesiosis in the 1890s. In 1900, South African scientists successfully infected a goat with heartwater, infested it with bont ticks, and then transferred the ticks to healthy goats, which soon developed heartwater. Koch's Postulates were fulfilled, and the mode of transmission was known, although the heartwater agent remained elusive until 1925 because of its small size.

Prevention and Treatment

As of 2009, no commercial heartwater vaccine is available. Studies of attenuated and recombinant vaccines are in progress, but these vaccines have not been fully validated under field conditions. Antibiotics such as tetracycline may be effective, but only at the earliest stages of the disease.

In some African countries, farmers induce a controlled heartwater infection by exposing cattle to infected sheep blood, then monitor the animals' body temperature and give them antibiotics if necessary. Corticosteroids have also been used as supportive therapy. Control of tick infestation (by dipping animals or spraying brush) may help prevent infection, but excessive reduction of tick numbers may backfire by interfering with maintenance of adequate immunity.

A 1989 study showed that bont ticks prefer cows that male bont ticks have already visited, because the previous tick deposits a pheromone that labels the cow as a suitable meal. The downside is that ticks may avoid cattle that have been treated with acaricides. Also, treated cattle may gradually lose their immunity and later develop severe disease when exposed again.

Popular Culture

The Xhosa people of South Africa use several herbal remedies to treat their livestock for heartwater, including the bark of Cape lancewood (*Curtisia dentata*), red beech (*Protorhus longifolia*),

Cape beech (*Rapanea melanophloeos*), and false horsewood (*Hippobromus pauciflorus*); the root bark of kerriebossie (*Gnidia capitata*); the tuber of geranium (*Pelargonium reniforme*); and the climbing stem of swart teebossie (*Vernonia mespilifolia*).

Traditional methods of tick removal have evolved into urban legends that are best ignored. If you burn the tick with a match, poison it with gasoline, or cover it with nail polish or liquid soap or petroleum jelly, it may regurgitate infected material into the wound. It may also die, leaving its mouthparts embedded in the skin, where an infection or granuloma may develop. The best way to remove an attached tick (from a person or animal) is to grasp the tick lightly, just behind its head—using tweezers, a loop of thread, a tick extractor, or two fingernails—and pull it straight out, *slowly*. Don't decapitate the tick or squeeze its body, and don't stop pulling until the tick lets go. Then kill it.

The Future

Birds called cattle egrets often carry tropical bont ticks, the vectors of heartwater. Cattle egrets are already common in North America, but bont ticks are not; the question is whether infested egrets can make the journey. And the answer is yes—otherwise, we would not have chosen this example. In 1990, researchers marked cattle egrets on Caribbean islands where heartwater exists, and tracked one of these birds to the Florida Keys. Tortoises and other reptiles imported from Africa as pets are also a source of worry, since they, too, can carry bont ticks. These facts suggest that heartwater will reach North America sooner or later.

References and Recommended Reading

Allan, S. A., et al. "Ixodid Ticks on White-Tailed Deer and Feral Swine in Florida." *Journal of Vector Ecology*, Vol. 26, 2001, pp. 93–102.

Allsopp, M. T., et al. "Novel *Ehrlichia* Genotype Detected in Dogs in South Africa." *Journal of Clinical Microbiology*, Vol. 39, 2001, pp. 4204–4207.

Allsopp, M. T., et al. "*Ehrlichia ruminantium*: An Emerging Human Pathogen?" *Annals of the New York Academy of Sciences*, Vol. 1063, 2005, pp. 358–360.

Anderson, P. G. 2007. "Edmund Vincent Cowdry (1888–1975)." St. Louis, MO: Bernard Becker Medical Library, Washington University School of Medicine.

APHIS Veterinary Services. 2002. "Heartwater Factsheet." U.S. Department of Agriculture, July 2002.

Burridge, M. J. "Ticks (Acari: Ixodidae) Spread by the International Trade in Reptiles and Their Potential Roles in Dissemination of Diseases." *Bulletin of Entomological Research*, Vol. 91, 2001, pp. 3–23.

Burridge, M. J., et al. "Increasing Risks of Introduction of Heartwater onto the American Mainland Associated with Animal Movements." *Annals of the New York Academy of Sciences*, Vol. 969, 2002, pp. 269–274.

Center for Food Security and Public Health. "Heartwater." Factsheet, Iowa State University College of Veterinary Medicine, updated 28 September 2007.

Cocks, M. L., and A. P. Dold. "Cultural Significance of Biodiversity: The Role of Medicinal Plants in Urban African Cultural Practices in the Eastern Cape, South Africa. *Journal of Ethnobiology*, Vol. 26, 2006, pp. 60–82.

Gilfoyle, D. "The Heartwater Mystery: Veterinary and Popular Ideas about Tick-Borne Animal Diseases at the Cape, c. 1877–1910." *Kronos: The Journal of Cape History*, Vol. 29, 2003, pp. 139–160.

"Imported Ticks Can Pose Threat to Florida Herds." Associated Press, 29 October 1989.

Jancin, B. "Plucking at Myths Surrounding Tick Removal." *Family Practice News*, 1 November 2000.

Loftis, A. D., et al. "Infection of a Goat with a Tick-Transmitted *Ehrlichia* from Georgia, U.S.A., That Is Closely Related to *Ehrlichia ruminantium*." *Journal of Vector Ecology*, Vol. 31, 2006, pp. 213–223.

Loftis, A. D., et al. "Geographic Distribution and Genetic Diversity of the *Ehrlichia* sp. from Panola Mountain in *Amblyomma americanum*." *BMC Infectious Diseases*, 23 April 2008.

Louw, M., et al. 2005. "*Ehrlichia ruminantium*, an Emerging Human Pathogen—A Further Report." *South African Medical Journal*, Vol. 95, 2005, pp. 948–949.

Mahan, S. M., et al. "Development of Improved Vaccines for Heartwater." *Developments in Biologicals* (Basel), Vol. 114, 2003, pp. 137–145.

"Ostriches Barred Entry." Associated Press, 6 July 1989.

Pedregal, A., et al. "Toward Prevention of Cowdriosis." *Annals of the New York Academy of Sciences*, Vol. 1149, 2008, pp. 286–291.

Peter, T. F., et al. "*Ehrlichia ruminantium* Infection (Heartwater) in Wild Animals." *Trends in Parasitology*, Vol. 18, 2002, pp. 214–218.

Provost, A., and J. D. Bezuidenhout. "The Historical Background and Global Importance of Heartwater." *Onderstepoort Journal of Veterinary Research*, Vol. 54, 1987, pp. 165–169.

Reeves, W. K., et al. "The First Report of Human Illness Associated with the Panola Mountain *Ehrlichia* Species: a Case Report." *Journal of Medical Case Reports*, 30 April 2008.

Shkap, V., et al. "Attenuated Vaccines for Tropical Theileriosis, Babesiosis and Heartwater: The Continuing Necessity." *Trends in Parasitology*, Vol. 23, 2007, pp. 420–426.

"Specialist Warns Against Foreign Ticks." *Gazette-Enterprise* (Texas), 23 July 2000.

"Ticks Follow Fellows to Tastiest Cows." *Science News*, 25 February 1989.

Wagner, G. G., et al. "Babesiosis and Heartwater: Threats Without Boundaries." *Veterinary Clinics of North America Food Animal Practice*, Vol. 18, 2002, pp. 417–430.

Yabsley, M. J., et al. "Natural and Experimental Infection of White-Tailed Deer (*Odocoileus virginianus*) from the United States with an *Ehrlichia* sp. Closely Related to *Ehrlichia ruminantium*." *Journal of Wildlife Diseases*, Vol. 44, 2008, pp. 381–387.

Yunker, C. E. "Heartwater in Sheep and Goats: a Review." *Onderstepoort Journal of Veterinary Research*, Vol. 63, 1996, pp. 159–170.

CLASSICAL SWINE FEVER

Summary of Threat

Classical swine fever (CSF) is a highly contagious viral disease that has caused large epidemics and major financial losses. Hosts include domestic pigs and wild boars. The actual death rate is low for most modern CSF virus strains, but controlling outbreaks may require culling large number of pigs. This disease is classified as a biosecurity threat under the Bioterrorism Protection Act of 2002.

Other Names

Classical swine fever is also called hog cholera, and the classical swine fever virus (CSFV) is also called hog cholera virus (HCV). Less common names are pig typhoid, pig plague, swine plague, Billings' swine plague, swine pest, pneumoenteritis of swine, swine diphtheritis, and infectious pneumonia of swine. Names in other languages include *peste porcina clásica* or *cólera porcina* (Spanish), *peste suína clássica* (Portuguese), *peste du porc* (French), *sikarutto* (Finnish), *Virusschweinepest* (German), and *klassieke varkenspest* (Dutch).

Media references to "swine fever" may lead to confusion with African swine fever, a different disease that (as of 2009) has never been found in the United States. The name "hog cholera" also refers to at least two other diseases of swine.

Description

The agent is a small RNA virus in the family Flaviviridae, a group that also includes the agents of dengue fever and hepatitis C in humans. CSF spreads by contact with infected animals or contaminated surfaces, insemination or other exchange of body fluids, close-range airborne or aerosol transmission, and possibly by flying insects, lice, and various small animals. Symptoms include high fever, convulsions, loss of appetite, a depressed appearance, and the tendency of pigs to huddle together (Figure 4.7). Diarrhea, breathing difficulty, and eye irritation may occur. Pigs infected with a virulent strain often die within a week or two. Those with chronic CSF may survive for a few months, or else recover and then suffer a fatal relapse. Survivors often have smaller litters, stillbirths, or high mortality at weaning.

As of 2009, classical swine fever remains a deadly disease partly because of international policy. To export pigs, a nation must prove that its pigs are free of CSF antibodies. But since vaccination produces antibodies that are indistinguishable from those resulting from infection, the only economically feasible way to stop an outbreak is by slaughtering pigs.

Figure 4.7 Depression is an early symptom of classical swine fever.

Source: Frank Filippi, CSIRO Australian Animal Health Laboratory.

Which Animals Are at Risk?

The only known hosts for classical swine fever are pigs (including wild boars), although other animal species can be infected experimentally. If exposed, all unvaccinated pigs are at risk.

The Numbers

In 1997–1998, a classical swine fever epidemic in the Netherlands resulted in the deaths of about 11 million pigs. Only about 10 percent of those pigs died from the disease itself; the rest were slaughtered to stop the outbreak while maintaining disease-free status according to European Union policy. Estimates of the resulting losses to the swine industry range from $2.3 billion to $5 billion in U.S. dollars.

History

Classical swine fever first appeared in Indiana sometime before 1833 and was a major threat to swine production until 1962, when Public Law 87-209 started a vaccination campaign that eliminated CSF from the United States by 1977. As of 2009, the United States remains free of this disease.

A compulsory slaughter program in the United Kingdom eliminated the disease in 1966, but minor outbreaks occurred in 1971, 1986, and 2000. The last outbreak in England closely followed the epidemics of mad cow disease and foot-and-mouth disease, and the result was a temporary reduction of consumer confidence in British food.

Prevention and Treatment

Live attenuated vaccines are commercially available, but they make it hard to detect circulating viruses. Maternal antibodies may also interfere with the induction of immunity. "Marker" vaccines allow discrimination between infected and vaccinated pigs, but these vaccines are less effective than live vaccines, and the diagnostic test also has limitations. As a result, most developed nations prohibit vaccination in CSF-free areas, and they control outbreaks by slaughtering confirmed cases and contacts.

Wild boars and feral pigs might serve as a reservoir for classical swine fever, but it would not be feasible to test or kill all of them. The EU has used a combination of oral vaccines and hunting to control disease in wild boars.

Popular Culture

American journalist Stephen Bonsal's 1937 memoir *Heyday in a Vanished World* relates how the Dual Monarchy of Austria and Hungary barred Serbian pigs after some were found with hog cholera. But since Serbian pigs were not considered edible until they had fattened on Hungarian pastures, the export ban resulted in starvation for swine and swineherds alike, despite the press of swine in the streets of Belgrade.

Some people have claimed that the USDA inspection marks stamped on supermarket meat represent the "number of the beast," or that the indelible purple dye is poisonous. In fact, this harmless vegetable dye was the invention of Dr. Marion Dorset, the same USDA scientist who identified the agent of classical swine fever and developed the vaccine (Case Study 4-6).

A book on Native American herbal medicine states that the Meskwaki people fed their pigs the root of the feathery false Solomon's seal plant (*Smilacina racemosa*) to prevent hog cholera or classical swine fever. Since this disease first reached North America in the early 1800s, the practice was not an ancient one.

German farmers once believed that pigs would not contract hog cholera if they were treated for seven days with water containing a plant called asphodel (probably *Asphodelus ramosus*). In ancient Europe, this plant was sacred to the goddess Persephone, whose worship involved the sacrifice of pigs.

Case Study 4-6: The Legacy of Dr. Marion Dorset

USDA researcher Marion Dorset (1872–1935) was responsible for the conquest of hog cholera in the United States. He also studied bovine TB and other animal diseases, and colleagues compared his work to that of Pasteur and Koch, yet Dr. Dorset was renowned for his dedication and modesty. His obituary states, in part:

> On the occasion of his discovery of anti-hog cholera serum, Doctor Dorset had the opportunity to acquire wealth through the manufacture and sale of this product for which a large demand promptly developed. But after applying for and receiving a patent [No. 1,784,928], he gave it to the government and to the public so that any person in the United States might use the method without payment of royalty.*

Nor did his legacy end with his death in 1935. His son Virgil "Jack" Dorset became a physician and volunteered for service in Manila in World War II, only to be captured at the fall of Bataan. He continued his role as medical officer in the prison camps and, in 1944, was moved to a camp in Japan, where he witnessed the cloud from the atomic bomb at Nagasaki. After the camp was liberated, Col. Dorset became a surgeon in the U.S. Marine Hospital. He died in Texas in 1999, a hero in the best tradition.

**Wyoming Stockman-Farmer,* 1 August 1935.

The Future

Regardless of how the reader (or author) may feel about animal rights issues, recent events have made it clear that significant numbers of people worldwide oppose the mass slaughter of pigs or cattle for disease control purposes. Thus, an alternative course of action will most likely present itself in the near future. An improved marker vaccine might satisfy both public health and ethical concerns.

References and Recommended Reading

Artois, M., et al. "Classical Swine Fever (Hog Cholera) in Wild Boar in Europe." *Revue Scientifique et Technique*, Vol. 21, 2002, pp. 287–303.

Blome, S., et al. "Assessment of Classical Swine Fever Diagnostics and Vaccine Performance." *Revue Scientifique et Technique*, Vol. 25, 2006, pp. 1025–1038.

Edwards, S., et al. "Classical Swine Fever: the Global Situation." *Veterinary Microbiology*, Vol. 73, 2000, pp. 103–119.

Greiser-Wilke, I., et al. "Diagnostic Methods for Detection of Classical Swine Fever Virus—Status Quo and New Developments." *Vaccine*, Vol. 25, 2007, pp. 5524–5530.

Luy, J., and K. R. Depner. "The Need for a Paradigm Shift in the Control of Classical Swine Fever." *EurSafe News*, Vol. 8, 2006, pp. 3–6.

Moennig, V., et al. "Clinical Signs and Epidemiology of Classical Swine Fever: A Review of New Knowledge." *Veterinary Journal*, Vol. 165, 2003, pp. 11–20.

Morilla, A., et al. (Eds.) 2002. *Trends in Emerging Viral Infections of Swine*. Ames: University of Iowa Press.

Paton, D. J., and I. Greiser-Wilke. "Classical Swine Fever—An Update." *Research in Veterinary Science*, Vol. 75, 2003, pp. 169–178.

Reynolds, D. "Vigilance for Classical Swine Fever and FMD." *Veterinary Record*, Vol. 158, 2006, p. 383.

Ribbens, S., et al. "Transmission of Classical Swine Fever. A Review." *Veterinary Quarterly*, Vol. 26, 2004, pp. 146–155.

Ribbens, S., et al. "Evidence of Indirect Transmission of Classical Swine Fever Virus through Contacts with People." *Veterinary Record*, Vol. 160, 2007, pp. 687–690.

Suradhat, S., et al. "Factors Critical for Successful Vaccination against Classical Swine Fever in Endemic Areas." *Veterinary Microbiology*, Vol. 119, 2007, pp. 1–9.

Terpstra, C., and A. J. de Smit. "The 1997/1998 Epizootic of Swine Fever in the Netherlands: Control Strategies Under a Non-Vaccination Regimen." *Veterinary Microbiology*, Vol. 77, 2000, pp. 3–15.

U.S. Department of Agriculture, Animal and Plant Health Inspection Service. "Procedure Manual for Classical Swine Fever (CSF) Surveillance." Version 2.0, 1 April 2007.

van Oirschot, J. T. "Diva Vaccines that Reduce Virus Transmission." *Journal of Biotechnology*, Vol. 73, 1999, pp. 195–205.

van Oirschot, J. T. "Emergency Vaccination against Classical Swine Fever." *Developments in Biologicals*, Vol. 114, 2003, pp. 259–267.

BLUE-EAR PIG DISEASE

Summary of Threat

A mysterious viral disease caused great concern in 2007 when it killed an estimated 10 million pigs in China, but it later turned out to be the same as a disease first seen in North America in 1987. It mainly affects the respiratory and reproductive systems. As of 2009, there are no known hosts other than swine, although some wild animals have been infected experimentally. A commercial vaccine is available.

Other Names

In the early 1990s, the media referred to this disease as "mystery swine disease" or MSD. It is now called blue-ear (or blue-eared) pig disease, porcine reproductive and respiratory syndrome (PRRS), porcine epidemic abortion and respiratory syndrome (PEARS), or porcine high fever disease (PHFD). Older names include swine infertility and respiratory syndrome (SIRS), new pig disease, pig plague of 1989, Wabash syndrome, and swine plague. The agent is Lelystad virus, porcine arterivirus, or PRRSV.

Blue-ear pig disease is *abortus blauw* in Dutch, *seuchenhafter Spätabort der Schweine* or *rätselhafte Schweinekrankheit* in German, *syndrom reproductif et respiratoire du porc* in French, *síndrome disgenésico y respiratorio del cerdo* or *enfermedad misteriosa del cerdo* in Spanish, *heko-heko* in Japanese, and *lan er bing* in Chinese.

Description

The agent is a small RNA virus in the family Arteriviridae. Symptoms may include encephalitis, heart problems, infertility, stillbirth, and late-term abortion in adult pigs, or severe pneumonia in piglets. The blue or purple discoloration that gives the disease its popular name occurs in only 1 to 2 percent of pigs, but the name has stuck (Figure 4.8).

Figure 4.8 Pigs suffering from porcine reproductive and respiratory syndrome (PRRS), also called blue-ear pig disease.

Source: CSIRO Australian Animal Health Laboratory.

PRRS spreads mainly by direct contact with body fluids or contaminated surfaces, including the hands, clothing, or vehicles of swine workers. Flies and mosquitoes can also serve as mechanical vectors. The disease is present in North and South America, Asia, and most of Europe. As of 2009, only three countries (Australia, New Zealand, and Switzerland) are officially free of PRRS.

Which Animals Are at Risk?

All domestic and wild pigs appear to be susceptible, but the death rate is highest among recently weaned pigs from infected litters. Older pigs often recover, but long-term effects may include immune suppression and secondary bacterial infections.

The Numbers

The case fatality rate for PRRS is usually 10 to 20 percent for all age groups or 40 percent for piglets, not including losses from culling. The 2006–2007 epidemic in China killed an estimated 10 million pigs, but China is a big country; to put this number in perspective, China reportedly loses about 25 million pigs to all diseases in a typical year (see also Case Study 4-7, page XXX). The epidemic contributed to the economic impact from pork prices, which reportedly rose by about 75 percent in 2007.

Case Study 4-7: Strong Pig

After the 2008 earthquake that devastated southwest China, a pig was found that had survived for 36 days buried under rubble, sustained only by rainwater and a bag of charcoal. The pig became China's animal of the year and now lives in a museum. This event is noteworthy in a nation that lost some 10 million pigs to blue-ear pig disease the year before. Like Phoenix the Calf (Case Study 4-2), the Strong Pig (Zhu Jianqiang) became a symbol of hope and renewal that transcends cultural boundaries.

As of 2007–2008, PRRS infected about 60 percent of swine herds in the United States each year, and the annual economic impact of PRRS on pork producers in the United States alone was estimated to be $560 million to $762 million.

In 2008, the Vietnamese government reported that three PRRS outbreaks had infected a total of 338,736 pigs and necessitated the slaughter of 288,000. Vietnam requested international help and imported vaccine from China.

History

First reported in the United States in 1987, "mystery swine disease" caused a series of epidemics in North America and Europe. Researchers in South Dakota, Minnesota, and Missouri identified the cause as a viral infection and began work on a vaccine. PRRS was pandemic by the early 1990s, and the first of several live virus vaccines became available in 1994.

A swine disease that appeared in China in 2006 later turned out to be a highly virulent form of PRRS. Changes in swine production in the second half of the twentieth century, including increased transportation of live animals and sperm, may have facilitated the spread of this virus.

Prevention and Treatment

As of 2009, no consistently effective treatment is available, but commercial vaccines (both live and killed) have helped to control outbreaks and reduce losses. These vaccines are not entirely effective because of the high genetic diversity of the virus. Other preventive measures are the same as for any disease: breed resistant strains, maintain surveillance programs, and keep pig farms reasonably clean. Swine workers should wash their hands and change clothes after handling infected pigs.

Popular Culture

Since this disease was virtually unknown to the public until about 1991, it is too new to have permeated popular culture to any great extent. We know of no motion pictures or songs about this disease. In 2009, the website of an American man living in Vietnam displayed a photograph of a local bakery's cake, topped with a miniature blue-eared pig sculpted in frosting, but this phenomenon is too recent to evaluate.

The 2006–2007 PRRS epidemic did, however, start at least one Internet rumor. In 2008, after the press reported that the Chinese government tried to cover up the epidemic, suspicion grew that China might be endangering global health by its secrecy on issues such as SARS, bird flu, contaminated pet food, toxic drywall, hair bands made from recycled condoms, and lead-based paint. For some, the next logical step was to point out the bruised appearance of "blue-eared pigs" and suggest that the Chinese were concealing an outbreak of hemorrhagic fever, possibly Ebola. The timing was good; in 2008, Ebola Reston and PRRS were found to coexist in Philippine pigs.

Next, someone speculated that a recombinant Ebola virus had escaped from a Chinese lab, causing the human disease outbreak later blamed on *Streptococcus suis* (Chapter 3). In the next few years, these suspicions may coalesce into an elaborate conspiracy theory.

The Future

After the 2007 outbreak, colloquium participants predicted that improved PRRS vaccines would be available within 5 to 10 years. Meanwhile, the Chinese government instituted a nationwide program to prevent future outbreaks. If other countries follow China's example, blue-eared pigs may soon exist only in history books.

References and Recommended Reading

Barboza, D. "Virus Spreading Alarm and Pig Disease in China." *New York Times*, 16 August 2007.

Beilage, E. G., and H. J. Batza. "PRRSV-Eradication: An Option for Pig Herds in Germany?" *Berliner und Münchener Tierärztliche Wochenschrift*, Vol. 120, 2007, pp. 470–479. [German]

"Blue-Ear Pig Disease Discovered Nationwide." *China Daily*, 12 June 2007.

Chae, C. "A Review of Porcine Circovirus 2-Associated Syndromes and Diseases." *Veterinary Journal*, Vol. 169, 2005, pp. 326–336.

"China Arrests Makers of Fake Vaccine for Pig Disease." Associated Press, 29 October 2007.

Cho, J. G., and S. A. Dee. "Porcine Reproductive and Respiratory Syndrome Virus." *Theriogenology*, Vol. 66, 2006, pp. 655–662.

Collins, J. E., et al. "Isolation of Swine Infertility and Respiratory Syndrome Virus (Isolate ATCC VR-2332) in North America and Experimental Reproduction of the Disease in Gnotobiotic Pigs." *Journal of Veterinary Diagnostic Investigation*, Vol. 4, 1992, pp. 117–126.

Goyal, S. M. "Porcine Reproductive and Respiratory Syndrome." *Journal of Veterinary Diagnostic Investigation*, Vol. 5, 1993, pp. 656–664.

Hornby, L. "China Official Fired over Blue Ear Disease Outbreak." Reuters, 11 February 2009.

Keffaber, K. K. "Reproductive Failure of Unknown Etiology." *American Association of Swine Practitioners Newsletter*, Vol. 1, 1989, pp. 1–10.

Mateu, E., and I. Diaz. "The Challenge of PRRS Immunology." *Veterinary Journal*, Vol. 177, 2008, pp. 345–351.

Plagemann, P. G. "Porcine Reproductive and Respiratory Syndrome: Origin Hypothesis." *Emerging Infectious Diseases*, Vol. 9, 2003, pp. 903–908.

"Quake Zone Hero Pig Named China's Animal of the Year." *Taipei Times*, 29 December 2008.

"Scientists Solve Part of Mystery Swine Disease." Associated Press, 13 November 1991.

"Swine Study Looks at Mutating Pig Virus." United Press International, 21 February 2008.

Terpstra, C., et al. "Experimental Reproduction of Porcine Epidemic Abortion and Respiratory Syndrome (Mystery Swine Disease) by Infection with Lelystad Virus: Koch's Postulates Fulfilled." *Veterinary Quarterly*, Vol. 13, 1991, pp. 131–136.

Tong, G.-Z., et al. "Highly Pathogenic Porcine Reproductive and Respiratory Syndrome, China." *Emerging Infectious Diseases*, Vol. 13, 2007, pp. 1434–1436.

"USDA Renews Program to Fight Pig Syndrome." United Press International, 24 July 2008.

"Vietnam Hit by Outbreak of Blue-Eared Pig Disease." *China Post*, 8 April 2006.

Zimmerman, J. (Ed.). 2003. *PRRS Compendium Producer Edition*. Des Moines, IA: National Pork Board.

NEWCASTLE DISEASE

Summary of Threat

Newcastle disease (ND) is a deadly viral infection of chickens, turkeys, and other domestic and wild birds. The most virulent strains can quickly kill entire flocks. ND can also cause illness (usually mild) in humans. Vaccines are available for some strains. This highly contagious disease is classified as a biosecurity threat under the Bioterrorism Protection Act of 2002.

Other Names

Newcastle disease, or ND, is also called New Castle disease (NCD), fowlpest, Doyle's disease, Egyptian fowl-plague, Korean fowl-plague, Asiatic fowl-plague, pseudoplague, pseudo-poultry plague, avian distemper, Ranikhet disease, Tetelo disease, avian pneumoencephalitis, atypical Geflugelpest, or respiratory nervous disorder. Names of specific ND strains include virulent Newcastle disease (VND), Asiatic Newcastle disease (AND), velogenic neurotropic ND (VNND), and velogenic viscerotropic Newcastle disease (VVND). Exotic Newcastle disease (END) consists of both VVND and VNND. The Newcastle disease virus itself (NDV) is also called avian paramyxovirus-1 or APMV-1. Newcastle disease is *peste avícola de Egipto* in Spanish, *Newcastle-Krankheit* in German, *hoensepest* in Danish, and *ayam tetélo* in Indonesian.

Description

As of 2009, Newcastle disease is the most important poultry disease (and perhaps the one with the most acronyms). It is endemic in the Middle East, Africa, and Central and South America, but outbreaks occur worldwide. The agent is a paramyxovirus, like the agents of measles and rinderpest. ND has four forms, or pathotypes:

- Asymptomatic enteric: birds do not appear sick.
- Lentogenic: birds have mild respiratory illness.
- Mesogenic: birds have major respiratory or neurological symptoms, with mortality up to about 10 percent.
- Velogenic: either neurotropic, with respiratory or neurological symptoms (VNND), or viscerotropic, with hemorrhagic lesions in the intestine (VVND); mortality may exceed 90 percent.

"Virulent" ND includes the mesogenic, VNND, and VVND pathotypes, but "exotic" ND means VNND or VVND. In some virulent ND outbreaks, 100 percent of birds in a flock show symptoms, and 90 percent die within days after exposure. All forms spread mainly by aerosol transmission or direct contact with other birds or infected surfaces. Outbreaks often result from illegal importation of birds for the pet trade or cockfighting. Pigeons that scavenge feed from poultry operations may spread the disease.

NDV can also cause illness in humans. Most of these infections are mild, but there is one report of fatal NDV pneumonia after peripheral blood stem cell transplant surgery. Humans can also act as carriers and transmit the virus to other people or birds. Many other animal species serve as hosts or mechanical vectors. The virus remains active in moist soil for 22 days, on feathers at room temperature for 123 days, and in lake water for 19 days.

Which Animals Are at Risk?

Chickens, pigeons, ducks, parrots, and many wild migratory birds are at risk for ND, even after vaccination. Chickens are highly susceptible, but some birds, including turkeys and budgies, may harbor the infection without showing obvious signs. Virulent strains are endemic in wild cormorants, and there is concern that gulls might transfer the disease between cormorant colonies and farms. ND poses no serious risk to most humans, but poultry workers sometimes develop malaise or eye irritation.

The Numbers

A 1971–1973 outbreak caused the deaths of about 12 million chickens in southern California, with estimated losses of $56 million. Some chickens died of the disease itself, whereas others were killed to stop the outbreak. An outbreak in Italy in 1999–2000 necessitated the culling of 13 million chickens and guinea fowl. In 2000, an outbreak in Mexico destroyed nearly 14 million chickens. The most recent U.S. outbreak, in 2003, cost $160 million and claimed the lives of 4 million chickens in southern California, Arizona, and Nevada.

History

The first ND outbreak in the British Isles took place at Newcastle-upon-Tyne in 1927, the year after the disease made its global debut in Indonesia. In 1956, however, researchers found records of a similar disease that killed all the chickens in the Western Isles of Scotland in 1896. Surviving farmers remembered that the sick chickens wheezed and staggered. But since other bird diseases (such as avian influenza) can cause similar symptoms and high mortality, the 1896 outbreak may never be positively identified.

Virulent forms of ND reached the United States in about 1945, possibly with infected partridges and pheasants imported from Hong Kong. Major outbreaks occurred in southern California in 1971 and 2002. As of 2009, the United States has been free of Newcastle disease outbreaks since 2003.

Several recent sources claim that Newcastle disease drove the North American passenger pigeon (*Ectopistes migratorius*) to extinction. But the last wild passenger pigeons died in 1899, and the last known passenger pigeon died in a zoo in 1914, long before there was any evidence of Newcastle disease in North America. ND was not even discovered until 1926, and the Scottish outbreak of 1896 (whatever it was) did not affect New World poultry. Thus, the claim is dubious at best. The source most often cited is IUCN 2008, which cites BirdLife International, which cites several review papers—which simply mention Newcastle disease as one of several factors that *might* have contributed to the extinction of the passenger pigeon. More likely factors include hunting, deforestation, and an unrelated parasitic disease called trichomoniasis.

Prevention and Treatment

ND is controllable, but the global cost of vaccination, surveillance, and testing may exceed that of any other animal disease. Commercial vaccines for some strains have been available since

Case Study 4-8: Taking the Heat

Many rural villages in the tropics have no refrigeration facilities for transport and storage of vaccines. In 2005–2007, three agricultural organizations in Tanzania collaborated on a project to increase rural poultry productivity by making a heat-stable Newcastle disease vaccine available to farmers and teaching them how to use it. By the end of 2006, the vaccination program had decreased chicken mortality by 80 percent. Average household poultry production increased from 16 to 74 chickens, with concomitant improvement in family nutrition and financial security. Many participants invested their surplus income in other livestock or facilities.

1944, and vaccination is routine in some countries (Case Study 4-8). Methods include spraying entire flocks, adding vaccines to feed or water, or vaccinating individual birds (Figure 4.9).

The live ND vaccine is effective, but it is unpopular because of its potential ability to revert to full virulence. Since 1970, inactivated vaccines have also been available. In 2006, researchers in Mexico reported the invention of genetically modified maize that induces ND antibody production when fed to poultry.

If chicks are not vaccinated during the first four days of life, vaccination should be postponed until the third week to avoid maternal antibody interference with active immune response. Some countries prohibit ND vaccination because of the problem of distinguishing between infected and vaccinated flocks.

Other control measures include quarantine and disinfection of contaminated areas. Heat, bleach, or ultraviolent light deactivates the virus on environmental surfaces.

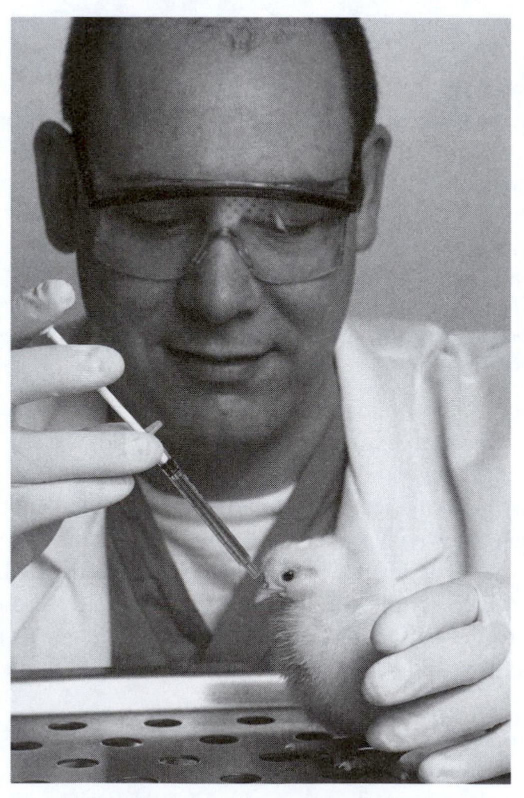

Figure 4.9 A baby chick receives a virosome vaccine to protect it from exotic Newcastle disease.
Source: U.S. Department of Agriculture, Agricultural Research Service.

Popular Culture

In 1962, Cuban Premier Fidel Castro claimed that the United States had deliberately infected his country's flocks with Newcastle disease. A Canadian poultry expert who visited Cuba reportedly told the press that the U.S. Central Intelligence Agency (CIA) had paid him $5,000 to introduce ND to Cuban turkeys. This expert further claimed that he discarded the viral cultures after accepting the money and never carried out his mission, yet an outbreak occurred in Cuba soon after his visit. ND is highly infectious, and the true source of the 1962 outbreak is unknown.

It is illegal to import cascarones—confetti-filled eggshells used in Easter celebrations—from Mexico to the United States because of the danger of spreading Newcastle disease to poultry.

Goosey Fair was an annual event in the town of Tavistock in southern England for over 800 years, where people bought their Christmas geese at an open market. After the mid-1960s, however, laws enacted to limit the spread of Newcastle disease made it illegal to sell live poultry at markets.

According to a field manual, African folk cures for ND include any of the following added to water: white vinegar, ground chili peppers, mango or tamarind bark, detergent, potassium permanganate, ground garlic, cactus sap, or car battery acid. Some African farmers associate outbreaks with the Harmattan (a seasonal dry wind), whereas others believe the disease strikes at Christmas time, or when mango trees are flowering—or when diseased birds are nearby. Popular control measures include selling sick birds, or eating them as quickly as possible and burying the leftovers.

The Future

Since about 1964, doctors have administered vaccine strains of Newcastle disease virus to human cancer patients as an oncolytic (tumor-destroying) agent or immune enhancer. Some clinical trials with NDV have yielded positive results, but the results are controversial. As of 2009, research in this area is continuing.

References and Recommended Reading

Aldous, E. W., and D. J. Alexander. "Newcastle Disease in Pheasants (*Phasianus colchicus*): A Review." *Veterinary Journal*, Vol. 175, 2008, pp. 181–185.

Alexander, D. J., et al. "Characterization of Paramyxoviruses Isolated from Penguins in Antarctica and Sub-Antarctica During 1976–1979." *Archives of Virology*, Vol. 109, 1989, pp. 135–143.

Alexander, D. J. "Newcastle Disease and Other Avian Paramyxoviruses." *Revue Scientifique et Technique*, Vol. 19, 2000, pp. 443–462.

Blockstein, D. E., and H. B. Tordoff. "Gone Forever: A Contemporary Look at the Extinction of the Passenger Pigeon." *American Birds*, Vol. 39, 1985, pp. 845–851.

Broder, J. M. "8,000 California Birds Killed in Bid to Stop Virus." *New York Times*, 26 October 2002.

Chakrabarti, S., et al. "Detection and Isolation of Exotic Newcastle Disease Virus from Field-Collected Flies." *Journal of Medical Entomology*, Vol. 44, 2007, pp. 840–844.

El Saawi, N. "Crying Fowl Play." Associated Press, 23 February 2003.

"Exotic Newcastle Disease." Factsheet, USDA Animal and Plant Health Inspection Service (APHIS) Veterinary Services, January 2003.

Goebel, S. J., et al. "Isolation of Avian Paramyxovirus 2 from a Patient with a Lethal Case of Pneumonia." *Journal of Virology*, Vol. 81, 2007, pp. 12709–12714.

Guerrero-Andrade, O., et al. "Expression of the Newcastle Disease Virus Fusion Protein in Transgenic Maize and Immunological Studies." *Transgenic Research*, Vol. 15, 2006, pp. 455–463.
Hanson, R. P. "The Possible Role of Infectious Agents in the Extinctions of Species." Pages 439–454 in Hickey, J. J. (Ed.). 1969. *Peregrine Falcon Populations*. Madison: University of Wisconsin Press.
Huang, Z., et al. "Recombinant Newcastle Disease Virus as a Vaccine Vector." *Poultry Science*, Vol. 82, 2003, pp. 899–906.
"Illegal Bird Smuggling Could Cause Disease." United Press International, 8 January 2005.
MacPherson, L. W. "Some Observations on the Epizootiology of Newcastle Disease." *Canadian Journal of Comparative Medicine*, Vol. 20, 1956, pp. 155–168.
Maendeleo Agricultural Technology Fund. "Dissemination of Thermo-Stable New Castle Disease Vaccine in Rural Chickens of Mwanza Region, Tanzania." *MATF Newsletter*, Vol. 6, 2007, pp. 14–15.
Romero, S. "Virus Takes a Toll on Texas Poultry Business." *New York Times*, 16 May 2003.
Sawahel, W. "GM Maize Protects Chickens from Deadly Virus." *Science and Development Network*, 18 August 2006.
Schorger, A. W. 1955. *The Passenger Pigeon: Its Natural History and Extinction*. Madison: University of Wisconsin Press.
Senne, D. A., et al. "Control of Newcastle Disease by Vaccination." *Developments in Biologicals* (Basel), Vol. 119, 2004, pp. 165–170.
Sinkovics, J. G., and J. C. Horvath. "Newcastle Disease Virus (NDV): Brief History of Its Oncolytic Strains." *Journal of Clinical Virology*, Vol. 16, 2000, pp. 1–15.
Spradbrow, P. B. "Epidemiology of Newcastle Disease and the Economics of Its Control." Workshop Proceedings, Danish Agricultural and Rural Development Advisers' Forum, 1999.
Verwoerd, D. J. "Ostrich Diseases." *Revue Scientifique et Technique*, Vol. 19, 2000, pp. 638–661.
Zulkifli, M. M., et al. "Newcastle Diseases Virus Strain V4UPM Displayed Oncolytic Ability against Experimental Human Malignant Glioma." *Neurological Research*, 18 October 2008.

AVIAN INFLUENZA

Summary of Threat

Avian influenza or bird flu is the same disease as influenza A (Chapter 2), but most strains that infect birds appear harmless to humans and vice versa. Some bird flu strains, including the highly publicized H5N1, can cause severe illness in both birds and humans and might have the potential to cause deadly pandemics. Avian influenza is classified as a biosecurity threat under the Bioterrorism Protection Act of 2002.

Other Names

The names of influenza A antigenic subtypes refer to glycoproteins on the surface of the virus (see Figure 2-7, page 33). The letter H followed by a number identifies a glycoprotein called a hemagglutinin, and N followed by a number identifies another glycoprotein called a neuraminidase. Thus, H5N1 is the influenza A virus subtype that has hemagglutinin 5 and neuraminidase 1. Within each subtype are various strains, some more dangerous than others. Some strains or subtypes also have common names based on the animal species they usually infect, such as avian (bird) flu, equine (horse) flu, canine (dog) flu, and porcine (swine) flu.

The acronym HPAI stands for "highly pathogenic avian influenza." (In this case, the H does not stand for hemagglutinin.) HPAI was once called fowl plague, a name that now usually refers to Newcastle disease. Less dangerous strains are called LPAI (low pathogenic avian influenza) or

NPAI (non-pathogenic avian influenza). Words for influenza in most languages refer to the human disease (Chapter 2).

Description

Whether the patient is a bird, a human, or a horse, the avian influenza virus (Figure 4.10) causes high fever, breathing difficulty, nasal discharge, and the appearance of discomfort. In birds, egg production may drop and feathers may appear ruffled. Although most strains are relatively harmless, HPAI is an exception. It often causes additional symptoms, such as swelling, greenish diarrhea, hemorrhaging under the skin, and neurological complications. HPAI can rapidly kill large numbers of birds.

Poultry workers and other exposed humans can also contract bird flu, and in a few cases it has spread directly from one human to another. In 2006, a person in Indonesia transmitted H5N1

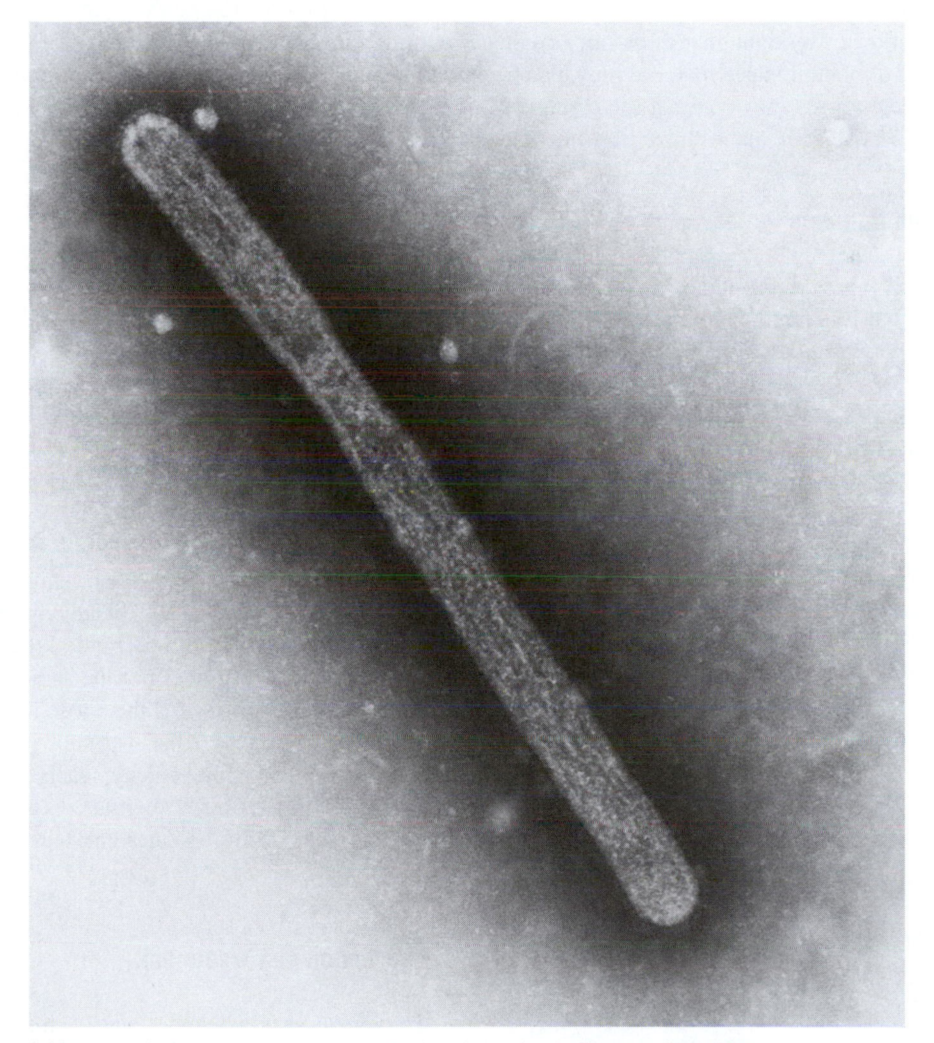

Figure 4.10 Transmission electron micrograph showing avian influenza virus H5N1.
Source: U.S. Centers for Disease Control and Prevention, Public Health Image Library.

to six household members, and one infected an eighth person. But bird flu remains largely a disease of birds, and the feared human pandemic has not yet happened. As of 2009, avian influenza occurs worldwide, but the strains known as HPAI (H5N1 and others) have not been reported in the United States since 2004.

Transmission of avian influenza is usually by close-range airborne inhalation, direct contact with infected animals, ingestion of infected birds, or contact with contaminated surfaces. The incubation period ranges from a few days to one week.

Which Animals Are at Risk?

Domestic chickens, turkeys, geese, parrots, ostriches, and other birds are at risk for avian influenza. It spreads rapidly and causes high mortality in confined poultry feeding operations, but migratory wild ducks, gulls, and shorebirds sometimes carry the virus in their intestines without becoming sick.

Humans have contracted bird flu by prolonged contact with infected poultry (or people), but apparently not by handling, cooking, or consumption of dressed poultry meat. Sewer workers exposed to contaminated material may also be at risk. Domestic dogs, cats, and other animals that eat infected birds may contract the disease (Case Study 4-9). Alligators and crocodiles are also potential hosts, but studies have been inconclusive.

The Numbers

H5N1 has killed hundreds of millions of birds in Russia, the Middle East, Europe, and Africa, but it is not the only potentially lethal subtype. The 1983 outbreak of H5N2 in the northeastern United States killed an estimated 17 million chickens.

As of 2008, there were 16 known hemagglutinin (H) subtypes and 9 known neuraminidase (N) subtypes, for a total of 144 (16×9) possible influenza subtypes, all with possibly different host susceptibilities.

Case Study 4-9: Tweety Nails Sylvester

In 2004, the media reported that HPAI-H5N1 avian influenza killed a large number of tigers at two zoological parks in Thailand. In both cases, the tigers apparently contracted the disease by eating raw or inadequately cooked chicken or pork. One zoo had 441 tigers, of which 45 died of influenza and 102 were euthanized. Antiviral drugs may have saved some of them. At the other zoo, two tigers and two leopards died of the same disease. The tiger is an endangered species, and the 149 tigers that died of bird flu represented about 3 percent of all tigers in the world.

History

Avian influenza has existed for centuries (or longer), but the first known cases of subtype H5N1 were in Scotland in 1959. It reportedly killed two flocks of chickens, but it was not the same as the highly pathogenic H5N1 virus that arose in China in 1997 and resurfaced in 2004. The subtype alone does not determine pathogenicity; each subtype has various strains. Table 4.2 summarizes the most common subtypes with the dates and locations of recent outbreaks.

Prevention and Treatment

Influenza vaccines can control losses by reducing viral replication and shedding. In the

Table 4.2 Recent Avian Influenza Outbreaks

Subtype	Outbreak	Location
H5N1	1997	Europe, Asia, Africa
H5N2	1983	Pennsylvania
H5N2	1994	Mexico
H5N2	1997	Italy
H5N2	2004	Texas
H7N1	1999	Italy
H7N3	1994	Australia, Pakistan
H7N3	2002	Chile
H7N4	1997	Australia
H7N7	2003	Netherlands

United States, the use of H5 and H7 avian influenza vaccines on animals requires USDA approval. In 2008, U.S. researchers reported a new vaccine that could protect chickens, cats, and humans in the event of a bird flu pandemic. Infected flocks can also be treated with antibiotics to control secondary infections. In 2006, the FDA prohibited the use of antiviral drugs to treat poultry because of the danger of drug resistance.

As of 2009, there is no proof that humans have caught avian influenza by handling dressed poultry, but reasonable precautions are in order. Eating raw chicken (or feeding it to the dog, cat, or pig) is clearly out. Consumers should cook poultry thoroughly and wash anything that comes in contact with raw poultry or eggs.

Popular Culture

The 2006 made-for-TV movie *Fatal Contact: Bird Flu in America* is not a documentary but a work of fiction, in which the H5N1 influenza virus mutates into a pandemic form that spreads easily from person to person. An American businessman visiting China contracts the virus in a Hong Kong market and takes it home. The movie highlights the importance of cooperation and planning, and even mentions Tamiflu® resistance. As expected, the pandemic swamps the health-care system and disrupts other public services.

During the 2004 avian influenza outbreak, investigators tracked rumors of outbreaks in both humans and animals. Most rumors appeared within the first seven weeks after the public health alert; even in a crisis, people quickly adjust and go on with their lives. Rumors regarding human cases were both more numerous and less likely to be true than those regarding animal outbreaks. The findings were consistent with the basic law of rumor, which states that the number of rumors in circulation is roughly equal to the importance of the rumor times the uncertainty surrounding the rumor. As more information becomes available, uncertainty decreases, and fewer rumors circulate even if the importance remains high.

In 2005, the media reported that British scientists quarantined a parrot imported from Surinam that was suspected of having avian influenza. Eventually, someone noticed that the parrot was already dead.

The Future

For years, some scientists have warned that H5N1 or a related virus could mutate to a form resembling the Spanish Flu on steroids and wipe out as many as 1 billion people. Others, however, have concluded that the high death rate in the 1918 pandemic resulted mainly from secondary bacterial infections and not from unique properties of the flu itself. Thus, to mix a metaphor, bird flu (like the 2009 swine flu) may be a paper tiger. The real problem that won't go away might be antibiotic resistance (Chapter 3), which could make it nearly as hard to fight bacterial infections in 2018 as in 1918.

References and Recommended Reading

Alexander, D. J. "An Overview of the Epidemiology of Avian Influenza." *Vaccine*, Vol. 25, 2005, pp. 5637–5644.

Causey, D., and S. V. Edwards. "Ecology of Avian Influenza in Birds." *Journal of Infectious Disease*, Vol. 197, Supplement 1, 2008, pp. S29–S33.

Davis, L. M., and E. Spackman. "Do Crocodilians Get the Flu? Looking for Influenza A in Captive Crocodilians." *Journal of Experimental Zoology, Part A, Ecological Genetics and Physiology*, 31 March 2008.

Fish, D. "What About the Ducks? An Alternative Vaccination Strategy." *Yale Journal of Biology and Medicine*, Vol. 78, 2005, pp. 301–308.

Fitzpatrick, S. "Jakarta Bird Flu Theory 'Nutty.'" *The Australian*, 27 February 2008.

Gharaibeh, S. "Pathogenicity of an Avian Influenza Virus Serotype H9N2 in Chickens." *Avian Diseases*, Vol. 51, 2008, pp. 106–110.

Gilbert, M. "Climate Change and Avian Influenza." *Revue Scientifique et Technique*, Vol. 27, 2008, pp. 459–466.

Lipatev, A. S., et al. "Domestic Pigs have Low Susceptibility to H5N1 Highly Pathogenic Avian Influenza Viruses." *PLoS Pathogens*, Vol. 4, 2008, pp. e1000102.

Londt, B. Z., et al. "Highly Pathogenic Avian Influenza Viruses with Low Virulence for Chickens in *In Vivo* Tests." *Avian Pathology*, Vol. 36, 2007, pp. 347–350.

Marangon, S., and L. Busani. "The Use of Vaccination in Poultry Production." *Revue Scientifique et Technique*, Vol. 26, 2007, pp. 265–274.

McDowell, R. "US Controls Bird Flu Vaccines over Bioweapon Fears." Associated Press, 11 October 2008.

Mumford, E. L., and U. Kihm. "Integrated Risk Reduction along the Food Chain." *Annals of the New York Academy of Sciences*, Vol. 1081, 2006, pp. 147–152.

Mumford, E. L., et al. 2007. "Avian Influenza H5N1: Risks at the Human-Animal Interface." *Food and Nutrition Bulletin*, Vol. 28 (Supplement), 2007, pp. S357–S363.

Olsen, B., et al. "Global Patterns of Influenza A Virus in Wild Birds." *Science*, Vol. 312, 2006, pp. 384–388.

Rosenthal, E. "Bird Flu Reports Multiply in Turkey, Faster than Expected." *New York Times*, 9 January 2006.

Samaan, G., et al. "Rumor Surveillance and Avian Influenza H5N1." *Emerging Infectious Diseases*, Vol. 11, 2005, pp. 463–466.

Sims, L. D., et al. "Origin and Evolution of Highly Pathogenic H5N1 Avian Influenza in Asia." *Veterinary Record*, Vol. 157, 2005, pp. 159–164.

Song, D., et al. "Transmission of Avian Influenza Virus (H3N2) to Dogs." *Emerging Infectious Diseases*, Vol. 14, 2008, pp. 741–746.

Spackman, E. "A Brief Introduction to the Avian Influenza Virus." *Methods in Molecular Biology*, Vol. 436, 2008, pp. 1–6.

Specter, M. "Nature's Bioterrorist." *The New Yorker*, 5 February 2005.

Srikantiah, S. "Mass Culling for Avian Influenza: Rational Strategy or Needless Destruction?" *Indian Journal of Medical Ethics*, Vol. 5, 2008, pp. 52–54.

Thanawongnuwech, R., et al. "Probable Tiger-to-Tiger Transmission of Avian Influenza H5N1." *Emerging Infectious Diseases*, Vol. 11, 2005, pp. 699–701.

"Vaccine Protects Against Bird Flu." United Press International, 20 October 2008.
Walsh, B. "Living Cheek by Beak in Indonesia." *Time*, 14 June 2007.
Weber, T. P., and N. I. Stilianakis. "Ecologic Immunology of Avian Influenza (H5N1) in Migratory Birds." *Emerging Infectious Diseases*, Vol. 13, 2007, pp. 1139–1143.
Yamada, T., et al. "Ready for Avian Flu?" *Nature*, Vol. 454, 2008, p. 162.
Zamiska, N. "How Academic Flap Hurt World Effort on Chinese Bird Flu." *Wall Street Journal*, 27 February 2006.

HONEYBEE COLONY COLLAPSE DISORDER

Summary of Threat

In colony collapse disorder, large numbers of bees die or fail to return to the hive. As of 2009, scientists have not yet determined the exact reason for this phenomenon. Parasites, bacteria, viruses, fungi, stress, and pesticides all may be contributing factors. There is evidence that periodic die-offs of bee colonies have plagued humanity for thousands of years.

Other Names

Colony collapse disorder is also called CCD, BCCD (bee colony collapse disorder), fall dwindle, spring dwindle, May disease, autumn collapse, bee die-off, hive death, disappearing disease, vanishing bee syndrome, or *Mary Celeste* syndrome. Most of these names are self-explanatory; the last refers to a ship that was found sailing the Atlantic in 1872 without its passengers or crew.

One website author facetiously identifies colony collapse disorder as a psychiatric problem called ADHBee, with reference to the contemporary buzzword—sorry, now we're doing it—ADHD, or attention deficit hyperactivity disorder. In other words, the bees lose interest in the things we want them to do (visiting flowers or making wax or whatever) and just wander off. Facetious or not, it's as good an explanation as any to date.

Description

As of 2009, colony collapse disorder is a symptom rather than a disease in the usual sense. Worker bees (Figure 4.11) fail to return to the hive, and the colony dies, but experts are not sure why. There is some evidence that such die-offs are simply a characteristic of honeybee colonies, and that the crisis is largely an illusion.

Various studies have identified the culprit as Israeli acute paralysis virus, deformed wing virus, an Asian bee parasite (*Nosema ceranae*), or the familiar varroa mites (Figure 4.12) and tracheal mites that decimated American bee colonies in the mid-1980s. Other studies have implicated clothianidin and other pesticides, electromagnetic radiation from cell phones, inbreeding, climate change, habitat loss, immune suppression, or stress (Case Study 4-10).

But while some farmers complain about a shortage of bees, others claim that there are too many bees in the wrong places. For example, almond growers in California's San Joaquin Valley need bees for pollination, but tangerine growers on adjacent farms need to keep bees away. If pollinated, tangerines develop unwanted seeds (pips) that reduce their market value. Similar problems affect seedless grapes and watermelons. The law protects bees by requiring growers to avoid

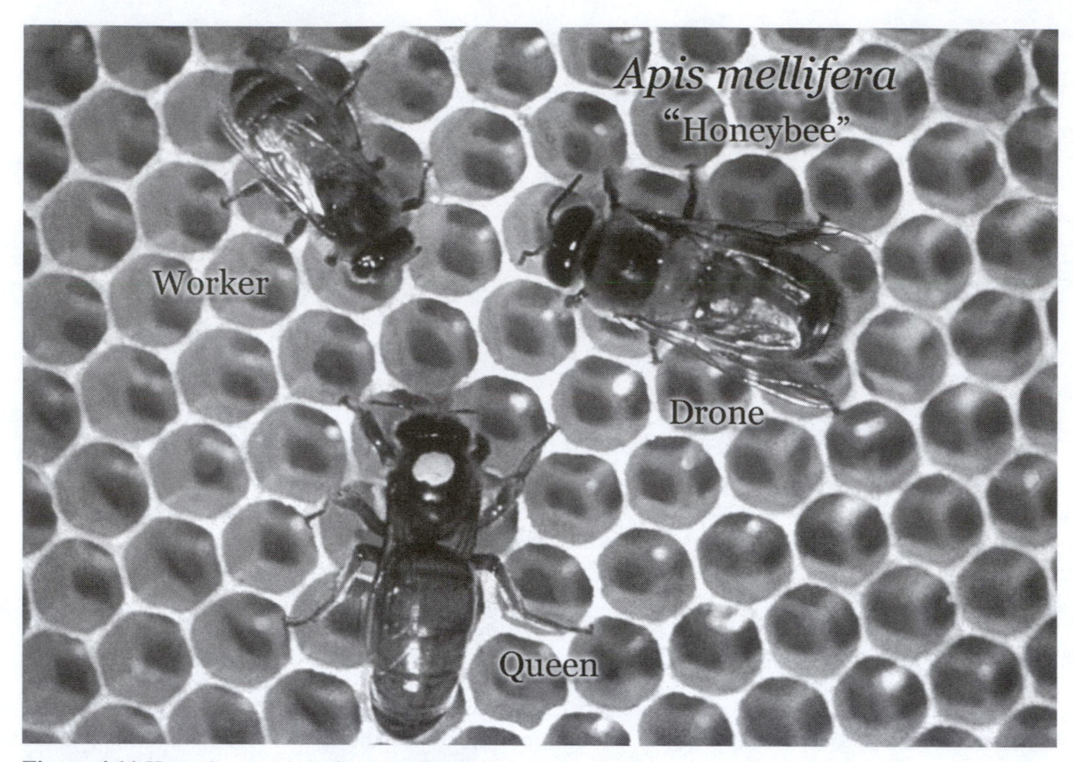

Figure 4.11 Honeybee worker, drone, and queen.

Source: U.S. Centers for Disease Control and Prevention, Public Health Image Library.

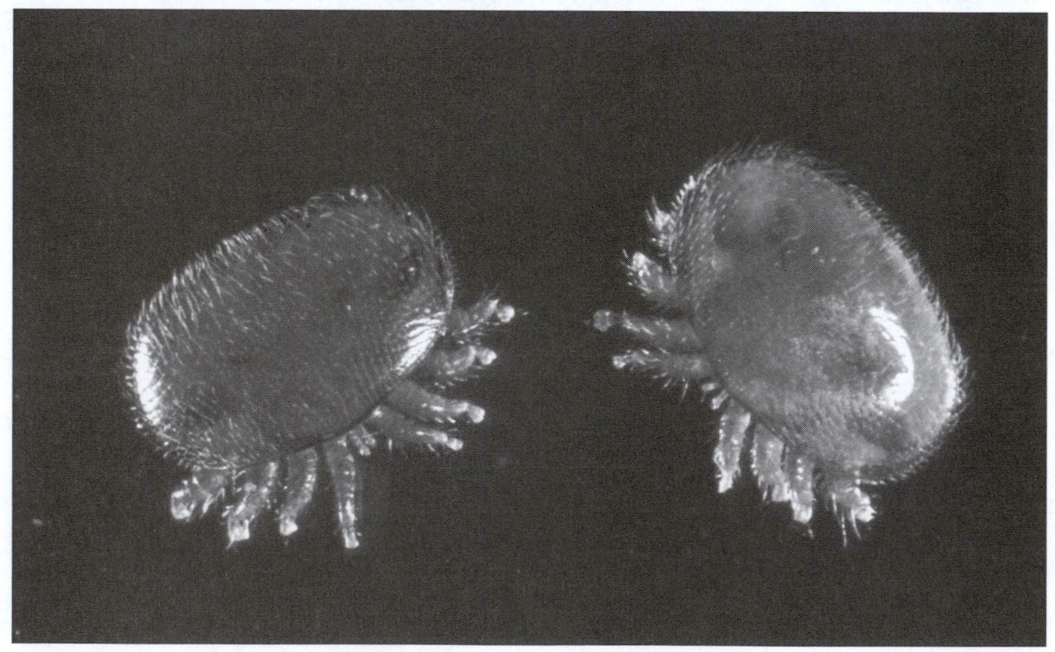

Figure 4.12 *Varroa jacobsoni*, a mite that lives as a parasite on honeybees.

Source: U.S. Department of Agriculture, Agricultural Research Service.

spraying for pests when bees are present, but scheduling conflicts are inevitable.

Which Animals (and Plants) Are at Risk?

Bees extract nectar from flowers, modify it by digestion and evaporation, and store it in the form that we call honey. But in the process of gathering nectar, the bees also collect pollen on their bodies and transfer it to other flowers, more efficiently than wind alone could do. This is the plant world's equivalent of sex. Thus, anything that threatens bees also threatens many plants (Table 4.3).

Colony collapse disorder has affected managed honeybee colonies in North America, Europe, and Asia at least since 2006. Most wild bee species are not at risk for colony collapse because they do not live in colonies, but they might be individually susceptible to whatever disease, pesticide, or other stressors may be causing colony collapse. As of 2009, the cause and risk factors remain unknown.

Case Study 4-10: Give Bees a Chance

Stress is more than a human malady. To the extent that the term encompasses a range of biological responses to perceived threats or unfavorable conditions, it is probably fair to say that bees can experience stress. The question is whether a commercial bee lot subjects bees to a level of stress for which evolution has not prepared them. Wild honeybees usually build three or four hives per square mile, for example, whereas managed colonies are packed together in rows a few feet apart. When forage is scarce, farmers keep bees alive by giving them high-fructose corn syrup. And wild honeybees normally have a winter rest period, but winter-blooming crops won't wait. In 2007, journalists described how beekeepers from all over the United States converged on California in a mad dash to pollinate the almond crop and collect their money before the bees dropped dead. Perhaps the majority of bees simply cannot adjust to fast food and a commuter lifestyle.

Table 4.3 Food Crops That Require Bees for Pollination

Name	Pollinators	Product
Almond	Honeybees, bumblebees, solitary bees	Nut
Apple	Honeybees, bumblebees, solitary bees	Fruit
Apricot	Honeybees, bumblebees, solitary bees	Fruit
Avocado	Honeybees, stingless bees, solitary bees	Fruit
Blackberry	Honeybees, bumblebees, solitary bees	Fruit
Blueberry	Honeybees, bumblebees, solitary bees	Fruit
Brazil nut	Bumblebees, orchid bees, carpenter bees	Nut
Broccoli	Honeybees, solitary bees	Flower, stem
Brussels sprouts	Honeybees, solitary bees	Leaves
Buckwheat	Honeybees, solitary bees	Seed
Cabbage	Honeybees, solitary bees	Leaves
Cacao	Flies and unidentified wild bees	Seed (cocoa)
Canola	Honeybees, solitary bees	Seed, oil
Cantaloupe	Honeybees, bumblebees, solitary bees	Fruit
Casaba melon	Honeybees, bumblebees, solitary bees	Fruit
Cashew	Honeybees, bumblebees, stingless & solitary bees	Nut
Cauliflower	Honeybees, solitary bees	Flower
Chayote	Honeybees, stingless bees	Fruit
Cherry	Honeybees, bumblebees, solitary bees	Fruit
Citron	Honeybees, bumblebees, solitary bees	Fruit

(Continued)

Table 4.3 (*Continued*)

Name	Pollinators	Product
Cranberry	Honeybees, bumblebees, solitary bees	Fruit
Crenshaw melon	Honeybees, bumblebees, solitary bees	Fruit
Cucumber	Honeybees, bumblebees, solitary bees	Fruit
Currant	Honeybees, bumblebees, solitary bees	Fruit
Dewberry	Honeybees, solitary bees	Fruit
Eggplant	Honeybees, bumblebees, solitary bees	Fruit
Feijoa	Honeybees, solitary bees	Fruit
Fennel	Honeybees, solitary bees	Seed, bulb
Gooseberry	Honeybees, solitary bees	Fruit
Hazelnut	Honeybees, solitary bees	Nut
Honeydew melon	Honeybees, bumblebees, solitary bees	Fruit
Huckleberry	Honeybees, solitary bees	Fruit
Jujube	Honeybees, solitary bees	Fruit
Kiwi fruit	Honeybees, bumblebees, solitary bees	Fruit
Kohlrabi	Honeybees, solitary bees	Leaves
Loquat	Honeybees, bumblebees	Fruit
Macadamia	Honeybees, stingless bees, solitary bees	Nut
Mango	Honeybees, stingless bees	Fruit
Mustard	Honeybees, solitary bees	Seed, oil
Nectarine	Honeybees, bumblebees, solitary bees	Fruit
Onion	Honeybees, solitary bees	Bulb
Parsley	Honeybees, flies	Leaves
Parsnip	Honeybees, solitary bees	Root
Passionfruit	Carpenter bees, solitary bees, bumblebees	Fruit
Peach	Honeybees, bumblebees, solitary bees	Fruit
Pear	Honeybees, bumblebees, solitary bees	Fruit
Persian melon	Honeybees, bumblebees, solitary bees	Fruit
Persimmon	Honeybees, bumblebees, solitary bees	Fruit
Plum	Honeybees, bumblebees, solitary bees	Fruit
Pumpkin	Honeybees, bumblebees, solitary bees	Fruit
Radish	Honeybees, solitary bees	Root
Raspberry	Honeybees, bumblebees, solitary bees	Fruit
Rutabaga	Honeybees, solitary bees	Root
Squash	Honeybees, bumblebees, solitary bees	Fruit
Starfruit	Honeybees, stingless bees	Fruit
Sunflower	Honeybees, wild bees	Seed, oil
Tea	Wild bees	Leaves
Turnip	Honeybees, solitary bees	Root, leaves
Watermelon	Honeybees, bumblebees, solitary bees	Fruit

Note: Many other crops also benefit from bee pollination but do not absolutely require it. Sources differ as to the primary pollinators for certain plants.
Source: USDA Agricultural Research Service.

The Numbers

In the United States, an estimated 2.4 million to 3 million managed honeybee colonies pollinate $14 billion worth of crops every year. In addition to the value of the crops themselves, beekeepers earn a living by supplying farmers with beehives for pollination. In 2005, beekeepers in California alone received over $121 million for this service, in addition to sales of 10,000 tons of honey and 200 tons of beeswax.

History

In the fall of 2006, American beekeepers reported that approximately one-third of all managed colonies were suddenly vacant for no apparent reason. By the end of 2007, the problem was worse; according to a report in *Science*, between 50 percent and 90 percent of all honeybee colonies in the United States were lost.

Bee diseases and pests may spread more rapidly nowadays because of the frequent long-distance transport of beehives, but the myth of Aristaeus (see Popular Culture) suggests that beekeepers have had similar problems for thousands of years. When colonists brought the honeybee to America in the 1600s, its diseases and pests came along for the ride. Beekeepers reported large die-offs in the 1890s, and something called "disappearing disease" affected many colonies in the late 1970s. Honeybees, like all living things, coexist with a host of microbes, parasites, predators, and competitors whose precise interactions may never be fully understood, although the effort continues.

Prevention and Treatment

Agricultural extension offices distribute literature on methods of protecting bees from pesticides and other known or suspected stressors. There are also online and traditional college courses in beekeeping management. Unfortunately, no one knows exactly how to prevent colony collapse disorder.

Chemical pesticides are available for treatment of mite infestations, but some of these chemicals also harm bees, and others have become less effective on mites due to resistance.

Popular Culture

Archaeological evidence shows that people in Egypt and the Middle East have raised honeybees in portable hives for at least 5,000 years. According to ancient Greek mythology, the first beekeeper was Aristaeus, who also dabbled in the nascent olive oil and cheesemaking industries. He was doing quite well, until one day he chased after a married woman—the wife of the divine Orpheus, worse luck—and she stepped on a venomous snake and died. Thus, the gods punished Aristaeus by killing all his bees. Olives are wind-pollinated, but honey and wax were also valuable commodities, and Aristaeus was devastated. But instead of apologizing to the widower, or trying to figure out what went wrong, Aristaeus scolded his *mother* for not protecting him. (The story is reminiscent of many a dispute between the farming industry and the USDA.) Eventually, on the advice of a shape-shifting prophet, Aristaeus made amends by sacrificing some cattle in a sacred grove. Nine days later, he returned to the grove and found that a swarm of wild bees had landed on one of the carcasses, thus putting him back in business.

In ancient times, many European societies regarded bees as messengers who notified the gods when a person died. This tradition may explain the rural European and American custom of "telling the bees" about the death of a household member. If not thus informed, the bees may die or stop making honey. Other popular beliefs held that bees would die if purchased on a Friday, or if purchased for cash instead of barter, or if people quarreled about their ownership. In A.D. 77, Pliny the Elder wrote that a swarm of bees would die if a menstruating woman looked at them. These beliefs suggest that bee colonies have always tended to die unexpectedly.

The Future

Honeybee colonies will probably continue to suffer massive die-offs, because they have always done so. One possible solution is to encourage more pollination of crops by wild native bees. Many wild bees nest in the ground, for example, so farmers need to watch where they plow. These insects are also vulnerable to pesticides, and they need flowers during the breeding season. But with a bit of effort and planning, some farms near native habitats can receive full pollination from wild bees alone.

References and Recommended Reading

Aliouane, Y., et al. "Subchronic Exposure of Honeybees to Sublethal Doses of Pesticides: Effects on Behavior." *Environmental Toxicology and Chemistry*, 13 August 2008.

Amdam, G. V., and S.-C. Seehuu. "Order, Disorder, Death: Lessons from a Superorganism." *Advances in Cancer Research,* Vol. 95, 2006, pp. 31–60.

Anderson, D., and I. J. East. "The Latest Buzz about Colony Collapse Disorder." *Science*, Vol. 319, 2008, pp. 724–725.

Antúnez, K., et al. "Honeybee Viruses in Uruguay." *Journal of Invertebrate Pathology*, Vol. 93, 2006, pp. 67–70.

Ariana, A., et al. "Laboratory Analysis of Some Plant Essences to Control *Varroa destructor* (Acari: Varroidae)." *Experimental and Applied Acarology*, Vol. 27, 2002, pp. 319–327.

Barbassa, J. "Honeybee Deaths Increase." Associated Press, 7 May 2008.

Benjamin, A. "Last Flight of the Honeybee?" *The Guardian*, 31 May 2008.

"Clues Sought in Honey Bees' Demise." United Press International, 1 July 2008.

Cone, T. "Tangerine Growers Tell Beekeepers to Buzz Off." Associated Press, 9 January 2009.

Cox-Foster, D. L., et al. "A Metagenomic Survey of Microbes in Honey Bee Colony Collapse Disorder." *Science*, Vol. 318, 2007, pp. 283–287.

Delaplane, K. S., and D. F. Mayer. 2000. *Crop Pollination by Bees*. New York: CABI Publishing.

Farley, J.D. "Bee by Bee." *New York Times*, 30 June 2008.

" Honeybee Disorder Still Stumps Researchers." United Press International, 5 January 2009.

Kay, J. "Lawsuit Seeks EPA Pesticide Data." *San Francisco Chronicle*, 18 August 2008.

Kuchinskas, S. "Are the Bees Dying Off Because They're Too Busy?" *East Bay Express*, 11 August 2007.

Muz, M. N. "Sudden Die-Off of Honeybee Colonies." *Türkiye Parazitolojii Dergisi,* Vol. 32, 2008, pp. 271–275. [Turkish]

Pinto, M. A., et al. "Temporal Pattern of Africanization in a Feral Honeybee Population from Texas Inferred from Mitochondrial DNA." *Evolution*, Vol. 58, 2004, pp. 1047–1055.

Revkin, A. C. "Bees Dying: Is It a Crisis or a Phase?" *New York Times*, 17 July 2007.

Roulston, T. H. "Practices That Encourage Native Bees in Vine Crops." Great Lakes Fruit Vegetable and Farm Market Expo, Grand Rapids, MI, 9–11 December 2008.

Russell, S. "UCSF Scientist Tracks Down Suspect in Honeybee Deaths." *San Francisco Chronicle*, 26 April 2007.

Seeley, T. D., and D. R. Tarpy. "Queen Promiscuity Lowers Disease Within Honeybee Colonies." *Proceedings Biological Sciences*, Vol. 274, 2007, pp. 67–72.

Shen, M., et al. "Intricate Transmission Routes and Interactions between Picorna-like Viruses (Kashmir Bee Virus and Sacbrood Virus) with the Honeybee Host and the Parasitic Varroa Mite." *Journal of General Virology*, Vol. 86, 2005, pp. 2281–2289.

CONCLUSION

Infectious disease outbreaks represent a major threat for species that live in densely packed communities. This statement applies to wild bee colonies, bat caves, and human cities, as well as cattle feedlots and crowded piggeries. New vaccines and genetically modified livestock strains may alleviate this problem, but new diseases will continue to appear. Populations that pass through a bottleneck of disease will theoretically emerge with enhanced resistance, but that reward offers small consolation in the short term.

5

Food Insecurity, Continued

The great peril of our existence lies in the fact that our diet consists entirely of souls.

—Attributed to Aua, Inuit shaman

A major die-off of livestock might plunge world markets into chaos and exacerbate international tensions, but global famine would not be automatic. Meat and milk represent only about 9 percent of total caloric intake in low-income nations and 24 percent in high-income nations. The balance of the human diet consists of grains and other plant products. Nor is this a modern aberration; many hunter-gatherer societies obtained an estimated 15 percent of their diet from animals and 85 percent from plants.

In other words, plants are what we really need. If even one staple crop fails, such as rice, wheat, or maize, widespread starvation may result. Rice alone represents about 20 percent of the world's food, and wheat and maize together account for another 40 percent. So how safe are these resources? Table 5.1 lists examples of past outbreaks, and Table 5.2 summarizes major diseases of staple crops.

Before dismissing the ten examples in this chapter as inconsequential brown spots or green bugs, consider the fact that the U.S. Department of Homeland Security lists some of them as biosecurity threats. These tiny adversaries (and hundreds of others) have caused more death and economic devastation than any human army, and in many cases we don't know how to stop them.

CITRUS TRISTEZA VIRUS

Summary of Threat

Citrus tristeza virus (CTV) is a catastrophic disease of orange, grapefruit, and other citrus fruits, which are among the most important cash crops in tropical and subtropical regions worldwide. Transmission is mainly by aphid vector, grafting, and mechanical damage. Outbreaks have destroyed millions of trees and caused great economic damage. Control of tristeza is difficult, and new strains continue to arise.

Table 5.1 Some Major Outbreaks of Plant Diseases and Pests

Disease	Year	Location	Estimated Losses
Citrus tristeza virus	1940s	Brazil	9 million citrus trees
Citrus tristeza virus	1984–1986	Florida	50% of citrus crop
Citrus tristeza virus	2002	Italy	10,000 citrus trees
Bacterial wilt (tomato)	1966–1968	Philippines	15% of tomato crop
Bacterial wilt (banana)	2006	Uganda	Up to 90% of banana crop
Southern corn leaf blight	1970	United States	15% of corn crop
Citrus canker	1984–1989	Florida	20 million citrus trees[1]
Late blight of potato	1845–1849	Ireland	Potato crop + 1 million humans
Late blight of potato	1997	Peru	80% of potato crop
Soybean rust	2003	Brazil	$1.3 billion
Witches' broom disease	1989	Brazil	90% of cacao crop
Phoma stem canker	2004	Argentina	32% of canola in affected fields
Wheat rust	1954	United States	40% of wheat crop
Asian soybean aphid	2000–2002	United States	$2.2 billion
Desert locust	2004	West Africa	$2.5 billion

[1]Burned as part of an unsuccessful campaign to eradicate the disease.

Other Names

Citrus tristeza virus is also called CTV, citrus tristeza closterovirus, citrus quick decline virus, tristeza-quick decline (T-QD) virus, grapevine A virus, grapefruit stem pitting virus, grapefruit stunt bush virus, Ellendale mandarin decline virus, hassaku dwarf virus, citrus seedling yellows virus, and lime dieback virus. Names in other languages include *tristeza dos citros* (Portuguese, "sadness of citrus") and *podredumbre de las raicillas* (Spanish, "corruption of citrus").

Description

The agent is an RNA virus that infects citrus trees, causing loss or discoloration of foliage or stem pitting (Figure 5.1). Infected trees may die within a few months or survive for years in poor condition, producing small fruit. This disease may occur wherever citrus trees are present. Specific effects depend on scion varieties and rootstock combinations, but there are three distinct syndromes:

- Tristeza, the "quick decline" form of the disease, first seen on citrus grafted on sour orange, with wilting and dieback
- Stem pitting or honeycombing, visible under the bark on small branches
- Seedling yellows, with stunting, curling, and yellowing of leaves

The most efficient vectors for this disease are brown citrus aphids (*Toxoptera citricida*, Case Study 5-1), but other aphids get the job done. Trees also become infected by grafting and mechanical inoculation, or by way of a parasitic plant called a bridging dodder.

Table 5.2 The World's Staple Crops and Their Principal Diseases

Crop	Fungi	Viruses	Nematodes	Bacteria
Barley	Mildew (*Erysiphe graminis*) Spot blotch (*Cochliobolus sativus*) Scald (*Rhynchosporium secalis*) Scab (*Gibberella zeae*) Rusts (*Puccinia*) Net blotch (*Pyrenophora teres*) Barley stripe (*Pyrenophora graminea*) Smuts (*Ustilago*)	Barley yellow dwarf luteovirus Barley stripe mosaic hordeivirus	Root-knot nematode (*Meloidogyne*) Cyst nematode (*Heterodera*) Root-lesion nematode (*Pratylenchus*)	
Cassava	Anthracnose (*Colletotrichum gloeosporioides*) Root rot (*Polyporus sulphureus*)	African cassava mosaic geminivirus East African cassava mosaic geminivirus Indian cassava mosaic geminivirus		Bacterial blight (*Xanthomonas axonopodis* pv. *manihotis*)
Lentil	Wilt (*Fusarium oxysporum* f.sp. *lentis*) Blight (*Ascochyta lentis*) Rust (*Uromyces viciae-fabae*) Vascular wilt (*Fusarium oxysporum* f.sp. *lentis*) Anthracnose (*Colletotrichum truncatum*)			
Maize	Northern corn leaf blight (*Helminthosporium turcicum*) Downy mildew (*Sclerospora* and others) Southern corn leaf blight (*H. maydis*) Rust (*Puccinia*) Smut (*Ustilago zeae*) Stalk and ear rots (*Gibberella zeae*, *Diplodia*, others)	Chlorotic dwarf machlovirus Streak: geminivirus Yellow dwarf luteovirus		Stewart's wilt (*Erwinia stewartii*) Corn stunt disease (*Spiroplasma kunkelii*)
Common millet	Downy mildew (*Sclerospora graminicola*)			
Finger millet	Blast (*Pyricularia setariae*) Leaf blight (*Cochliobolus nodulosus*)			

(Continued)

Table 5.2 (*Continued*)

Crop	Fungi	Viruses	Nematodes	Bacteria
Foxtail millet	Blast (*Pyricularia setariae)* Rust (*Uromyces setariae-italicae)* Smut (*Ustilago crameri)* Downy mildew (*Sclerospora graminicola)*			
Pearl millet	Ergot (*Claviceps fusiformis)* Downy mildew (*Sclerospora graminicola)*			
Teff	Rust (*Uromyces eragrostidis)* Head smudge (*Helminthosporium miyakei)*			
Oats	Crown rust (*Puccinia coronata)* Stem rust (*Puccinia graminis)* Powdery mildew (*Erysiphe graminis*) Smut (*Ustilago avenae* and *U. hordei*) Leaf blight (*Phaeosphaeria avenaria*) Root rot and crown rot (*Fusarium*) Seedling blight (*Glomerella graminicola*) Snow mold (*Monographella nivalis*) Leaf blotch (*Pyrenophera avenae*) Groat blackening (*Alternaria* and *Cladosporium*)	Barley yellow dwarf luteovirus Oat mosaic potyvirus Oat golden stripe furovirus		Halo blight (*Pseudomonas syringae* pv. *coronafaciens)*
Potato	Early blight (*Alternaria solani*) Black scurf(*Rhizoctonia solani*) Late blight (*Phytophthora infestans*) Pink rot (*Phytophthora erythroseptica*)	Potato leafroll luteovirus Potato X potexvirus Potato Y potyvirus		Bacterial wilt (*Ralstonia solanacearum*) Bacterial soft rot (*Erwinia carotovora*) Common scab (*Streptomyces scabies*) Bacterial ring rot (*Clavibacter michiganensus* subsp. *sepedonicus*)
Rice	Blast (*Magnaporthe grisea*) Brown spot (*Cochliobolus miyabeanus*) Sheath blight (*Rhizoctonia solani*)	Rice tungro spherical machlovirus Rice tungro bacilliform badnavirus Barley yellow dwarf luteovirus		Bacterial leaf blight (*Xanthomonas oryzae* pv. *oryzae*)

Rye	Snow mold (*Monographella nivalis*) Brown rust (*Puccinia recondita*) Ergot (*Claviceps purpurea*) Eyespot (*Tapesia yallundae*) Sharp eyespot (*Rhizoctonia solani*) Powdery mildew (*Erysiphe graminis*) Stem rust (*Puccinia graminis*) Glume blotch (*Phaeosphaeria nodorum*)	Barley yellow dwarf luteovirus	Eelworm (*Ditylenchus dipsaci*)	
Sorghum	Grain molds (*Cochliobolus, Fusarium, Mycosphaerella*, others) Anthracnose (*Glomerella graminicola*) Leaf blight (*Setosphaeria turcica*) Zonate leaf spot (*Gloeocercospora sorghi*) Tar spot (*Phyllachora sorghi*) Charcoal rot (*Macrophomina phaseolina*) Rust (*Puccinia purpurea*) Ergot (*Claviceps sorghi*) Downy mildew (*Peronosclerospora sorghi*)	Maize streak geminivirus		
Soybean	Rust (*Phakopsora pachyrhizi*) Downy mildew (*Peronospora manshurica*) Anthracnose (*Colletotrichum truncatum* and *Glomerella glycines*) Purple seed stain (*Cercospora kikuchii*) Pod and stem blight (*Diaporthe phaseolorum* var. *sojae*)	Soybean mosaic potyvirus Bean yellow mosaic potyvirus		Bacterial pustule (*Xanthomonas axonopodis* pv. *phaseoli*)
Sweet Potato	Scab (*Sphaceloma batatas*) Fusarium wilt (*Fusarium oxysporum*) Black rot (*Ceratocystis fimbriata*) Java black rot (*Botryodiplodia theobromae*) Scurf (*Monilochaetes infuscans*)	Sweet potato feathery mottle potyvirus	Root-knot nematode (*Meloidogyne*)	Soil rot (*Streptomyces ipomoea*) Sweet potato little leaf phytoplasma

(Continued)

Table 5.2 *(Continued)*

Crop	Fungi	Viruses	Nematodes	Bacteria
Wheat	Stem rust (*Puccinia graminis* f.sp. *tritici*) Leaf rust (*Puccinia recondita* f.sp. *tritici*) Stripe or yellow rust (*Puccinia striiformis*) Spot blotch (*Cochliobolus sativus*) Head scab and foot/root rot (*Fusarium*) Sclerotium foot rot (*Corticium rolfsii*) Tan spot (*Pyrenophora tritici-repentis*) Powdery mildew (*Erysiphe graminis*) Speckled leaf blotch (*Mycosphaerella graminicola*) Glume blotch (*Phaeosphaeria nodorum*) Alternaria leaf blight (*Alternaria*) Loose smut (*Ustilago nuda* f.sp. *tritici*) Rhizoctonia root rot (*Rhizoctonia*)	Barley yellow dwarf luteovirus		Bacterial leaf streak or black chaff (*Xanthomonas translucens* pv. *undulosa*)
Yam	Anthracnose (*Colletotrichum gloeosporioides*) Tuber rots (*Fusarium, Penicillium, Rosellinia*)	Yam mosaic potyvirus Yam mild mosaic potyvirus		
Banana	Panama disease (*Fusarium oxysporum*) Black sigatoka (*Mycosphaerella fijiensis*)	Banana bunchy top virus (BBTV)		Bacterial wilt (*Ralstonia solanacearum*)

Sources: International Society for Plant Pathology; United Nations Food and Agriculture Organization (FAO).

Figure 5.1 Effects of citrus tristeza virus.

Source: Photo by S. M. Garnsey. Reproduced with permission from *Compendium of Citrus Diseases*, 2nd edition, 2000, American Phytopathological Society.

Case Study 5-1: The Secret Life of Aphids

The main vector of citrus tristeza is the brown citrus aphid. Unlike most aphids, this species rarely produces a sexual generation. Nearly all brown citrus aphids in North America are females, and they give birth to more females by parthenogenesis (virgin birth). These babies mature in about a week at room temperature and start having more babies. In three weeks, in a world without pesticides or predators, a single aphid could produce at least 4,400 descendants. This lifestyle helps the aphids disperse from one location to another and start new colonies. If the wind blows just one female brown citrus aphid to a new orchard, she can get right down to business without having to look for a mate. And if she happens to be carrying the tristeza virus, so will her thousands of descendants, and they will quickly spread this disease to all the trees where they feed. Even if a farmer manages to kill 90 percent of all aphids in a field, the remaining 10 percent can repopulate the field in a week.

Which Crops Are at Risk?

Natural hosts of CTV include orange, mandarin, grapefruit, lemon, and lime trees, plus passionfruit and several less famous tropical fruits (such as Nigerian powder flask, pamburas, and aeglopsis). The most susceptible trees are those grafted on sour orange rootstocks. Other rootstocks, such as Rangpur lime, appear to be resistant.

The Numbers

The annual worldwide citrus yield is on the order of 105 million tons, or over 2 billion boxes. In 1981, global losses from CTV alone amounted to 50 million trees. Brazil is the world's leading producer of citrus crops, with an estimated output of 360 million boxes of fruit in 2007–2008 and about 256 million citrus trees on 2 million acres.

The United States is the second largest citrus producer, and Florida produces about 70 percent of the U.S. crop. The Florida citrus harvest varies from year to year, but total production in the 2005–2006 season was about 175 million boxes or 8 million tons, valued at $1.1 billion. Florida has about 621,000 acres planted in citrus. In a normal year, Florida loses about 15 percent of its citrus trees to the citrus tristeza virus. Losses in a major outbreak can exceed 25 percent.

History

The first citrus trees traveled from Asia to the New World, Europe, and South Africa in the form of seed. Since CTV is not seed-transmitted, the disease did not arrive until later, when growers exported rootstocks in an effort to control other diseases. In the late 1890s, citrus growers in South Africa noticed that trees of sweet orange or mandarin on sour orange rootstock usually died or declined in two or three years. In the 1930s, scientists identified the problem as a viral disease that originated in Japan and followed citrus growers around the world.

The new disease reached Brazil in about 1937, and the name *tristeza* ("sadness") immediately stuck. By 1950, CTV had destroyed millions of trees in Brazil, Argentina, and southern California. It reached Florida by 1951, where its most efficient vector (the brown citrus aphid) joined it in 1995, thus increasing the potential for damage. As of 2009, farmers in California, Arizona, and Texas anxiously await the arrival of this aphid, which not only transmits CTV but also threatens trees by eating foliage.

Prevention and Treatment

Once trees are infected, the only known control measure is to destroy the trees. Prevention is complicated, since it requires virus-free budwood and rootstock.

In California, for example, the Citrus Tristeza Virus Interior Quarantine requires any citrus buds, cuttings, or scions to be tested for CTV before a nursery can use them. The California Department of Food and Agriculture (CDFA) maintains an inventory of certified tristeza-free citrus trees, each with a metal tag and identification number. CDFA requires annual tristeza testing; if any registered trees test positive, their registrations are cancelled, but the trees stay in the database. Laws passed in 2008 will further tighten these testing and quarantine requirements by 2010.

Popular Culture

As oranges gained popularity, scholars of Greek mythology proposed that the life-giving Golden Apples of the Hesperides were really oranges. Thus we have the words *hesperidium*, meaning any citrus fruit with a thick rind, and *hesperidin*, a nutrient found in citrus. But the Hesperides were the daughters of Hesperus, the evening star—the planet Venus, seen in the west at sunset—so the fabled orchard should be in the mythic lands beyond the Pillars of Hercules, perhaps in California. In fact, oranges are native to Asia, and Columbus brought them to the New World on his second voyage in 1493.

Famous last words are an aspect of popular culture. In 1945, a Texas agronomist wrote: "Standard sour orange has given universally good results as an understock for most varieties of citrus which are produced in this region. There is no reason for suggesting a change at this time, except for the possibility that a destructive disease which destroys trees on sour orange stock may eventually invade this area."[1]

The Future

There is some evidence that hot weather suppresses the citrus tristeza virus, but it is unlikely that global climate change will help the citrus industry in the long run, because hurricanes and other extreme weather events appear to spread a second disease called citrus canker, discussed later in this chapter.

References and Recommended Reading

Bandyopadhyay, R., and P. A. Frederiksen. "Contemporary Global Movement of Emerging Plant Diseases." *Annals of the New York Academy of Sciences*, Vol. 894, 1999, pp. 28–36.

Bar-Joseph, M., et al. "The Continuous Challenge of Citrus Tristeza Virus Control." *Annual Review of Phytopathology*, Vol. 27, 1989, pp. 291–316.

Bennett, L. "5,000 Trees Destroyed to Halt Virus." *Los Angeles Times*, 26 December 1979.

Brlansky, R. H., et al. "Tristeza Quick Decline Epidemic in South Florida." *Proceedings of the Florida State Horticultural Society*, Vol. 99, 1986, pp. 66–69.

Cambra, M., et al. "Incidence and Epidemiology of Citrus Tristeza Virus in the Valencian Community of Spain." *Virus Research*, Vol. 71, 2000, pp. 85–95.

Cox, W. "Tristeza Reigns as No. 1 Citrus Foe: Scientists Push Valley Virus Vectors Search." *Fresno Bee*, 6 February 1977.

Djelouah, K., and A. M. D'Onghia. 2001. "Occurrence and Spread of Citrus Tristeza in the Mediterranean Area." In Myrta, A., et al. (Eds.), *Production and Exchange of Virus-Free Plant Propagating Material in the Mediterranean Region*. Bari, Italy: International Center for Advanced Mediterranean Agronomic Studies (CIHEAM).

Futch, S. H., and R. H. Brlansky. "Field Diagnosis of Citrus Tristeza Virus." Gainesville, FL: University of Florida, Institute of Food and Agricultural Sciences, 2005.

Kallsen, C. "Controlling the Citrus Tristeza Virus in the San Joaquin Valley of California." University of California Cooperative Extension, 2 May 2002.

---

1. *Brownsville Herald*, 11 November 1945.

Karp, D. "An Orange Whose Season Has Come." *New York Times*, 22 January 2003.
Marroquin, C., et al. "Estimation of the Number of Aphids Carrying Citrus Tristeza Virus That Visit Adult Citrus Trees." *Virus Research*, Vol. 100, 2004, pp. 101–108.
Moreno, P., et al. "Citrus Tristeza Virus: A Pathogen That Changed the Course of the Citrus Industry." *Molecular and Plant Pathology*, Vol. 9, 2008, pp. 251–268.
Nawaz, M. A., et al. "Tristeza Virus: A Threat to Citrus Fruits." *DAWN*, 14 July 2008.
Reuther, W., et al. (Eds.) 1989. *The Citrus Industry*. Berkeley: University of California.
Rocha-Peña, M. A., et al. "Citrus Tristeza Virus and Its Aphid Vector *Toxoptera citricida*." *Plant Disease*, Vol. 79, 1995, pp. 437–445.

BACTERIAL WILT

Summary of Threat

Bacterial wilt affects nearly 200 plant species, including potatoes, bananas, tomatoes, peppers, and other important crops. As of 2009, there is no effective treatment other than quarantine and sanitation procedures. The agent (*Ralstonia solanacearum*) has several strains that affect different plants. One strain (Race 3, Biovar 2) is classified as a biosecurity threat under the Bioterrorism Protection Act of 2002.

Other Names

Names include southern bacterial wilt, southern wilt, spur canker, blast, or brown rot (of any affected crop); potato slime disease, jammy eye or sore eye of potato, banana slime disease, banana blood wilt, banana moko disease, tobacco granville wilt, and granville wilt of tobacco. Related bacteria cause Sumatra disease of cloves and bugtok and blood disease in bananas. Bacterial wilt is *flétrissement bactérien* in French, *mörk ringröta* in Swedish, and *Kartoffel-Schleimkrankheit* in German. Most names in other languages are direct translations, but there are interesting exceptions, such as *vaquita de la papa* (Spanish, "little cow of the potato").

Description

The agent of this disease is a bacterium (*Ralstonia solanacearum*) that occurs worldwide and can survive for at least three years in water or soil. It infects over 200 plant species, including staples, cash crops, garden flowers, and wild plants. There are three races with different host ranges:

- Race 1 (potato, tomato, pepper, eggplant, legumes, tobacco, other crops)
- Race 2 (triploid banana and plantain)
- Race 3 (potato, tomato, eggplant, geranium, nettle)

Symptoms depend on the host plant and race but usually include discoloration, drooping leaves, and decayed roots, tubers, or fruits (Figure 5.2). Race 1 is endemic in the southeastern United States and many other countries, but it is not considered a quarantine pest. The serious threat is Race 3, biovar 2, which can cause more extensive damage.

Besides infecting many different plants, bacterial wilt has several modes of transmission. The agent can enter a plant through the roots or damaged parts, or farm workers can spread it

Figure 5.2 Tobacco plant showing effects of bacterial wilt (*Ralstonia solanacearum*).
Source: Clemson University USDA Cooperative Extension Slide Series, Bugwood.org.

by transporting infected soil on shoes, hands, or tools. The Colorado potato beetle is a mechanical vector for this disease on potatoes. Other insects, birds, and animals that visit host plants may also serve as vectors.

Which Crops Are at Risk?

Common host plants for bacterial wilt include potato, tomato, banana, plantain, tobacco, red ginger, taro, chili pepper, eggplant, olive, mulberry, peanut, and clove, plus a number of economically important nonfood plants such as castor bean, heliconia, and geranium (Case Study 5-2). Overwatering or heavy rainfall may increase risk.

The Numbers

Bacterial wilt is one of the world's most expensive plant diseases. Global losses to the potato industry alone exceed $1 billion per year. In Indonesia, bacterial wilt costs the ginger industry an estimated 75 billion rupiah (U.S. $750,000) per year. This disease is also a major threat to the tomato industry. Worldwide production of tomatoes is about 104 million tons per year, but bacterial wilt can reduce yield by up to 90 percent in India and other endemic areas.

History

Bacterial wilt disease was first described in 1896 and may be native to the Andean highlands of Peru. Race 3, biovar 2 has not yet become established in the United States or Canada, but it

Case Study 5-2: Geranium Wars

In 2003, a greenhouse in Kenya accidentally sent several geranium cuttings infected with *Ralstonia solanacearum* to a flower producer in California. Later in the year, the same flower producer received a second shipment of infected cuttings, this time from a greenhouse in Guatemala. As a result, more than 900 of the firm's customers in 47 states received potentially infected plants and were subject to quarantine. The U.S. Department of Agriculture (USDA) requires the destruction of all plants shipped with infected geraniums, so this was an expensive mistake. Worse, since the same disease affects potato, tomato, tobacco, and many other crops, there was a possibility that the bacterium could escape and cost American farmers billions of dollars. About 2 million plants were destroyed, and the eradication was successful. But numerous greenhouses sued the company that made the slip, and the company, in turn, blamed the government for overzealous regulation. This bacterium is on a short list of pathogens that are subject to special reporting requirements under the Bioterrorism Protection Act of 2002, but apparently that fact was unrelated to the USDA's controversial action, which reflected its mission to block the importation of all plant diseases.

is present in most other potato-growing regions, including Africa, Asia, eastern Australia, Central and South America, and Mexico. It reached western Europe in the 1970s and apparently spread through infected seed potatoes and contaminated irrigation water. This form of bacterial wilt reached the United States in 1999 and 2003 but was successfully eradicated on both occasions.

Prevention and Treatment

As of 2009, no known chemical treatment is consistently effective. Some investigators have reported successful control of this disease by inoculating plants with strains of bacteria that inhibit the growth of *Ralstonia solancearum*. Most control measures focus on the principles of integrated pest management (IPM): Use healthy seed in clean soil, choose disease-resistant or tolerant varieties, graft on resistant rootstocks, and rotate with crops that are not susceptible to bacterial wilt.

Popular Culture

In 2003, the Internet and other media circulated a rumor to the effect that bananas may become extinct within the next ten years because of disease. The source appears to be a controversial report in the journal *New Scientist*, which focused on two fungal diseases of bananas: black sigatoka and Panama disease. As the rumor grew and improved, moko disease (bacterial wilt) and others were also cited as potential agents of this banana Armageddon. But according to the American Phytopathological Society, the rumor is unfounded. There are many banana strains, and only the Cavendish banana is currently at risk. Also, it would not become extinct, but commercially unviable. (The Cavendish has been America's favorite banana since the 1960s, when another disease wiped out the much-loved Gros Michel banana. The fact that people were trying to get high by smoking banana peel probably did not help matters either.) Bananas are, however, at greater risk than most fruits, because they produce no seeds and are grown from cuttings.

In 2000, an American company claimed that its biopesticide would prevent *Ralstonia solanacearum* wilt. The product itself was the result of legitimate research at Cornell University, but the aftermath belongs to popular culture. The active ingredient was a bacterial protein that was intended to activate a plant's immune response. The manufacturer won the EPA's Green Chemistry Award in 2001—but EPA approval requires only proof that a product is safe, not that it works. When put to the test, the biopesticide was ineffective. The media reported that the manufacturer sold the technology in 2007, executed a "pump and dump" in 2008, and voted to dissolve. This disappointing outcome should not discourage other companies from trying to develop the effective green technologies that the world so desperately needs.

The Future

Bacterial wilt is one of several diseases that may become worse, or more widespread, as a result of global warming (Chapter 6). As a result, the economic impact of bacterial wilt will continue to increase, particularly in those countries least able to afford it. Subsistence farmers have been hit particularly hard in recent years, and better control methods are urgently needed. Since 1992, plant pathologists have organized the International Bacterial Wilt Symposium (IBWS) series to disseminate research findings.

References and Recommended Reading

Allen, C., et al. (Eds.) 2005. *Bacterial Wilt Disease and the* Ralstonia solanacearum *Species Complex*. St. Paul, MN: APS Press.

Aranowski, A. "Plant Pathologists Unpeel Rumors of Banana Extinction." Press Release, American Phytopathological Society, 14 February 2003.

Chalker-Scott, L. "The Myth of the Magic Bullet." Washington State University, Puyallup Research and Extension Center, April 2005.

Daughtry, M. 2003. "Southern Bacterial Wilt, Caused by *Ralstonia solanacearum*." Cornell University, Long Island Horticultural Research and Extension Center.

Davies, J. C., and B. K. Rubin. "Emerging and Unusual Gram-Negative Infections in Cystic Fibrosis." *Seminars in Respiratory and Critical Care Medicine*, Vol. 28, 2007, pp. 312–321.

"Detection of *Ralstonia solanacearum* Race 3 Biovar 2 in New York Greenhouse." Phytosanitary Alert System, North American Plant Protection Association, 5 January 2004.

Eyres, N., and N. Hammond. "Factsheet: Moko Disease *Ralstonia solanacearum* (Race 2, Biovar 1)." Western Australia Department of Agriculture and Food, 2006.

"First Report of Brown Rot on Potato in Ireland." *SeedQuest*, 18 October 2007.

Floyd, J. 2008. "New Pest Response Guidelines, *Ralstonia solanacearum* Race 3 Biovar 2." Riverdale, MD: USDA/APHIS/PPQ.

Genin, S., and C. Boucher. "Lessons Learned from the Genome Analysis of *Ralstonia solanacearum*." *Annual Review of Phytopathology*, Vol. 42, 2004, pp. 107–134.

Ji, X., et al. "Biological Control Against Bacterial Wilt and Colonization of Mulberry by an Endophytic *Bacillus subtilis* Strain." *FEMS Microbiology and Ecology*, 10 July 2008.

Lemay, A., et al. 2003. Pest Data Sheet: "*Ralstonia solanacearum* Race 3 Biovar 2." Raleigh, NC: USDA/APHIS/PPQ Center for Plant Health Science and Technology, Plant Epidemiology and Risk Analysis Laboratory.

Moorman, G. W. "Bacterial Wilt—*Ralstonia solanacearum*." Plant Disease Facts, Pennsylvania State University, Department of Plant Pathology, updated 2008.

Prior, P., et al. (Eds.) 1998. *Bacterial Wilt Disease: Molecular and Ecological Aspects*. New York: Springer.

Priou, S., et al. 1999. *Integrated Control of Bacterial Wilt of Potato*. Geneva: United Nations Food and Agriculture Organization, 16 pp.

Smith, E. F. "A Bacterial Disease of Tomato, Pepper, Eggplant and Irish Potato (*Bacillus solanacearum* nov. sp.)." *USDA Division of Vegetable Physiology and Pathology Bulletin*, Vol. 12, 1896, pp. 1–28.

Stansbury, C., et al. "Bacterial Wilt: *Ralstonia solanacearum* – Race 3. Exotic threat to Western Australia." Agriculture Western Australia, Fact Sheet No. 7/2001.

"USDA Defends Action." *Gilroy Dispatch*, 25 April 2003.

SOUTHERN CORN LEAF BLIGHT

Summary of Threat

Southern corn leaf blight (SCLB) is a fungal disease that destroyed about 15 percent of the United States maize crop in 1970. Some strains of maize are more susceptible than others, and farmers have largely defeated this disease—for now—by planting resistant strains. The disease has become a favorite of economists and sociologists because of the various principles that its history appears to illustrate.

Other Names

English names include southern corn leaf blight (SCLB), southern leaf blight of corn, corn leaf blight, southern leaf blight (SLB), southern corn blight, stalk rot, and ear rot. In Spanish, SCLB is *la marchitez foliar del maíz del sur* or *mancha de la hoja del maíz.* The French name is *helminthosporiose du maïs,* and the German equivalent is *Blattfleckenkrankheit an Mais.* Names for the principal host include Indian corn and maize (Case Study 5-3).

Description

The infectious agent is a fungus called *Cochliobolus heterostrophus* (formerly *Bipolaris maydis, Helminthosporum maydis, or Drechslera maydis*). Its only known hosts are maize or corn (*Zea mays*) and its relatives teosinte (wild grasses in the genus *Zea*) and sorghum (*Sorghum bicolor* and related species). The fungus produces a toxin that prevents the plant from capturing energy from its metabolism, but the virulence of this toxin varies, depending on the race of fungus and the host. The race that caused the 1970 outbreak was called Race T because it produced the T toxin, so named because it attacked specific varieties of hybrid corn with T (Texas) cytoplasm. The predominant strain in the United States now is Race O.

Outbreaks have occurred in the Philippines, Asia, Africa, Europe, the United States, and Latin America. Spores disperse by wind or water droplets and can survive in soil or plant debris during the winter. After landing on corn leaves, the spores germinate and enter the plant through stomata. Symptoms include tan lesions on leaves, sometimes with dark red or purple edges (Figure 5.3). A black substance may be present on the ears, and rot may extend into the interior of the cob. Stalks may also be damaged.

Case Study 5-3: We Call it Corn

In the United States, the word "corn" usually refers to a plant (*Zea mays*) that the rest of the world calls maize or Indian corn. "Corn" is a generic term for the most important cereal crop in any region: wheat in Britain, oats in Scotland, and maize in North America. So whenever this book refers to corn, it really means maize. Mind you, we aren't knocking barley, rye, millet, or quadrotriticale. But studies of human hair samples have shown that a startling 70 percent of all the carbon in the body of the average American comes from just one plant: maize. Even people who eat very little "visible" corn absorb a great deal of it in the form of corn syrup and other food and beverage additives.

Which Crops Are at Risk?

Only corn with a specific mutation is highly susceptible to southern corn leaf blight. The fungus also attacks other strains of corn, and a few related plants such as teosinte and sorghum, but the effects are less severe. Outbreaks are more likely in hot, wet weather.

Figure 5.3 Cornfield showing effects of southern corn leaf blight.
Source: J. K. Pataky, University of Illinois.

The Numbers

The southern corn leaf blight outbreak in 1970 destroyed about 15 percent of the U.S. corn crop, causing economic losses estimated at $1 billion or more, depending on the source. At the time, 85 percent of all corn plants in the United States were said to be descendants of *one* corn plant discovered in Texas in 1944.

History

Southern corn leaf blight first appeared in the Philippines in about 1962, and its effect on hybrid corn was known by 1969, but it appears that the risk was not taken seriously. After the 1970 disaster, the USDA expected the disease to return in 1971, so they advised farmers to plant more corn than usual to compensate. Some did, others didn't. Some were unable to get enough resistant seed, so they planted other crops such as milo instead. But the blight caused only minor losses in 1971, and the resulting record corn crop drove down prices. Farmers also wasted money on fungicides.

Some analysts later decided that the USDA had acted without sufficient data, but others put a different spin on the same events:

> In the 1970's [*sic*], an unheard-of disease, the southern corn leaf blight, swept through the fields of the Midwest. In a few days, the tall, green, tasseled corn was devastated, as if someone had taken a blowtorch to it. Over that winter, scientists and farmers developed resistant corn varieties in time for the next

> spring planting. A national food disaster was stopped dead in its tracks—a triumph of faith, science, and inventiveness.[2]

The southern corn leaf epidemic of 1970, and the nonepidemic that followed in 1971, are among the few events on record that are regularly cited as both (a) an example of government incompetence and failure, and (b) an example of a magnificent triumph over disease. In 1971, the Associated Press quoted USDA officials to the effect that the corn blight could even dash President Nixon's hopes for re-election. And in 1972, a number of farmers filed class action suits against major seed companies, alleging that they should have known that T hybrid corn was susceptible to southern corn leaf blight. Apparently, none of these suits ever went to trial.

Prevention and Treatment

The most effective preventive measure is to plant resistant strains of corn, which are now available worldwide. This strategy has prevented significant outbreaks for many years. Some fungicides are also approved for use, but they are expensive and not always effective, even with repeated applications.

Popular Culture

In 1970, when southern corn leaf blight wiped out entire cornfields and threatened to devastate the American crop, the message from more than one church pulpit was clear: the blight was yet another tribulation, foretold by the Old Testament story of the seven thin ears blasted by the east wind (Genesis 41). The same vision featured seven lean cows, and some observers noted the spread of the cattle disease rinderpest to the Middle East at about the same time. Other people, however, interpreted the southern corn blight outbreak in less spiritual terms as the result of agro-scientists meddling with things best left alone—in this case, the genetic diversity of maize. Yet others found confirmation of their longstanding distrust of the government's farm policy. And, inevitably, there were those who claimed that Fidel Castro's legions had sabotaged the U.S. corn crop—just as Castro blamed the United States for outbreaks of dengue fever and Newcastle disease in Cuba during the same era. A natural disaster often serves as a Rorschach test to identify what people are really worried about.

The Future

This specific disease poses no known risk for the immediate future, but the story of southern corn blight stands as a parable on the dangers inherent in monoculture—in this case, the practice of planting large areas with genetically similar or identical plants that are all vulnerable to the same diseases. This danger is as great today as it was in 1970.

References and Recommended Reading

"Corn Blight May Affect Nixon's Election Hopes for '72." Associated Press, 24 January 1971.
Gupta, S. "If We Are What We Eat, Americans Are Corn and Soy." CNN, 22 September 2002.

---

2. President George (H. W.) Bush, excerpt from remarks to the American Farm Bureau Federation in Orlando, Florida, 8 January 1990.

Hansen, S. F., et al. "Categorizing Mistaken False Positives in Regulation of Human and Environmental Health." *Risk Analysis*, Vol. 27, 2007, pp. 255–269.

Hooker, A. L. "Studies Related to the Development and Control of the Southern Leaf Blight of Corn Caused by *Helminthosporium maydis*: Final Report." Illinois Agricultural Experiment Station, 1976.

Kloppenburg, J. R. 2004. *First the Seed: The Political Economy of Plant Biotechnology*. Madison: University of Wisconsin Press.

Levings, C. S. "Thoughts on Cytoplasmic Male Sterility in cms-T Maize." *The Plant Cell*, Vol. 5, 1993, pp. 1285–1290.

Martinson, C. "Essays on the College of Agriculture's History: Southern Corn Leaf Blight Epidemic." Iowa State University, College of Agriculture and Life Sciences, 2007.

McGee, D. C. 1988. *Maize Diseases: A Reference Source for Seed Technologists*. St. Paul, MN: American Phytopathological Society, 150 pp.

Miller, R. J., and D. E. Koeppe. "Southern Corn Leaf Blight: Susceptible and Resistant Mitochondria." *Science*, Vol. 173, 1971, pp. 67–69.

Sindhu, A., et al. 2008. "A Guardian of Grasses: Specific Origin and Conservation of a Unique Disease-Resistance Gene in the Grass Lineage." *Proceedings of the National Academy of Sciences*, Vol. 105, 2008, pp. 1762–1767.

Smith, N. A. 1972. "Corn Leaf Blights." Michigan State University, *Cooperative Extension Service Bulletin* E-832.

Tatum, L. A. "The Southern Corn Leaf Blight Epidemic." *Science*, Vol. 171, 1971, pp. 1113–1116.

Ullstrup, A. J. "The Impacts of the Southern Corn Leaf Blight Epidemics of 1970–1971." *Annual Review of Phytopathology*, Vol. 10, 1972, pp. 37–50.

Waggoner, P. E. "Epimay: a Simulator of Southern Corn Leaf Blight." *Bulletin, Connecticut Agricultural Experiment Station*, 1972, 84 pp.

Warren, H. L., et al. "Morphological and Physiological Differences between *Bipolaris maydis* Races O and T." *Mycologia*, Vol. 69, 1977, pp. 773–782.

CITRUS CANKER

Summary of Threat

Citrus canker may be the world's most feared disease of citrus crops. The agent is a bacterium that infects citrus trees, discoloring the fruit and making it drop prematurely. It spreads by wind and rain or by contact with infected or contaminated material. Although citrus canker does not usually kill trees outright, it makes the fruit unmarketable and forces removal of infected trees as a control measure.

Other Names

Citrus canker (CC) is also called citrus bacterial canker (CBC), citrus bacteriosus, or citrus cancrosis. Strains include Asiatic or oriental citrus canker, or cancrosis A; false citrus canker, or cancrosis B; Mexican lime cancrosis, or cancrosis C; citrus bacteriosus, also called Mexican bacteriosus or cancrosis D; and citrus bacterial spot, or cancrosis E. Citrus canker is *el chancro de los agrios* or *úlcera de los cítricos* in Spanish, *chancre des citrus* in French, *cancro degli agrumi* in Italian, *citruskräfta* in Swedish, and *Zitruskrebs* in German.

Description

The infectious agent is the bacterium *Xanthomonas axonopodis* (formerly called *Xanthomonas campestris, X. citri, Bacillus campestris, Bacterium campestris, Phytomonas campestris,* or

Figure 5.4 Grapefruit and leaves showing effects of citrus canker.
Source: Timothy Schubert, Florida Department of Agriculture and Consumer Services.

Pseudomonas campestris). It infects most commercially valuable citrus species, and its range is worldwide, although some local areas (such as Queensland, Australia) have eradicated the disease. The agent disperses in water droplets with the help of wind-driven rain. The strains vary by level of virulence and host range. Early symptoms include scablike lesions on fruit (Figure 5.4), often followed by premature fruit drop, loss of leaves, and fatal weakening of the entire tree.

Fruit infected with canker is safe to eat but unappetizing in appearance. The worst effects are economic rather than biological. When citrus growers cannot export their fruit because of quarantine restrictions, more fruit ends up in juice, and the price of juice falls. Eventually, citrus canker could force many growers out of business.

Which Crops Are at Risk?

Grapefruit, lime, tangerine, orange, lemon, and most other citrus fruits are susceptible to citrus canker. So is the Sichuan pepper tree. Some of the less popular citrus fruits are resistant, such as citron, calamondin, pomelo, and kumquat. An unrelated plant called goatweed, which grows in citrus orchards in India, can serve as a host.

Citrus canker causes more damage if an insect called citrus leaf miner (*Phyllocnistis citrella*) is also present, because its feeding creates breaks in the foliage where bacteria can enter. Other risk factors include exposure to infected fruit or any contaminated object, such as farm workers' hands, clothes, tools, or equipment.

The Numbers

Between 1984 and 1989, agricultural inspectors in Florida burned an estimated 20 million citrus trees—including many that showed no signs of disease—in an effort to stop a citrus canker outbreak. The cost was about $27 million to the state of Florida and $14 million to the federal government. In 1989, however, officials announced that the disease was not citrus canker after all and budgeted another $20 million to settle lawsuits filed by growers whose stock was destroyed. Interpretations of these events vary.

The discovery that citrus canker can spread 580 meters (1,900 feet) during one storm resulted in the "1,900-foot rule," which required the destruction of all citrus within 1,900 feet of an infected tree. Four hurricanes in 2005 caused $2.2 billion in damage to Florida's crops and farming infrastructure—including the loss of citrus valued at $180 million—and also spread citrus canker, necessitating the removal of another 10 percent of the state's citrus groves.

By 2005, growers had received an estimated $125 million in compensation. But the biggest number relevant to this discussion is $9 billion, the estimated annual value of Florida's citrus crop.

History

The first reports of citrus canker appeared in Java and India in the mid-nineteenth century. The disease made its debut in Florida in 1911 and spread throughout the Gulf states, but a 20-year program eradicated it. The disease turned up again in Florida in the 1980s and was declared eradicated again in 1994.

The most recent outbreak, in 1999, set off the most expensive government program ever devoted to a single plant disease (Case Study 5-4). In 2005, following a series of setbacks, the USDA determined that the eradication of citrus canker was essentially impossible, and the effort was formally abandoned in 2006. Management of this disease now focuses on control rather than eradication.

Prevention and Treatment

As of 2009, it appears that there is no effective way to treat citrus canker except by removing infected trees. Multiple applications of copper bactericides are sometimes effective, if expensive. The use of windbreaks may also help reduce the dispersal of this disease. Another approach is the use of bacteriophages (viruses that infect bacteria), but the results to date have been disappointing. There are recent reports of improved screening methods that will make it easier to detect bacteria on infected fruit.

Case Study 5-4: Payback Time

More than 40,000 residents of Palm Beach County, Florida, reportedly filed a class action suit after state and federal agricultural agencies destroyed their citrus trees in an ultimately futile effort to stop the spread of citrus canker. In 2007 the court awarded them $100 for the first tree lost and $55 for each additional tree. More class action suits were also filed. It is unclear whether this outcome vindicates the people who insisted that the eradication campaign was pointless in the first place.

Popular Culture

In 2005, a Florida citrus farmer offered the following prayer during a citrus canker outbreak: "We know that not a sparrow falls to the ground that you don't know about. We respectfully ask you to take your eyes off the dadburn sparrows and put them on the citrus industry."[3]

In 2000, Florida officials agreed to evaluate a product called Celestial Drops or Kabbalah Water as a cure for citrus canker. After six months of testing at taxpayer expense, the product turned out to be plain water, a chemical that Florida already possessed in sufficient quantity. In 2006 and 2008, other entrepreneurs made headlines with similar claims that (as of 2009) have not borne fruit.

In the 2005 novel *Predator*, by Patricia Cornwell, a Florida woman is found murdered after complaining about harassment from a citrus canker inspector. In the words of one reviewer, "You will learn more about citrus canker than you ever wanted to know." We won't spoil the ending, but a character named Hog speaks for many Florida residents: "I've seen entire orchards burned because of the canker. People's lives ruined."

In Nancy Cohen's 2003 novel *Highlights to Heaven*, some characters discuss issues related to the Florida citrus canker eradication program. Citrus canker is also a theme in Steve Glassman's 2001 mystery novel *The Near Death Experiment,* which reportedly is "set against the background of Florida's orange juice industry."

The Future

In the immortal words of Yoda, "Always in motion is the future." Citrus canker is one of several diseases that are expected to become worse, or more widespread, as a result of global warming (Chapter 6).

References and Recommended Reading

Balogh, B., et al. "Control of Citrus Canker and Citrus Bacterial Spot with Bacteriophages." *Plant Disease*, Vol. 92, 2008, pp. 2048–2052.

Bronson, C. H., and R. Gaskalla. "Comprehensive Report on Citrus Canker in Florida." Division of Plant Industry, Florida Department of Agriculture and Consumer Services, 15 October 2007.

Brown, K. "Florida Fights to Stop Citrus Canker." *Science*, Vol. 292, 2001, pp. 2275–2276.

"Citrus Canker Outbreak Reported in Australia." United Press International, 8 July 2004.

"Citrus Crop in Danger Due to Disease." Associated Press, 4 February 2000.

"Florida Braces for Small Orange Harvest." United Press International, 13 July 2006.

Golmohammadi, M., et al. "Diagnosis of *Xanthomonas axonopodis* pv. Citri, Causal Agent of Citrus Canker, in Commercial Fruits by Isolation and PCR-Based Methods." *Journal of Applied Microbiology*, Vol. 103, 2007, pp. 2309–2315.

Gottwald, T. R. "Citrus Canker: The Pathogen and Its Impact." *Plant Health Progress*, August/September 2002.

Irey, M., and T. R. Gottwald. "Post-Hurricane Analysis of Citrus Canker Spread and Progress toward the Development of a Predictive Model to Estimate Disease Spread Due to Catastrophic Weather Events." *Plant Health Progress*, 22 August 2006.

Kennedy, S. "Florida Still Struggles Over the Citrus Canker." *New York Times*, 2 July 1989.

Layden, L. "Storms Carried Citrus Canker to New Areas Across Florida." Scripps Howard News Service, 2 March 2005.

Li, W., et al. "Genetic Diversity of Citrus Bacterial Canker Pathogens Preserved in Herbarium Specimens." *Proceedings of the National Academy of Sciences USA*, Vol. 104, 2007, pp. 18427–18432.

---

3. *Miami Herald*, 25 January 2005.

Munson, S. "Loss of Lifestyle after the Hurricane." Letter to the Editor, *Charlotte Sun* (Port Charlotte, Florida), 15 January 2009.

Salisbury, S. "Citrus Canker Spreads Across State." *Miami Herald*, 25 January 2005.

Spreen, T. H., and M. L. Zansler. "The Costs and Value Loss Associated with Florida Citrus Groves Exposed to Citrus Canker." *Proceedings of the Florida State Horticultural Society*, Vol. 116, 2003, pp. 289–294.

Stall, R. E., and E. L. Civerolo. "Research Relating to the Recent Outbreak of Citrus Canker in Florida." *Annual Review of Phytopathology*, Vol. 29, 1991, pp. 399–420.

Stratton, J. "'Celestial Drops' Failed on Canker." *Knight-Ridder/Tribune Business News*, 5 July 2005.

Sun, M. "The Mystery of Florida's Citrus Canker." *Science*, Vol. 226, 1984, pp. 322–323.

Timmer, L. W., et al. (Eds.) 2000. *Compendium of Citrus Diseases*. 2nd ed. St. Paul, MN: American Phytopathological Society, 92 pp.

LATE BLIGHT OF POTATO

Summary of Threat

Late blight is a fungus-like organism (an oömycete) that destroys potatoes and related crops. More than one authority has called late blight the worst crop disease in the world. By forcing large-scale use of fungicides, this disease makes potato farming expensive and drives up prices. Late blight caused about 1 million human deaths in Ireland in the 1840s, and now costs the potato industry $3 billion per year.

Other Names

Names for late blight include potato blight or tomato blight, depending on the crop. In Ireland, late blight was often called *mí-adh* (pronounced mee-aw, literally "bad luck"). Names that refer specifically to the 1845–1849 late blight epidemic in Ireland include *an gorta mór* ("the great hunger") or the potato murrain. It is important not to confuse late blight with early blight, a completely different disease.

Description

Strictly speaking, the agent of late blight is not a fungus but an oömycete (*Phytophthora infestans*). Its relatives include the agents of sudden oak death and rhododendron root rot. Until recently, biologists regarded the oömycetes as fungi, members of the same group that includes mushrooms and yeast. Most books still describe late blight as a fungus, and the chemicals used to kill it are usually called fungicides rather than oömyceticides.

Late blight is arguably the worst crop disease in the world, infecting potato crops in cool, humid regions of every continent. Although it is now controllable with high levels of fungicide, the cost of treatment has driven up the price of potatoes. If untreated, late blight can still destroy an entire crop in a few weeks. Symptoms include pale green water-soaked lesions that grow into brown or black areas on leaves, stems, and tubers (Figure 5.5). Later, white mold may appear on the underside of leaves, and the potatoes eventually rot. A severely infected field has a distinctive odor.

Which Crops Are at Risk?

Potatoes are the main crop at risk, although the same disease also affects tomatoes, eggplant, peppers, and a few other crops. The agent will not sporulate (grow spores) if the air is too cold,

Figure 5.5 Potato tuber showing effects of late blight.
Source: United Nations Economic Commission for Europe (UNECE).

too hot, or too dry, so farmers use the "temperature-humidity rule" to predict outbreaks. Blight often develops within two or three weeks after a period when the temperature is 50°–80°F (10°–27°C) and humidity is 75 to 80 percent or higher for at least two days.

The Numbers

Late blight costs the global potato industry an estimated $3 billion per year. In the United States alone, annual losses amount to $200 million to $400 million per year, plus $100 to $200 per acre for fungicides.

History

The agent apparently originated in the South American Andes and spread from there to Mexico. It reached the northeastern United States by about 1840, and ships transported it to Europe as an accidental passenger. Once established, late blight destroyed potato crops, not only in Ireland but also in Belgium, the Netherlands, France, and England.

The late blight pandemic of 1845–1849 (Case Study 5-5) not only caused the deaths of an estimated 1 million Irish people but also forced millions more to emigrate. While many of these refugees tried to distance themselves from anything reminiscent of poverty, others started nationalist movements. Most historians would agree that the Irish Famine changed the world, and some have attributed the European revolutions of 1848 to the failure of the potato and other crops.

Ironically, a Belgian scientist named Charles Morren (1807–1858) found an effective control measure in 1845, the first year of the Famine: removing the upper parts of infected potato plants would save the tubers. But in those pre-Internet days, most farmers were probably unaware of Dr. Morren's findings.

For the past 150 years, the claim has periodically surfaced that the British government deliberately mismanaged the Famine as part of a campaign of genocide (or Malthusian adjustment) directed at the Irish. We won't touch this one—but in the words of Irish nationalist John Mitchel, "The Almighty, indeed, sent the potato blight, but the English created the Famine."[4]

From Europe the disease spread to Asia and Africa. It continued to be a major problem even after the 1885 invention of the copper fungicide Bordeaux mixture. Some new strains are highly virulent and resistant to fungicides; as a result, late blight is now a re-emerging disease. Major outbreaks occurred in Germany in 1916, in southern Alberta in 1992, and in New Brunswick in 2003.

Case Study 5-5: Hard Times

Imagine that you are a farmer in western Ireland, sitting down to dinner at the end of a hard day in 1845, 6 years after the Night of the Big Wind. Each person at the table receives 14 pounds of potatoes. No, this is not a joke or a typographic error; farming is strenuous, and farm workers eat a lot. Without butter, 14 pounds of cooked potatoes yield about 3,000 calories, not an unreasonable ration for a hardworking adult. But having become dependent on one crop to sustain your family for generations, now imagine the shock of seeing that crop disappear. You still have a few other foods to keep your spouse and nine children alive, such as seaweed, sour milk, salted herring, and congealed cow's blood mixed with strong butter. Yum! But these foods, once the rural Irish equivalent of antipasto, have never been available in sufficient quantity—except for seaweed, and once the famine starts in earnest, the shore is stripped bare. And so you hit the road in search of food or work, until the famine fever comes.

Prevention and Treatment

About 40 years after the Irish Famine, scientists found that a copper sulfate solution called Bordeaux mixture was effective against late blight. Resistant strains later appeared, and scientists countered with multiple applications and new fungicides. As of 2009, these chemicals remain the first line of defense against late blight, but they are expensive. In an effort to reduce cost, scientists are investigating the use of plant essential oils as late blight suppressors. Another, more controversial approach to this problem is the creation of genetically modified potatoes.

Popular Culture

The nineteenth-century Irish song "The White Potatoes" refers to late blight:

It's a thousand and eight hundred years
Forty and six years no lie
Since our Savior descended in human form
Until the potatoes of the world rotted.[5]

The germ theory of disease existed in the 1840s, but most people had never heard of it. Thus, when late blight struck Ireland and northern Europe, farmers blamed it on the Evil Eye, the sins of the world, or just *mí-adh*—bad luck. By Irish tradition, a kettle of boiled potatoes dumped in a neighbor's field would reduce his crop; in the early days of the potato murrain,

4. John Mitchel, *The Last Conquest of Ireland* (1861).

5. A. Gribben (Ed.), *The Great Famine and the Irish Diaspora in America* (University of Massachusetts Press, 1999), pp. 119–120.

many misunderstandings occurred before anyone recognized the extent of the disaster. In Scotland, the potato was already suspect because of its relationship to the toxic nightshades and because the Bible did not mention it.

Many celebrated novels refer to the Irish Famine, including Peter Behrens' *The Law of Dreams* (2006), Nuala O'Faolain's *My Dream of You* (2001), Liam O'Flaherty's *Famine* (1937), Anthony Trollope's *Castle Richmond* (1860), and William Carleton's *The Black Prophet* (1847). Motion pictures about the Famine include *The Field* (1991) and *Untamed* (1955).

The Future

It is not clear whether global climate change will increase or decrease the range and severity of late blight. In some parts of the world, higher temperatures might reduce opportunities for sporulation; in others, climate change might favor blight by increasing humidity. Public acceptance of genetically modified potatoes and other crops may increase as alternatives fail.

References and Recommended Reading

Anderson, P. K., et al. "Emerging Infectious Diseases of Plants: Pathogen Pollution, Climate Change and Agrotechnology Drivers." *Trends in Ecology and Evolution*, Vol. 19, 2004, pp. 535–544.

Andreu, A. B., et al. "Enhancement of Natural Disease Resistance in Potatoes by Chemicals." *Pest Management Science*, Vol. 62, 2006, pp. 162–170.

Bandyopadhyay, R., and P. A. Frederiksen. "Contemporary Global Movement of Emerging Plant Diseases." *Annals of the New York Academy of Sciences*, Vol. 894, 1999, pp. 28–36.

Bhattacharjee, S., et al. "The Malarial Host-Targeting Signal is Conserved in the Irish Potato Famine Pathogen." *PLoS Pathogens*, Vol. 2, 2006, p. e50.

Cummins, J. "Genes from a Wild Plant *Solanum bulbocastanum* Used to Resist Potato Blight Fungus." News Release, GM-Free Ireland, 28 January 2006.

Egelko, B., and B. Tansey. "Engineered Alfalfa Ban Upheld on Appeal." *San Francisco Chronicle*, 3 September 2008.

Froyd, J. D. "Can Synthetic Pesticides be Replaced with Biologically-Based Alternatives? An Industry Perspective." *Journal of Industrial Microbiology and Biotechnology*, Vol. 19, 1997, pp. 193–195.

Gribben, A. (Ed.) 1999. *The Great Famine and the Irish Diaspora in America*. Amherst: University of Massachusetts Press.

Grunwald, N. J., and W. G. Flier. "The Biology of *Phytophthora infestans* at Its Center of Origin." *Annual Review of Phytopathology*, Vol. 43, 2005, pp. 171–190.

Judelson, H. S. "The Genetics and Biology of *Phytophthora infestans*: Modern Approaches to a Historical Challenge." *Fungal Genetics and Biology*, Vol. 22, 1997, pp. 65–76.

Kandell, J. "Building a Better Potato." *Los Angeles Times*, 11 August 2002.

Lee, M. R. "The Solanaceae: Foods and Poisons." *Journal of the Royal College of Physicians*, Edinburgh, Vol. 36, 2006, pp. 162–169.

O'Callaghan, M. "BASF Admits Defeat of GMO Potato Experiment." News Release, GM Free Ireland, 24 May 2006.

"Potato Blight Threatens New Brunswick Crop." United Press International, 11 August 2003.

Ristaino, J. B. "Tracking Historic Migrations of the Irish Potato Famine Pathogen, *Phytophthora infestans*." *Microbes and Infection*, Vol. 4, 2002, pp. 1369–1377.

Ristaino, J. B., et al. "PCR Amplification of the Irish Potato Famine Pathogen from Historic Specimens." *Nature*, Vol. 411, 2001, pp. 695–697.

Shattock, R. C. "*Phytophthora infestans*: Populations, Pathogenicity and Phenylamides." *Pest Management Science*, Vol. 58, 2002, pp. 944–950.

Staples, R. C. "Race Nonspecific Resistance for Potato Late Blight." *Trends in Plant Science*, Vol. 9, 2004, pp. 5–6.

Ulman, C., et al. "Zinc-Deficient Sprouting Blight Potatoes and Their Possible Relation with Neural Tube Defects." *Cell Biochemistry and Function*, Vol. 23, 2005, pp. 69–72.

Varzakas, T. H., et al. "The Politics and Science Behind GMO Acceptance." *Critical Reviews in Food Science and Nutrition*, Vol. 47, 2007, pp. 335–361.

SOYBEAN RUST

Summary of Threat

Soybean rust is a fungal disease that affects soybeans and other legume crops. It can cause defoliation, with losses of 70 to 80 percent. The fungus spreads by windborne spores. As of 2009, there are no resistant soybean strains, and control requires fungicide use. The U.S. government classified soybean rust as a biosecurity threat under the Bioterrorism Protection Act of 2002, but then delisted it again in 2005 to make it easier to study.

Other Names

English names for this disease include Asian soy or soybean rust (ASR), Australasian soybean rust, Asian rust, and soy rust. Most names in other languages are direct translations: *la roya de la soya del Asia* or *la roya de la habichuela soya* in Spanish, *la rouille de la soja* in French, *Asiatischen Sojarosts* in German, and *Sojabohnenrostes* in Swedish.

Description

"Soybean rust" actually refers to two related diseases. The agent of the more virulent disease is the fungus *Phakopsora pachyrhizi*, which can cause premature defoliation of soybean plants. Symptoms include small lesions on leaves that turn tan or reddish (Figure 5.6). Older names for this fungus include *Phakopsora sojae*, *P. calothea*, *Malupa sojae*, and *Uredo sojae*. Another fungus, *Phakopsora meibomiae*, causes a similar but milder disease. In this book, "soybean rust" means the virulent form unless otherwise specified.

Wind, rain, and contaminated objects transmit soybean rust. The disease is most likely to develop in rainy or humid weather, with temperatures between 59°F and 82°F (15°–28°C). The range of soybean rust changes from year to year, but in 2009 it is well established in the United States (16+ states), Mexico, and parts of South America, Asia, Africa, and Australia. Since low winter temperatures and low humidity limit its range, the only part of Europe at risk is the southern Mediterranean region.

More than 150 plant species can serve as hosts for soybean rust, including lima beans, black-eyed peas, garden peas, kidney beans, green beans, yam bean, and many common wild plants such as coral bean, clover, locoweed, lupine, and kudzu (*Pueraria montana*). Kudzu, a huge Asian weed that overruns everything, is resistant to soybean rust and can also survive cold weather. Thus, by infecting kudzu, soybean rust can safely overwinter in the United States.

Which Crops Are at Risk?

As of 2009, no known soybean strains are resistant to soybean rust. The U.S. crop consists mainly of cultivars that are considered highly susceptible. This plant represents an important part

Figure 5.6 Lesions of soybean rust on underside of leaf.
Source: Renato Boff, Universidade Federal do Rio Grande do Sul, Brazil.

of the modern diet; on average, Americans get 10 percent of their daily calories from soybean oil, and the soybean crop is also a major export. Soybean oil turns up in a surprising range of products, such as soaps, paints, insect repellents, herbicides, newspaper ink, and breast implants.

The Numbers

In 2005, the United States produced 84 million metric tons of soybeans, of which it exported about 30 percent. The rest of the crop, a mind-numbing 50 million metric tons (over 100 billion pounds), represented domestic consumption. These numbers explain the anxiety that surrounded the 2004 arrival of soybean rust; the United States was the world's leading producer and consumer of soybeans and in 2005 nearly tied Brazil as the world's leading exporter. In 2008, the National Agricultural Statistics Service estimated the total U.S soybean crop at 80 million metric tons. Thus far, soybean rust has not had a major impact on production, but the century is young.

History

Soybean rust was discovered in Japan in 1902 (Case Study 5-6). It spread to Australia by 1934, India by 1951, Hawaii by 1994, Africa by 1996, and Paraguay by 2001. When it reached Brazil in 2002, scientists realized that its arrival in North America was inevitable. It finally made its debut in Louisiana and adjacent southeastern states in November 2004, probably with help from Hurricanes Frances and Ivan.

Prevention and Treatment

As of 2009, the only preventive measures are chemical control and destruction of weed hosts such as kudzu. Several fungicides are effective, but somewhat expensive. Growers are advised to plant early in the season, with preventive applications of fungicides. Since rust spores adhere to clothing and boots, workers in infected fields should wear disposable spray suits.

Case Study 5-6: The Discovery of Soybean Rust

In 1902, Japanese scientist Dr. Torama Yoshinaga (1871–1946) first discovered the soybean rust fungus in a soybean field in Kochi Prefecture on the island of Shikoku. Ironically, soybean rust has never caused serious damage to Japan's soybean crops, probably because of the country's cold winters. The soybean plant itself has a long history in Asia. In 2853 B.C., the Chinese emperor Sheng-Nung designated the soybean as one of five sacred plants (the others were rice, wheat, barley, and millet).

Popular Culture

According to a peculiar rumor found on the Internet in 2008, the U.S. government imported the South American nutria (a large rodent, plural nutria) to get rid of kudzu—an imported weed that serves as an alternate host for soybean rust. This story makes no sense for three reasons. First, although nutria eat plants, there is no evidence that they prefer kudzu. Second, nutria have a well-authenticated history: ranchers brought them to the United States (and other countries) from South America in the early 1900s, hoping to create a demand for the fur and meat. When these markets failed to materialize, many captive nutria were liberated and became pests. Third, kudzu was not widespread in the United States until the 1930s, and the U.S. Soil Conservation Service was still advising farmers to plant it for erosion control as recently as the 1950s. Thus, the harmful effects of free-roaming nutria were recognized before the harmful effects of kudzu became apparent. Both species are now the targets of multi-million-dollar eradication programs.

The soybean itself is the subject of many popular beliefs. Soy contains chemical compounds similar to the female hormone estrogen, and health-food advocates have claimed both risks and benefits on that basis. Studies have yielded inconsistent results. Some sources insist that soy can cure everything from hot flashes to high blood pressure, whereas others believe that soy depresses the male sex drive and causes breast cancer in women. To the extent that there is any consensus, it appears that soy products are harmless in moderation.

The Future

The soybean industry needs cultivars with durable rust resistance—that is, resistance genes that the rust fungus cannot quickly defeat. As of 2009, related studies are in progress in the United States, Brazil, and other countries. The net effect of global climate change on soybean rust is hard to predict, but its range may expand into areas that are presently inhospitable due to cold winters.

References and Recommended Reading

"Asian Soybean Rust Identified in Iowa." Associated Press, 14 March 2007.

Bromfield, K. R. "World Soybean Rust Situation." Pages 481–500 in Hill, L. D. (Ed.). 1976. *World Soybean Research*. Danville, IL: Interstate Printers and Publishers.

Bromfield, K. R. "Soybean Rust" (monograph). American Phytopathological Society, December 1984, 65 pp.
Coblentz, B. "Soybean Rust Battle Takes Look at Kudzu." Press release, Mississippi State University Office of Agricultural Communications, 1 May 2008.
Del Ponte, E. M., et al. "Predicting Severity of Asian Soybean Rust Epidemics with Empirical Rainfall Models." *Phytopathology*, Vol. 96, 2006, pp. 797–803.
Dorrance, A. E., et al. 2004. Soybean Rust. Ohio State University Extension Fact Sheet.
Dorrance, A. E., et al. 2007. *Using Foliar Fungicides to Manage Soybean Rust*. Columbus: Ohio State University, 112 pp.
Feng, P. C., et al. "The Control of Asian Rust by Glyphosate in Glyphosate-Resistant Soybeans." *Pest Management Science*, Vol. 64, 2008, pp. 353–359.
"Iowa Officials Question Origin of Asian Soybean Rust Sample." *Associated Press*, 30 May 2007.
Keller, R. "Sentinel Plots Are Key." *AgProfessional*, January 2008.
Miles, M.R., et al. "Soybean Rust: Is the U.S. Soybean Crop at Risk?" American Phytopathological Society, June 2003.
Mueller, D., and D. Engelbrecht. "Soybean Rust Found in an Iowa Field." *Integrated Crop Management*, 1 October 2007.
Pivonia, S., and X. B. Yang. "Relating Epidemic Progress from a General Disease Model to Seasonal Appearance Time of Rusts in the United States: Implications for Soybean Rust." *Phytopathology*, Vol. 96, 2006, pp. 400–407.
Ratcliffe, S. T. "Soybean Rust, *Phakopsora pachyrhizi* and *P. meibomiae*." North Central Pest Management Center, 2002.
Rytter, J. L. "Additional Alternative Hosts of *Phakopsora pachyrhizi*, Causal Agent of Soybean Rust." *Plant Disease*, Vol. 68, 1984, pp. 818–819.
Shaner, G. E., et al. "Preparing for Asian Soybean Rust." Purdue University Cooperative Extension Service, ID-324, 2005, 16 pp.
Skeeles, J. "Soybean Rust Hurting Crops." *Chronicle-Telegram* (Elyria, Ohio), 18 January 2005.
United States Department of Agriculture. "Feeding America: The Rust Invasion." Cooperative State Research, Education and Extension Service (CSREES), Partners Video Magazine 17 (on DVD), 2007.
Wright, D. "Dispelling Myths about Asian Soybean Rust." Plant Health Initiative, Summer 2004.

WITCHES' BROOM DISEASE

Summary of Threat

Witches' broom is one of several diseases that attack cacao trees, whose seed pods are the source of cocoa and chocolate. In 1989, this fungus destroyed an estimated 90 percent of Brazil's cacao crop. It is one of several factors driving the recent increase in chocolate prices. Fungicides are largely ineffective, so farmers must cut off the infected parts of the trees to prevent the release of spores.

Other Names

The only common English name for this disease and its agent appears to be witches' broom fungus or witches' broom disease—"the witch" for short. Brazilian sources call it *fungo vassoura-de-bruxa*, the Portuguese equivalent. One website calls it "disease of the brush of witch," probably the result of machine translation.

Description

The agent is the fungus *Crinipellis perniciosa* (formerly called *Moniliophthora perniciosa* or *Marasmius perniciosus*). Wind, rain, and human activities can disperse the spores; in one reported case, an outbreak occurred when workers dumped infected plant debris into a river that carried it downstream to other plantations. Related fungi cause other cacao diseases called frosty pod and black pod. Witches' broom (Figure 5.7) causes an infected tree to send up a spray of random shoots from its flower clusters and branch tips and also infects the pods, making them unusable and reducing bean production. As of 2009, this disease is limited to the New World tropics and has not reached West Africa or Indonesia, the other major cacao-growing areas.

Figure 5.7 Witches' Broom Fungus (*Crinipellis perniciosa*), a mushroom that infects cacao trees and reduces yield.

Source: Scott Bauer, USDA Agricultural Research Service.

Every time this interesting mushroom rears its head, the press exploits public fears that it will devastate the global chocolate supply. But why is a threat to chocolate somehow more frightening than a threat to soybeans or potatoes? That's like asking why Ebola attracts so much media attention and pneumonia so little, although Ebola kills few people and pneumonia kills many. Chocolate is news.

Which Crops Are at Risk?

Other fungal diseases called "witches' broom" attack other crops, but the only major crop at risk from *Crinipellis perniciosa* is the cacao plant. This fungus also infects liana (*Arrabidaea verrucosa* and others), nightshades (*Solanum* sp.), and annatto (*Bixa orellana*). The fungus produces spores at night, in areas with annual rainfall of 60–80 inches (150–200 centimeters), air temperatures in the range 75°F to 80°F (24°–27°C), and relative humidity of 80–90 percent.

As of 2009, no genetically resistant cacao strains are known or widely available, and all cacao trees in South America, Central America, and southern Mexico are at risk. In the 1930s, resistant cacao strains were found in Trinidad, but these plants succumbed to more aggressive strains of the fungus in other countries.

The Numbers

The total area planted with cacao trees worldwide is about 27,000 square miles (or 70,000 square kilometers). As of 2009, annual retail chocolate sales total about $50 billion. Sales in the United States alone represent more than one-quarter of this total (about $13 billion).

Total cacao production in 2003–2004 was about 3.5 million tons, up from 1.5 million tons in 1983–1984. This trend represents an increase in planted areas rather than higher yield. Witches' broom and other cacao diseases continue to destroy 30–40 percent of the global crop each year, causing annual losses estimated at $2 billion. As a result, chocolate prices increased by nearly 50 percent between 2006 and 2007.

About 70 percent of the world's cacao beans grow in West Africa. The Ivory Coast contributes about 40 percent of the total, mostly from small family farms. By some estimates, if witches' broom and related fungi ever become established in West Africa, the world will lose at least one-quarter of its chocolate.

History

Cacao and its parasites originated in the Amazon basin, where Native Americans cultivated this crop as early as the seventh century A.D. A statement often attributed to the Aztec Emperor Montezuma alleges that a cup of chocolate permits a man to walk for a whole day without food. Christopher Columbus supposedly was the first European to taste chocolate, during his fourth visit to the New World in 1502. Hernán Cortez later introduced chocolate to Europe, where it was an instant success. The use of chocolate spread to the United States in about 1765, and a Dutch inventor introduced the cocoa press in 1828. Witches' broom disease of cacao was first described in South America in the late 1700s and has been the subject of scientific investigations since the 1890s (Case Study 5-7).

Prevention and Treatment

As of 2009, all known fungicides appear to be ineffective against witches' broom, or else too expensive for large-scale use. Researchers have tried using another fungus as a biocontrol agent to reduce witches' broom spore production, but results have been inconsistent. Until a better method is found, the most practical way to manage the disease is by removing infected parts of trees. Growers keep the trees short to make pruning easier. Brooms left on the ground can be sprayed with petroleum oil, which makes the surface repellent to rain water and prevents the fungus from producing spores.

Case Study 5-7: Lessons from Chocolate

The history of witches' broom disease contains a bittersweet lesson. In 1930, an early environmentalist named Albert Stoll Jr. wrote, in the context of a brief review of witches' broom fungus and other major crop diseases,

> All plant life, so necessary to the welfare of man, appears to go through successive cycles of devastation and immunity from various forms of disease, keeping the scientists continually on the jump to uncover means of combating the menace. No sooner is this accomplished than up bobs an invading army of insects to wreak more vengeance and make the control task the more difficult. Life is a great struggle and the plant world demonstrates this without a doubt.[1]

Mr. Stoll was conservation editor of the *Detroit News from* 1923 to 1950, and his efforts were largely responsible for the creation of Isle Royale National Park in Lake Superior.

[1]*Ironwood Daily Globe*, 4 August 1930

Popular Culture

As of 2009, chocolate addiction is a more-or-less recognized phenomenon, yet its users can do little to stop the onslaught of witches' broom and other diseases that imperil the world's cocoa supply. One seemingly unavoidable joke is that the United States government should add chocolate to its growing list of controlled substances, so that the trafficking and use of chocolate would become illegal. Its market price would then skyrocket, and with it the incentive and budget for more aggressive disease control measures. (Yes, of course we're kidding.)

Is it true that burning cacao beans or plants can release lethal clouds of hydrogen cyanide? No. Many plants, including cacao, contain cyanide compounds in small quantities and release some cyanide when burned, but at levels far below OSHA permissible exposure limits.

The Future

One way to solve the problem of witches' broom fungus would be to reduce global demand for cacao beans, perhaps by improving the quality (and reducing the cost) of chocolate produced in cell culture. Such a breakthrough might impact Third World economies, but that issue has not hindered the development of synthetic vanilla.

References and Recommended Reading

Andebrhan, T. "Studies on the Epidemiology and Control of Witches' Broom Disease of Cacao in the Brazilian Amazon." Pages 395–402 in *Proceedings of the 9th International Cocoa Research Conference, Lome, Togo, 12–18 February 1984*. Lagos, Nigeria: Cocoa Producers' Alliance, 1985.

Becker, H. "Fighting a Fungal Siege on Cacao Farms." *Agricultural Research*, November 1999.

Bowers, J. H., et al. "The Impact of Plant Diseases on World Chocolate Production." *Plant Health Progress*, 9 July 2001.

Clarence-Smith, W. G. 2000. *Cocoa and Chocolate, 1765–1914*. New York: Routledge.

Cronshaw, D. K. "Fungicide Application together with Cultural Practices to Control Cocoa Diseases Caused by *Crinipellis perniciosa, Monilia roreri, Phytophthora palmivora* in Ecuador." *Tropical Agriculture*, Vol. 56, 1979, pp. 165–170.

Evans, H. C. "Witches' Broom Disease: A Case Study. Cocoa in South American Countries." *Cocoa Growers Bulletin*, Vol. 32, 1981, pp. 5–19.

Evans, H. C. "Cacao Diseases—The Trilogy Revisited." *Phytopathology*, Vol. 97, 2007, pp. 1640–1643.

Evans, H. C., and R. W. Barreto. "*Crinipellis perniciosa*: A Much Investigated but Little Understood Fungus." *Mycologist*, Vol. 10, 1996, pp. 58–61.

Fulton, R. H. "The Cacao Disease Trilogy: Black Pod, Monilia Pod Rot, and Witches'-Broom." *Plant Diseases*, Vol. 73, 1989, pp. 601–603.

Hebbar, P. K. "Cacao Diseases: A Global Perspective from an Industry Point of View." *Phytopathology*, Vol. 97, 2007, pp. 1658–1663.

"Mars Teams Up with USDA to Improve Cacao Genetics for Pest and Disease Resistance, Better Yields and Climatic Adaptation." *Food Industry News*, 18 July 2008.

Money, N. P. 2007. *The Triumph of the Fungi: A Rotten History*. New York: Oxford University Press.

Patterson, R. "Recovery from This Addiction was Sweet Indeed." *Canadian Medical Association Journal*, Vol. 148, 1993, pp. 1028–1032.

Pegler, D. N. "*Crinipellis perniciosa* (Agaricales)." *Kew Bulletin*, Vol. 32, 1978, pp. 731–736.

Pereira, J. L., et al. "Witches' Broom Disease of Cocoa in Bahia: Attempts at Eradication and Containment." *Crop Protection*, Vol. 15, 1996, pp. 743–752.

"Plant Diseases Threaten Chocolate Production Worldwide." *ScienceDaily*, 6 June 2006.

Ploetz, R. C. "Cacao Diseases: Important Threats to Chocolate Production Worldwide." *Phytopathology*, Vol. 97, 2007, pp. 1634–1639.

Purdy, L. H., and R. A. Schmidt. "Status of Cacao Witches' Broom: Biology, Epidemiology, and Management." *Annual Review of Phytopathology*, Vol. 34, 1996, pp. 573–594.

Samuels, G. J., et al. "*Trichoderma stromaticum* sp. nov., a Parasite of the Cacao Witches Broom Pathogen." *Mycological Research*, Vol. 104, 2000, pp. 760–764.

Stoll, A., Jr. "Our Plant Disease Loss." *Daily Globe*, 4 August 1930.

Tovar Rodriguez, G. "Witches' Broom in Cacao, Science and Technology." *Agronomia Colombiana*, Vol. 3(1/2), 1986, pp. 15–30.

van den Doel, K., and G. Junne. "Product Substitution through Biotechnology: Impact on the Third World." *Trends in Biotechnology*, Vol. 4, 1986, pp. 88–90.

Wheeler, B. E. J., and Mepsted, R. "Pathogenic Variability Amongst Isolates of *Crinipellis perniciosa* from Cocoa (*Theobroma cacao* L.)." Plant Pathology, Vol. 37, 1988, pp. 475–488.

PHOMA STEM CANKER

Summary of Threat

The agents of phoma stem canker are two related fungi that infect cruciferous vegetables (such as broccoli and cauliflower) and canola, which is a major source of vegetable oil and animal feed. These products may not sound as glamorous as chocolate or oranges, but losses from phoma stem canker amount to nearly $1 billion per year and will probably increase as a consequence of global climate change.

Other Names

Phoma stem canker is also called blackleg of cabbage or crucifers, black stem disease, phoma leaf spot, crucifer canker, or crucifer dry rot. Spanish texts describe it as *pie negro* ("black foot" or "black stem"), *chancro* ("canker"), *podredumbre seca* ("dry rottenness"), or *podredum-*

bre de raíz, pie y tallo ("rottenness of the root, stem and stalk"). French and German names have similar translations.

Description

The agent, a fungus called *Leptosphaeria maculans*, has an asexual growth form with a different name, *Phoma lingam* (formerly *Sphaeria lingam* or *Plenodomus lingam*). A less aggressive species, *Leptosphaeria biglobosa*, also causes phoma stem canker.

Like most fungal diseases of plants, phoma stem canker spreads by airborne and waterborne spores, as well as by infected seed and debris (Figure 5.8). The spores enter a host plant through wounds or leaf stomata and grow toward the stem, causing damage that reduces market value or kills the plant. Growth is faster at higher temperatures.

Phoma stem canker infects cruciferous plants, including canola (oilseed rape), broccoli, turnip, bok choy, rutabaga, cauliflower, Brussels sprouts, cabbage, collards, kale, and others. Some cruciferous weeds, such as mustard, are also susceptible. The world could survive without cruciferous vegetables, but it would be a sadder place. Canola is a major cash crop, the world's third largest source of vegetable oil (after soy and palm) and one of the few that can grow in colder climates and in winter. Byproducts are also a major source of animal feed.

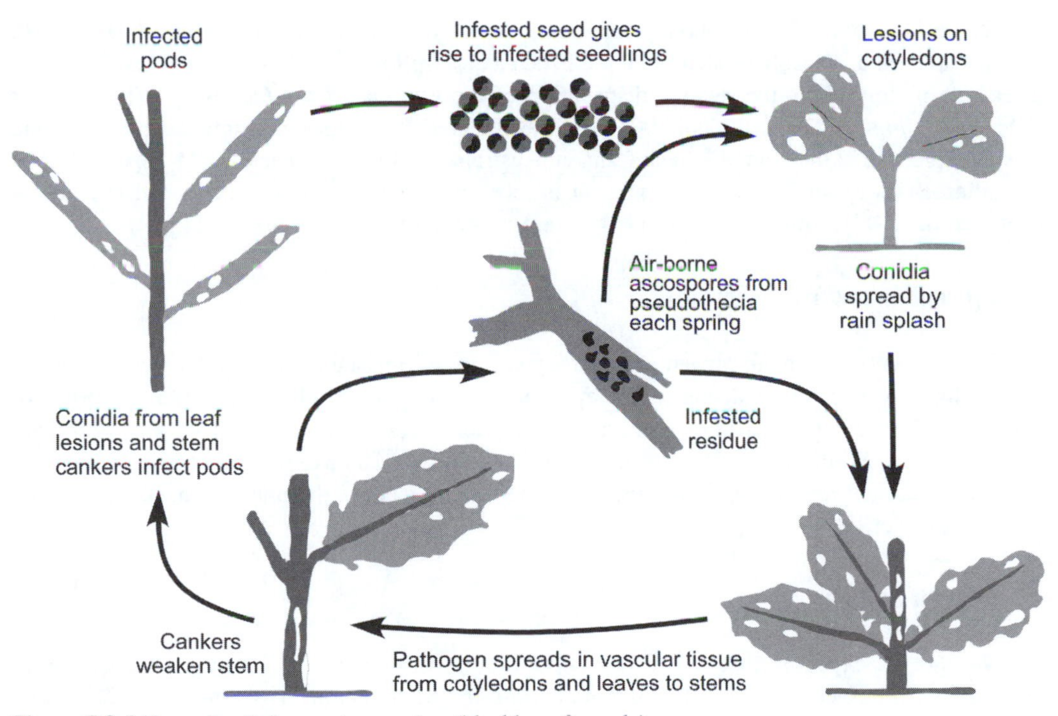

Figure 5.8 Life cycle of phoma stem canker (blackleg of canola).
Source: Canadian Phytopathological Society.

Case Study 5-8: Save the Crucifers

Turnips, mustard, and other crucifers were among the first vegetables cultivated by humans. The Chinese have grown these plants for at least 7,000 years. With such a history, it seems likely that the crucifers will survive the onslaught of phoma stem canker, even under conditions of global warming—but they might not survive the double-whammy of phoma stem canker *plus* the loss of honeybees (Chapter 4). Cruciferous vegetables, unlike many staple crops, depend on honeybees for effective cross-pollination.

Which Crops Are at Risk?

Most cruciferous plants are at risk (Case Study 5-8), and outbreaks are most likely in areas with high summer temperatures. Researchers have developed computer models to simulate the interaction of phoma stem canker development with cultural practices (sowing date, crop density, fungicide application, cultivar, nitrogen management) and physical variables such as temperature and rainfall.

The Numbers

In 2006–2007, the world produced about 18 million metric tons of canola oil. Six nations—China, Canada, India, Germany, France, and Britain—have historically accounted for about 80 percent of production. In addition to canola oil, Canada and the United States (mainly North Dakota) are major producers and exporters of canola seed. More than 80 percent of available seed has been genetically modified for increased resistance to herbicides and disease. As of 2008, worldwide annual losses of canola from phoma stem canker total about $900 million.

History

The official discoverer of phoma stem canker was German botanist H. J. Tode (1733–1797), who found it on dead cabbage stems and published a description in 1791. The recent history of this disease is mainly about the relative distribution of the two pathogens *Leptosphaeria maculans* (LM) and *L. biglobosa* (LB). As of 2005, LM was the predominant species in western Europe and was also present in the United States, Canada, Australia, and parts of Africa. LM appeared to be spreading eastward in 2005 and was present in most eastern European countries, but was absent from Russia. LB was present in North America, Europe, Australia, and part of China.

Prevention and Treatment

Until recently, the main prevention strategy was to use resistant cultivars. Unfortunately, the fungus has evolved, too, causing major epidemics in Australia and Europe. While studies of genetic resistance continue, fungicides are the main defense, combined with integrated pest management practices such as four-year crop rotation and the avoidance of infected sites. Early detection and treatment are essential, because once the fungus reaches the plant stem, fungicides are no longer effective.

Popular Culture

We did not expect to find any novels about cruciferous vegetables, but we were wrong. In Don Lee's acclaimed 2008 novel *Wrack and Ruin*, the main character is a northern California organic Brussels sprout farmer who must deal with the numerous diseases that afflict his crop.

According to a particularly silly urban legend, turnips and other cruciferous vegetables are never canned because they are treated with chlorine to kill fungus—possibly including *Leptosphaeria*—and the chlorine would interact with the lining of the can to release deadly chlorine gas. This is false. Many vegetables are washed in a chlorine solution, but it's rinsed off again before packaging. Gas bubbles sometimes form inside cans of food due to spoilage, but these gases are harmless, and the solution is to throw the cans away or take them back to the store. Turnips, cauliflower, and related vegetables are rarely canned because of their somewhat mushy texture, but we have talked to people who have seen and even eaten canned turnips.

The Future

Just when it seemed that global warming was responsible for everything, the press added one more negative consequence to the list. A 2007 study showed that a warmer (and possibly wetter) climate will probably increase the range and severity of phoma stem canker, thus reducing the world's supply of canola oil, and raising its price at the very time when renewable biodiesel fuels are likely to be in greater demand. The distribution of this disease will move northward into regions that are presently too cold, thus posing a serious risk to subsistence farmers in some parts of China and India.

References and Recommended Reading

Aubertot, J.-N., et al. "SimCanker: A Simulation Model for Containing Phoma Stem Canker of Oilseed Rape through Cultural Practices." Proceedings of the Fourth International Crop Science Congress, Brisbane, Australia, 2004.

Delaplane, K. S., and D. F. Mayer. 2000. *Crop Pollination by Bees*. New York: CABI Publishing.

Egan, J., et al. "Burning Canola Stubble May Not Control Blackleg." *Oilseed Outcomes*, Oilseed Industry and Agronomic Research Updates, July 2006, No. 1.

Evans, N., et al. "Range and Severity of a Plant Disease Increased by Global Warming." *Journal of the Royal Society Interface*, Vol. 5, 2008, pp. 525–531.

Fitt, B. D. L., et al. "World-Wide Importance of Phoma Stem Canker (*Leptosphaeria maculans* and *L. biglobosa*) on Oilseed Rape (*Brassica napus*). *European Journal of Plant Pathology*, Vol. 114, 2006, pp. 3–15.

Henderson, M. P. "The Black-Leg Disease of Cabbage Caused by *Phoma lingam* (Tode) Desmaz." *Phytopathology*, Vol. 8, 1918, pp. 379–431.

Howlett, B. J., et al. "*Leptosphaeria maculans*, the Causal Agent of Blackleg Disease of Brassicas." *Fungal Genetics and Biology*, Vol. 33, 2001, pp. 1–14.

Keri, M., et al. "Inheritance of Resistance to *Leptosphaeria maculans* in Brassica juncea." *Phytopathology*, Vol. 87, 1997, pp. 594–598.

Marcroft, S. "Australian Blackleg Management Guide." Canola Association of Australia, January 2005, 4 pp.

Markell, S., et al. "Blackleg of Canola." Fargo: North Dakota State University Extension Service, June 2008, 4 pp.

Moreno-Rico, O., et al. "Characterization and Pathogenicity of Isolates of *Leptosphaeria maculans* from Aguascalientes and Zacatecas, Mexico." *Canadian Journal of Plant Pathology*, Vol. 23, 2001, pp. 270–278.

Murdock, L., et al. 1991. "Canola Production and Management." University of Kentucky, College of Agriculture.

Pedras, M. S. C., and Y. Yu. "Stress-Driven Discovery of Metabolites from the Phytopathogenic Fungus *Leptosphaeria maculans*: Structure and Activity of Leptomaculins A-E." *Bioorganic & Medicinal Chemistry*, Vol. 16, 2008, pp. 8063–8071.

Rimmer, S. R., et al. (Eds.). *Compendium of Brassica Diseases*. St. Paul, MN: American Phytopathological Society Press.

Sosnowski, M., et al. 2001. "Symptoms of Blackleg (*Leptosphaeria maculans*) on the Roots of Canola in Australia." *New Disease Reports*, Volume 3, July 2001.

Sprague, S. J., et al. "Pathways of Infection of *Brassica napus* Roots by *Leptosphaeria maculans*. *New Phytologist*, Vol. 176, 2007, pp. 211–222.

Taylor, J. L. "A Simple, Sensitive, and Rapid Method for Detecting Seed Contaminated with Highly Virulent *Leptosphaeria maculans*." *Applied Environmental Microbiology*, Vol. 59, 1993, pp. 3681–3685.

West, J. S., et al. "Epidemiology and Management of *Leptosphaeria maculans* (Phoma Cell Canker) on Oilseed Rape in Australia, Canada and Europe." *Plant Pathology*, Vol. 50, 2001, pp. 10–27.

ASIAN SOYBEAN APHID

Summary of Threat

Asian soybean aphids (*Aphis glycines*) destroy plants by reproducing in large numbers and eating leaves. They also spread several viral plant diseases. Hosts for this aphid include soybean plants and other legumes, such as snap beans. Recent economic losses in the United States alone have approached $1 billion per year due to crop losses and the cost of pesticides.

Other Names

Common English names include exotic soybean aphid, soya bean aphid, SBA, or soy aphid. In other languages, this insect is *áfido de la soya* (Spanish), *puceron du soja* (French), *Sojabohnenblattlaus* (German, "soybean leaf louse"), *soijakirva* (Finnish), and *daizu-aburamusa* (Japanese).

Description

The soybean aphid (*Aphis glycines*) is an agricultural pest that has spread from Asia to all major soybean-producing regions of the world. It can reduce soybean yields by up to 50 percent by destroying plants directly, and it also spreads viral diseases such as soybean mosaic virus (SMV), soybean dwarf virus, cucumber mosaic virus, bean yellow mosaic virus, clover yellow vein virus, peanut mottle virus, and potato virus Y.

Soybean aphids are tiny greenish-yellow insects with reddish-brown eyes (Figure 5.9). They look similar to two other aphid species that live on the same host plants (*Aphis gossypii* and *A. nasturtii*), but the soybean aphid is the only one that forms large colonies. When conditions become crowded, soybean aphids produce a generation of winged adults that can fly to new fields. Wind and rain may also disperse these insects. In North America, winter hosts include a woody shrub called common buckthorn (*Rhamnus cathartica*) and the Asian weed kudzu (*Pueraria montana*).

Figure 5.9 Asian soybean aphid.

Source: Purdue University.

Unlike the brown citrus aphid (Case Study 5-1), the Asian soybean aphid has a sex life during part of the year. In summer, the entire population is female. In fall, both males and females are born, some of them with wings. The aphids migrate from soybeans to buckthorn shrubs, where they mate and lay their winter eggs. The eggs hatch in spring, and two or three generations of aphids live on buckthorn before returning to the soybean plants.

Which Crops Are at Risk?

Soybeans are the main crop at risk, but Asian soybean aphids can also infest snap beans and possibly alfalfa. In soybeans, yield loss depends partly on the stage of growth when the aphid infestation occurs. If aphid density is high when the soybean plants begin to flower, the damage is greatest, because the aphids interfere with pod development. Damage is also greater if plants are unhealthy due to dry soil or other unfavorable conditions. Aphids reproduce most rapidly at air temperatures of 75°–80°F (24°–27°C), and tend to stop reproducing when the mercury hits 90°F (32°C). Predators, parasites, some fungi, and thunderstorms can reduce aphid populations.

The Numbers

Between 2001 and 2004, the soybean aphid cost the U.S. economy an estimated $2.2 billion in crop losses, pesticides, and biological control programs. Each female soybean aphid has about

45 offspring, called nymphs, over a period of 15 days in summer. These nymphs become adults in about 5 or 6 days. The number of soybean aphids per plant required to cause economic damage may range from 2,500 to 4,000, depending on weather and other conditions. One Asian ladybird beetle can eat 160 aphids per day.

History

The Asian soybean aphid is native to eastern Asia. It arrived in the United States sometime before 2000, probably on airline passengers or horticultural cargo from Japan or China. The species was first identified in Wisconsin in the summer of 2000, and by 2003 it had spread to at least 20 states by active and wind-aided flight, and possibly also by clinging to vehicles. The reported distribution varies from year to year, but the species has occupied most soybean-growing areas.

Prevention and Treatment

Insecticides are the main control method, but these chemicals also kill beneficial insects and increase production costs. Studies of biological control methods have yielded some promising results (Case Study 5-9). For example, certain plants used as living mulches can increase populations of predators that eat aphids. Other control strategies include the use of resistant varieties, reflective mulches (because ultraviolet light deters winged aphids), trap cropping, weed control, rouging, and mineral oil sprays.

Since soybean aphids spend the winter on buckthorn shrubs, one obvious approach is to remove buckthorn near soybean fields. This is not as easy as it might sound, because buckthorn quickly regrows from a cut stump.

Case Study 5-9: Alien Versus Predators

When the Asian soybean aphid arrived in the United States in 2000, it became food for many predators, including the multicolored Asian lady beetle, which the USDA had already imported in the hope of controlling other aphids. By 2004, at least 22 predator species (native and nonnative) were eating Asian soybean aphids, yet outbreaks continued, often reducing soybean yield by as much as 40 percent. Scientists are now studying an Asian parasitic wasp called *Binodoxys communis* to see if it might serve as a form of biological control. It can kill Asian soybean aphids, but the problem is that it might compete with Asian lady beetles and other predators, instead of augmenting their efforts.

Popular Culture

Aphid folklore is surprisingly rich, perhaps because of the long association between aphids and farmers. In the Hebrides Islands of Scotland in the nineteenth century, children learned to extract honeydew by placing aphids on the backs of their hands and reciting a Gaelic poem, rendered in English as follows:

Carlin of the whey,
Whey-pail, whey-pail,
Give me of your whey
Or I will take head and feet off you.[6]

6. A. Goodrich-Freer, "More Folklore from the Hebrides" (*Folk-Lore*, Vol. 13, 1902, pp. 29–62).

In response to this ritual, the aphid would deposit a drop of sweet liquid on the child's hand. (This actually works.)

According to a legend of uncertain provenance, medieval European farmers once prayed to the Virgin Mary to rescue their crops from aphids. Swarms of small red beetles soon appeared and ate the aphids, thus saving the crops and preventing a famine. From that day forward, these insects were known as the beetles of Our Lady, Our Lady's bird, lady beetles, ladybirds, or ladybugs. Each lady beetle had seven black spots, representing the seven sorrows of the Blessed Virgin. In real life, the number of spots on a lady beetle may range from 0 to 24; a common European species has seven spots. But part of this story predates the Christian era, for the lady beetle was also associated with Freyja, the Norse mother goddess. The ladybug has inspired other legends, judging by its many unusual English names: fly-golding, goldie-bird, Bishop Barnaby, or God's almighty cow. Its Italian name means "the devil's chicken," whereas a more practical German name translates as "little hemispherical beetle."

The Future

Soybean aphid outbreaks are rare in the species' native China, thanks to the presence of parasitic wasps and other natural enemies, but it is impossible to recreate an entire foreign ecosystem piece by piece. Although further study may yield more effective biological controls, pesticides and the associated costs are likely to be necessary for the foreseeable future.

References and Recommended Reading

Aponte, W., and D. Calvin. "Fact Sheet: Soybean Aphid (*Aphis glycines*)." Penn State Entomology Department, July 2004.

"Bacteria Protect Soybeans from Aphids." United Press International, 14 April 2009.

Blackman, R. L, and V. F. Eastop. 1984. *Aphids on the World's Crops: An Identification and Information Guide.* New York: John Wiley.

Diaz-Montano, J., et al. "Chlorophyll Loss Caused by Soybean Aphid (Hemiptera: Aphididae) Feeding on Soybean." *Journal of Economic Entomology*, Vol. 100, 2007, pp. 1657–1662.

Griffiths, P. D. "Evaluation and Enhancement of Virus-Resistant Snap Beans." Final Project Report, NYS IPM Program, 2006.

Heimpel, G. E. "Soybean Aphid as Part of a Potential Four-Species Invasional Meltdown: Evaluation and Implications for Management." Abstract of paper presented at Ecological Society of America Annual Meeting, 16–19 November 2008.

Hill, C. B., et al. "Resistance of Glycine Species and Various Cultivated Legumes to the Soybean Aphid (Homoptera: Aphididae). *Journal of Economic Entomology*, Vol. 97, Vol. 2004, pp. 1071–1077.

Landis, D. A., et al. "Impact of Multicolored Asian Lady Beetle as a Biological Control Agent." *American Entomologist*, Vol. 50, 2004, pp. 153–154.

McCormack, B. P., et al. "Demography of Soybean Aphid (Homoptera: Aphididae) at Summer Temperatures." *Journal of Economic Entomology*, Vol. 97, 2004, pp. 854–861.

McGraw, L. "New Aphid Threatens U.S. Soybeans—*Aphis glycines*." *Agricultural Research*, May 2002.

Miao, J., et al. "Population Dynamics of Aphis glycines (Homoptera: Aphididae) and Impact of Natural Enemies in Northern China." *Environmental Entomology*, Vol. 36, 2007, pp. 840–848.

Myers, S. W., et al. "Effect of Soil Potassium Availability on Soybean Aphid (Hemiptera: Aphididae) Population Dynamics and Soybean Yield." *Journal of Economic Entomology*, Vol. 98, 2005, pp. 113–120.

Nault, B. A. "Aphid Ecology and Epidemiology of Cucumber Mosaic Virus in Snap Bean Fields: Implications for Management." Cornell University, March 2007.

O'Neil, R. "Biological Control of the Soybean Aphid." SABC Annual Report to North Central Soybean Research Program, April 2007.
"Reuniting Old Enemies: Releasing Aphid Enemies from Asia into Midwest Soybean Fields." Urbandale: Iowa Soybean Association, 2007.
Schmidt, N. P., et al. "Alfalfa Living Mulch Advances Biological Control of Soybean Aphid." *Environmental Entomology*, Vol. 36, 2007, pp. 416–424.
Shah, D. A., et al. "Incidence, Spatial Patterns, and Associations among Viruses in Snap Bean and Alfalfa in New York." *Plant Disease*, Vol. 90, 2006, pp. 203–210.
Sloderbeck, P. E., et al. "The Soybean Aphid." Kansas State University Agricultural Experiment Station and Cooperative Extension Service MF-2582, June 2003.
Takahashi, S., et al. "Life Cycle of the Soyabean Aphid *Aphis glycines* Matsumura, in Japan." *Japanese Journal of Applied Entomology and Zoology*, Vol. 37, 1993, pp. 207–212.
Wang, X. B., et al. "A Study on the Damage and Economic Threshold of the Soyabean Aphid at the Seedling Stage." *Plant Protection*, Vol. 20, 1994, pp. 12–13.
Wyckhuys, K. A. G., et al. "Parasitism of the Soybean Aphid *Aphis glycines* by *Binodoxys communis*: the Role of Aphid Defensive Behavior and Parasitoid Reproductive Performance." *Bulletin of Entomological Research*, Vol. 98, 2008, pp. 361–370.

LOCUSTS

Summary of Threat

Locusts are grasshoppers that migrate in huge swarms, destroying crops in their path. A single swarm may be several miles long and may devastate a field in minutes. Natural predators and parasites help reduce locust numbers. Other control measures include insecticide spraying from vehicles and plowing the soil to destroy locust eggs. Better methods are needed, particularly in Africa and the Middle East.

Other Names

English nicknames and euphemisms for locusts include "the countless," "darkener of the sun," "shrimp of the desert," "flying shrimp," or "hopper." In other languages, locusts are *anbeta* (Ethiopian), *saltamontes* (Spanish), *sauterelle* (French), *cavalletta* (Italian), *gafanhoto* (Portuguese), *si-khónyane* (Swazi), *adede* (Luo), *luzige* (Egyptian), *sprinkaan* (Afrikaans), *Woestijnsprinkhaan* (Dutch), or *Heuschrecke* (German).

Description

Locusts are grasshoppers that sometimes come together in huge migratory swarms. In some species, this swarming follows a physical transformation that results from crowding. Of about 8,000 grasshopper species, only 10 or 12 are known to have a locust phase. The most famous is the desert locust (*Schistocerca gregaria*), shown in Figure 5.10. This insect occurs throughout Africa, the Middle East, and western Asia—about 20 percent of the land surface of the Earth. China, South America, Australia, and other major agricultural areas have their own locust species. Some closely related grasshoppers occur in the United States, but they do not swarm, although they can cause severe crop damage.

Figure 5.10 Desert locust (*Schistocerca gregaria*).
Source: Spencer Behmer, Texas A&M University.

Which Crops Are at Risk?

All plants (and wooden objects and laundry) in the path of a migratory locust swarm are at risk. In some parts of the world, locust swarming is associated with high seasonal rainfall. The effects of temperature are less clear; warmer winters may enable locusts to survive, but hotter summers may be less favorable.

The Numbers

A single desert locust swarm can contain as many as 80 million locusts per square kilometer, or 31 million per square mile. The swarm may contain billions of locusts, and it can travel more than 81 miles (130 kilometers) per day, destroying more than 100 tons of vegetation every day.

One reported locust swarm in Africa was about 1 mile (1.6 km) wide at the front and 100 feet (about 30 meters) high, and took more than nine hours to pass overheard. But according to an 1880 report, a swarm of Rocky Mountain locusts in 1875 was much larger—an astonishing 110 miles (176 km) wide and 1,800 miles (2,880 km) long.

When China had a locust outbreak in 2008, a few weeks before the Olympic Games, the government sent 33,000 professional exterminators to contain the problem. But large-scale pesticide spraying causes other problems. By one estimate, there are over 25 million cases of work-related pesticide poisoning in developing countries every year.

History

The phenomenon of locust swarming is older than agriculture, but we will fast-forward for the sake of brevity. Examples of major outbreaks occurred in the United States in 1875; in Palestine in 1915; in South America in 48 of the 58 years between 1897 and 1954, but apparently none since; in Africa in 1986–1989 and 2004; in Australia in 2004; and in China in 2008.

Only once in the history of agriculture has mankind succeeded in driving a pest to extinction, and even that victory resulted from an accident. Rocky Mountain locusts (*Melanoplus spretus*) in the Great Plains aggregated in some of the largest swarms ever described, but by the early twentieth century they were extinct. Studies of frozen locusts in glaciers revealed the probable sequence of events: Although the invasion area of this locust was over 2 million square miles (5.5 million square kilometers), the areas where it laid its eggs were quite small, and were limited to river valleys where settlers grazed their cattle and plowed the soil. In so doing, they inadvertently wiped out the locust's underground nursery.

Prevention and Treatment

For more than a thousand years, the Chinese have kept records of locust activity and weather conditions in an effort to predict swarming. Other countries later adopted a similar approach using GIS technology and satellite imagery, but the result is the same. Tracking the movement of locust swarms and preparing for their arrival is an important component of prevention.

Farmers in Africa use vehicle-mounted and aerial insecticide sprayers to protect crops from locusts, which acquire the chemicals when they land on the plants and start eating them. Biopesticides—bacteria, fungi, pheromones, and plant extracts such as neem—work more slowly than chemical pesticides, but cause less harm to people and the environment. Natural enemies of locusts include birds, reptiles, and certain parasitic or predatory wasps, flies, and beetle larvae. Digging up locust eggs might work in theory, but in most regions this approach is impracticable because of the size of the area affected.

Popular Culture

The 1957 motion picture *Beginning of the End* is about giant man-eating grasshoppers, created by accident when ordinary grasshoppers get into a silo full of radioactive wheat at an experimental farm in Illinois. The scientists, who have tampered with forces best left alone, wring their hands until the Army comes to the rescue. Our favorite line, spoken by the General:

> Dr. Wainwright, you're a scientist, you know what grasshoppers can do. I'm a soldier, I know what guns can do.

In the 1950s, journalists speculated about the possibility of using radiation to create better crops and livestock, and this prospect inspired the same fears that resulted from genetic engineering experiments a generation later. But there is nothing unbelievable about grasshoppers suddenly becoming huge (at least collectively) and causing widespread destruction. Locusts do it all the time.

In the 2004 motion picture *Hidalgo*, desert locusts serve as food for the hero and his horse (see also Case Study 5-10, page 189). In many places where locusts occur, they are not the worst natural phenomena that farmers must deal with. An Ethiopian proverb holds that "It is better to have locusts than rain in November."[7] (Rain promotes mold, which can destroy grain stored after the fall harvest.)

In her 1937 memoir *On the Banks of Plum Creek*, Laura Ingalls Wilder (famous for *Little House on the Prairie*) vividly described a swarm of Rocky Mountain locusts attacking a wheat field. This species now appears to be extinct.

The Future

In 2009, researchers reported that a chemical called serotonin causes grasshoppers to transform into locusts. Someday it might be possible to develop chemical weapons that specifically block this transformation by inhibiting serotonin production in grasshoppers. But since serotonin is also a neurotransmitter in the brains of many other animal species—including humans—extensive testing will be required to ensure safety. Otherwise, the potential for urban legends, science-fiction novels, and real-life tragedy is immense.

Case Study 5-10: Deep-Fried Locust

In the fifth century B.C., the historian Herodotus wrote that people in the region we now call Libya ate powdered locusts mixed with milk. More recently, John the Baptist and the prophet Muhammad ate locusts. Dried locusts contain up to 75 percent protein and are rich in B vitamins. Here is a modern recipe, courtesy of the Peace Corps.

Ingredients: Vegetable oil, locusts, salt, and chili powder. Go out at night in locust season, wearing a headlamp, and grab handfuls of locusts that are attracted to the light. Put them in a bag and take them back to the kitchen. Pull off the wings, wash the locusts, and pat dry. Then fry the locusts for about two minutes in hot vegetable oil, and put them on a plate with a paper towel to absorb excess oil. Sprinkle with chili powder and salt. (The Peace Corps also recommends putting the plate out in the sun to make the locusts crunchy, presumably with a cover to keep flies from landing on it.)

In the United States, lubber grasshoppers are reportedly good to eat, although harvesting them is more labor-intensive because they do not swarm. Also, it is important to avoid eating grasshoppers that have been sprayed with pesticides. Other traditional locust or grasshopper recipes include boiling, roasting, stewing in butter, or mixing ground-up locusts with flour and water and baking them into cakes.

References and Recommended Reading

Bhattacharya, S. "Plague of Locusts Causes Mass Allergy Attack." *Sudan Tribune*, 12 December 2008.

Buhl, J., et al. "From Disorder to Order in Marching Locusts." *Science*, Vol. 312, 2006, pp. 1402–1406.

Bukkens, S. G. F. "The Nutritional Value of Edible Insects." *Ecology of Food and Nutrition*, Vol. 36, 1997, pp. 287–319.

Ceccato, P., et al. "The Desert Locust Upsurge in West Africa (2003–2005): Information on the Desert Locust Early Warning System and the Prospects for Seasonal Climate Forecasting." *International Journal of Pest Management*, Vol. 53, 2007, pp. 7–13.

Enserink, M. "Can the War on Locusts Be Won?" *Science*, Vol. 306, 2004, pp. 1880–1882.

Harmon, K. "When Grasshoppers Go Biblical: Serotonin Causes Locusts to Swarm." *Scientific American*, 30 January 2009.

Levy, S. "Last Days of the Locust." *New Scientist*, 21 February 2004.

Lockwood, J. 2004. *Locust: The Devastating Rise and Mysterious Disappearance of the Insect That Shaped the American Frontier*. New York: Basic Books, 304 pp.

"Locust Army Gathers in Africa." *New Scientist*, 11 September 2004.

"Locust Plague Sweeps South, Swarming Desert Capital." Associated Press, 6 August 2004.

Lomer, C. J., et al. "Biological Control of Locusts and Grasshoppers." *Annual Review of Entomology*, Vol. 46, 2001, pp. 667–702.

Lovejoy, N. R., et al. "Ancient Trans-Atlantic Flight Explains Locust Biogeography: Molecular Phylogenetics of *Schistocerca*." *Proceedings Biological Sciences*, Vol. 273, 2006, pp. 767–774.

7. J. McCann, *People of the Plow* (Univ. of Wisconsin Press, 1995).

Magnier, M. "Add Locusts to China's List of Calamities." *Los Angeles Times*, 3 July 2008.

"Mali Calls for Help to Fight Locust Swarms." Reuters, 25 July 2004.

Miller, G. A., et al. "Swarm Formation in the Desert Locust *Schistocerca gregaria*: Isolation and NMR Analysis of the Primary Gregarizing Agent." *Journal of Experimental Biology*, Vol. 211, 2008, pp. 370–376.

"North Africa Wars on Locusts." Associated Press, 22 April 1988.

Peatling, S. "Australia Braces for Locust Plague." *National Geographic News*, 30 November 2004.

Prior, C. "Locust and Grasshopper Biocontrol with Fungi." *Mycological Research*, Vol. 108, 2004, p. 724.

Ullman, M. "African Desert Locusts in Morocco in November 2004." *British Birds*, Vol. 99, 2006, pp. 489–491.

Whiting, J. D. "Jerusalem's Locust Plague." *National Geographic*, December 1915.

CONCLUSION: ONE TO GROW ON

There are hundreds of known plant diseases and pests, and nobody knows which ones will next emerge in a new and threatening form. Perhaps the most dreaded new candidate is a highly virulent, fungicide-resistant strain of a disease called black stem rust of wheat (Figure 5.11). This

Figure 5.11 Ug99 strain of wheat rust on wheat in Njoro, Kenya.

Source: Yue Jin, USDA Agricultural Research Service.

disease caused huge losses and famines in the first half of the twentieth century, but resistant wheat strains kept it under control. Now, it appears to be back.

The new strain is called Ug99 because it was discovered in Uganda in 1999. Reports indicate that Ug99, if not contained, is capable of destroying 10 percent of the world's wheat production. It spread to Kenya by 2002 and to Ethiopia by 2003. As predicted, it reached Sudan and Yemen in 2006, and Iran by 2007. Inevitably, it will spread to the rest of Asia and the Americas. Remedial action is in progress; a Global Rust Initiative has been established in Nairobi to monitor the progress of the disease and to develop resistant wheat.

References and Recommended Reading

"Bayer CropScience Fungicides Have Proven Control of Ug99 in Wheat Trials." Press release, Bayer CropScience, 19 June 2008.

Borlaug, N. E. "Stem Rust Never Sleeps." *New York Times*, 26 April 2008.

"Fungus Puts World Wheat Crop at Risk." United Press International, 27 March 2008.

Lacey, M. "New Strain of Wheat Rust Appears in Africa." *New York Times*, 9 September 2005.

Mackenzie, D. 2007. "Billions at Risk from Wheat Super-Blight." *New Scientist Environmental*, 3 April 2007.

"Scientists Fight Stem Rust UG99 before It Becomes a Threat." PhysOrg.com, 18 November 2008.

"Ug99 Now in Iran." Geneva: United Nations Food and Agriculture Organization, News Release, 5 March 2008.

6

Making Things Worse

> We have met the enemy and he is us.
>
> —Walt Kelly, 1970 Pogo cartoon

Biological threats do not exist in a vacuum. The topics in this chapter include physical, chemical, and sociological factors that increase the level of risk associated with some of the diseases discussed in Chapters 2 through 5. These accessory threats are often controversial, and most have no easy solutions.

TOO MANY BABIES: OVERPOPULATION

Summary of Threat

The human population has approximately tripled in the last 50 years. If the number continues to increase, it must eventually reach the point where the Earth can no longer produce enough food and other resources for everyone, and mass starvation or other so-called corrections will result. No one knows exactly when that day will arrive, how many people the Earth can support, or what quality of life the majority will accept.

So What?

After the invention of agriculture made it possible for large numbers of people to live near one another, the exchange of ideas favored proliferation of the arts and sciences. Unfortunately, not only ideas were exchanged; crowding also increases the opportunity for infectious disease transmission. Modern sanitation and public health policies have reduced the risk, but it remains one of the trade-offs of civilization.

Many airborne diseases, such as tuberculosis, influenza, and measles, spread rapidly under conditions of crowding and poverty (Figure 6.1). Aging or nonexistent sewage systems in densely

Figure 6.1 A public health worker visits a slum in northern India. Overcrowding, poverty, and famine promote the spread of many diseases. As the human population grows larger, these conditions are expected to increase.

Source: U.S. Centers for Disease Control and Prevention, Public Health Image Library.

populated urban slums and box cities promote outbreaks of waterborne disease. Population growth also has indirect effects; the already-faltering healthcare system in many countries will eventually be strained past the breaking point. History shows that deadly childhood diseases such as diphtheria tend to re-emerge when governments are unable to provide vaccination and other health services.

On Earth Day 2009, somebody finally said it (again). Dr. Charles A. Hall, a systems ecologist at the State University of New York, reportedly told the media that "Overpopulation is the only problem."

Population pressure also increases the likelihood of violence—possibly including biological warfare and bioterrorism, although we can't be certain, because there is no large-scale precedent for either. People under stress often increase their disease risk by turning to unsafe behaviors such as substance abuse, overeating, and indiscriminate sex. Finally, population growth is likely to exacerbate global warming, environmental pollution, and famine, all of which contribute to disease risk and a general reduction in quality of life.

Can't Scientists Do Something about It?

Reducing the population means lowering the birth rate or raising the death rate, and the latter is not an option. We can invent better birth control methods, but people must decide whether to use them. Another approach is to invent technologies that make limited resources go further. Farmers are learning to grow more food on less land, but efficiency has an upper limit. Engineers can build smaller and smaller cars, powered by solar cells or windup keys. Architects can design tiny cubicle apartments, stacked hundreds of stories high, in which humans can work online or raise their families or sit around. Generations accustomed to this type of life might find it satisfactory for a while, but its very complexity invites disaster. At some point, a key resource would fail. Ultimately, if population growth continues, all must fail.

Scientists are often better at describing problems than fixing them. For example, biologists recognize two reproductive strategies called *r* and *K* selection. The *r* stands for the rate of population growth, and *K* is carrying capacity—the density (a measure of crowding) at which population growth stops. When the density is low, and environmental conditions are unstable, an animal might have numerous babies and dump them into the world as quickly as possible. Most will die—often from random events such as weather—but in a good year they might get lucky. That is *r* selection. But when the density is high, and conditions are stable, it's more efficient to have fewer offspring and invest more energy in each one, to enable them to compete effectively for scarce resources. That's *K* selection. Strictly speaking, these terms do not apply to human societies, but the comparison seems irresistible. Having fifteen children and sending them out to beg on the street sounds like *r* selection, whereas having one child and sending her to Harvard sounds like *K* selection. But humans are not mice, and such behavior is not hardwired. Studies have shown, for example, that educating girls can reduce the birthrate (Chapter 7).

The Numbers

Some authorities believe that the human population will stabilize at somewhere between 8 and 10 billion people (Case Study 6-1). Others predict that it will crash before reaching that level, perhaps restabilizing at about 1 billion. Others claim that continued population growth will alternate with smaller corrections. (A crash or correction means a large number of human deaths, usually resulting from density-dependent factors such as famine, disease, or interpersonal violence.)

According to a 2008 study, the declining birthrate in the European Union will result in zero population growth by 2015. But immigration will continue, so the EU population is expected to increase from the present 495 million to a maximum of 521 million in 2035 and then decline to about 506 million in 2060. Meanwhile, the birthrate in the United States is higher than it has been in more than 40 years, with 4,265,555 births in 2006 and no upper limit in sight.

A total fertility rate (TFR) of 2.0 children per woman is the replacement rate necessary to maintain a population at the same level. In 1970, the TFR in China was about 5.8 children per woman, and the population was 820 million. By 1999, the population had risen to 1.3 billion, but the TFR had declined to 1.8 children per woman,

Case Study 6-1: The World of 2050

In 2050, the global population will probably exceed 9 billion—assuming that an asteroid or other cosmic "correction" doesn't hit us first—and most people in poorer countries will live in urban slums. This pattern could make the conquest of tuberculosis harder than ever, unless an effective vaccine is widely available by 2050. Water pollution and associated diseases may also increase. If present trends continue, about one-sixth of all humans will live in India, with nearly as many in China, and the United States coming in third with nearly 400 million people. About half of the Amazon rain forest will be gone, together with the Alpine glaciers and most of the world's ocean fisheries and coral reefs. Many vehicles may still run on fossil fuels, but not for long. The average global temperature will probably be about 7°F (4°C) warmer than in 2008, and some coastal areas will be under water.

thanks to China's controversial family planning policy. The Chinese government estimated that the first 30 years of this policy prevented 350 million to 400 million births that would otherwise have occurred.

During the same time period, India also made some headway in slowing population growth. Between the mid-1960s and 1997, India's TFR declined from 5.7 to 3.3 children per woman. But in 2008 the rate stood at about three children per woman. Infant mortality has also declined, and as a result, India's population continues to grow. In nearby Bangladesh, one of the world's poorest countries, TFR was 6.4 in 1970 and 3.6 in 2003, with further decline anticipated.

Rates of birth and death vary from one part of Africa to another. Tunisia has had remarkable success, reducing fertility from 7 children per woman in 1950 to 2 in 2006. Southern Africa, like India, reduced its TFR from about 5.7 children per woman in 1960 to 3.3 in 2000.

Some Middle Eastern countries have also joined in the global effort to slow the rate of population growth. Iran, for example, reduced its TFR from 6.5 children per woman in 1980 to 2.5 in 2002. Several others, including Yemen, Iraq, and the Palestinian territories, have seen little change in fertility, with an average of 5.5 children per woman in 2002.

Discussion

Human population control is a touchy subject. Some readers interpret the phrase as a call for abortion, euthanasia, or genocide. But population control really means any action (or inaction) that influences the rate of population increase. The word "control" implies that somebody is in charge, but in most cases the process is more like an old-fashioned Ouija board. Nobody consciously pushes the planchette in one direction or another, but the net efforts of the participants determine the outcome.

In everyday life, euphemisms tend to replace harsher terms for events that people in the midst of the action cannot ignore: the soldier's "collateral damage," the firefighter's "full involvement," the politician's "ethnic cleansing," the animal researcher's "humane endpoint," the demographer's "Malthusian adjustment," and the ecologist's "population correction." The latter term usually refers to something like mice or deer, not people. A population builds up to a high level, and then disease and famine take over. Most of the mice (or whatever) die, and everything is back to normal again.

So when a respected ecologist in 2006 told an audience that the world would be better off with 10 percent of the present population, it should have been obvious that he was not advocating mass murder. He assumed that the audience knew what he meant—that the human population has grown beyond a sustainable level, and the world would be a more comfortable place for everyone if there were fewer of us. But the wire services jumped on the speech, and columnists around the world redefined the ecoterrorist movement to include academic scientists and advocates of zero population growth. Some commentators have even claimed that the ultimate goal of the environmental movement is to annihilate the human race. (Note how easy it would be to quote the last part of that sentence out of context.)

At the other extreme, the media sometimes ignore outrageous statements about population control. In 1971, the Texas House of Representatives unanimously passed a resolution honoring Albert deSalvo (the Boston Strangler) for his unselfish service to mankind and the "unconventional" way he dealt with overpopulation. Rep. Tom Moore allegedly sponsored this resolution in order to prove that the legislators did not read carefully, and it would appear that he succeeded.

It's unlikely that any scientist, or anyone else, really wants to see billions of people die from famine and disease. On the contrary, most of us wish people would plan ahead and avoid that fate, but in some countries—including the United States—this just isn't happening. Any discussion of infectious disease comes back to population growth sooner or later, because a hungry, crowded world can expect frequent disease outbreaks.

Popular Culture

In the 1973 movie *Soylent Green*, an overcrowded future world resorts to a particularly creepy form of cannibalism. *Logan's Run* (1976) portrays a future with an easier solution: kill everyone over 30. In *Red Planet* (2000), humans deal with overpopulation by trying to establish a colony on Mars. But in 2006, the real fear finally surfaced in *Children of Men*, which depicts a future in which humans lose the ability to have children. It is probably a safe bet that most people find this outcome scarier than overcrowding.

In Alexei Panshin's 1968 novel *Rite of Passage*, "free birth" is among the most shocking of all socially irresponsible acts, and every child learns that war is a consequence of population pressure. Many science-fiction classics feature a depopulated Earth after a nuclear or biological holocaust. A more fanciful solution to crowding appears in Clifford Simak's 1952 novel *Ring Around the Sun*: human mutants arise who have the ability to step through dimensional barriers and enter an infinite number of worlds with fertile soil, clean air, and no people. If only the real problem were that easy to solve.

Other science-fiction treatments of overpopulation include Robert Silverberg's 1970 short story *A Happy Day in 2381* and his 1971 novel *The World Inside*; Ursula K. LeGuin's 1985 utopian novel *Always Coming Home*; John Brunner's classic 1968 *Stand on Zanzibar*; and James Blish's 1967 *A Torrent of Faces*. Jonathan Swift probably started the ball rolling in 1729 with his satirical essay *A Modest Proposal*, which proposed that the Irish poor should eat their children. (The Irish potato crop had recently failed for the first time, but not the last.)

The Old Testament (Judges 6:4–6) tells us how the Midianites multiplied in such numbers that they left no food or land for the Israelites. By the standards of the day, the solution was obvious: kill them! Gideon was a humble man who led a small but deceptively loud Israelite army to victory over a larger, less organized force that panicked and surrendered. The name Gideon originally meant "Destroyer," but it has acquired other connotations. In a 1969 *Star Trek* episode, for example, Gideon is an overpopulated alien world where birth control is forbidden. Recognizing that crowding is making life intolerable, the leaders introduce a deadly disease to kill many of their people. Thus, in both stories, one influential group imposes a violent solution on all.

Point/Counterpoint

P: Governments can't control population growth. Only free people can do that. Having children is a fundamental human right, and no government that claims otherwise will last long. In a free-market economy, population growth is self-correcting. Besides, the world isn't full yet. Huge areas are not being used for anything. Once the polar ice melts, we will have even more arable land. We can bring water to the deserts and farm the oceans, and there will be plenty of food for everyone.

CP: Yes, there was a nice "correction" in Ireland in 1847. The abundance of food enabled the population to grow rapidly for a few generations, and then a million people died horribly when one crop failed. Some parts of Africa are undergoing similar "corrections" today. Mass starvation and disease are fine ways to limit population growth if you are a rodent, but aren't humans ethically bound to do better? And for the sake of future generations, please don't wreck the deserts and oceans, too.

P: It always comes back to Ireland, doesn't it?

The Future

Human population growth cannot continue indefinitely, but recent numbers suggest that the global population might stabilize before the quality of life becomes unbearable. Meanwhile, the forces of *r* selection appear to be maintaining their lead while family planning advocates struggle to keep up.

According to a 2009 report, several countries (including China and the UK) use computer games to teach people about contraception. But just explaining the mechanical details of birth control may not be enough. Students also need to understand the long-term consequences of personal choice. Perhaps high school science teachers could make more effective use of computer games, such as Sim City, to dramatize the effects of overcrowding and resource depletion.

References and Recommended Reading

Aitken, R. J., et al. "As the World Grows: Contraception in the 21st Century." *Journal of Clinical Investigation*, Vol. 118, 2008, pp. 1330–1343.

Austin, L. "Prof. Criticized for Overpopulation View." Associated Press, 4 April 2006.

Bennett, J. "Overpopulation Is the Problem." *BioScience*, 1 February 2007.

"Combat Climate Change with Fewer Babies—OPT Report." News Release, Optimum Population Trust, 7 May 2007.

Dahl, R. "Population Equation: Balancing What We Have with What We Need." *Environmental Health Perspectives*, Vol. 113, 2005, pp. A599–A605.

Francis, D. R. "'Birth Dearth' Worries Pale in Comparison to Overpopulation." *Christian Science Monitor*, 14 July 2008.

Gillespie, D., et al. "Unwanted Fertility among the Poor: An Inequity?" *Bulletin of the World Health Organization*, Vol. 85, 2007, pp. 100–107.

"Highest U.S. Birth Rate in Four Decades." United Press International, 9 January 2009.

Hoffman, M. C. "Philippines in Struggle against Abortionist Population Control Initiative." LifeSiteNews.com, 22 July 2008.

Howard, G. "China's Population Control a Sensible Measure." *Pitt News*, 20 March 2008.

"Low-Cost Female Condom Popular in Britain." United Press International, 22 December 2008.

Morrison, P. "Who Will Heed the Warnings on the Population Bomb?" *Los Angeles Times*, 5 September 1999.

Mudie, L., et al. "Abuses under Population Policies." Radio Free Asia, 12 July 2008.

Murdock, D. "Extremists Want Better Living through Mass Death." *The Telegraph*, 17 June 2006.

Page, S. T., et al. "Advances in Male Contraception." *Endocrine Review*, Vol. 29, 2008, pp. 465–493.

"Population: End to Natural Demographic Growth in 2015." *European Social Policy*, 9 September 2008.

Prugh, T. "Women: Population's Once and Future Key." *World Watch*, 1 September 2008.

Richey, W. "Supreme Court Declines to Hear Asylum Case Involving Forced Abortion." *Christian Science Monitor*, 13 May 2008.

"Scientists Work on Garbage for Gas." United Press International, 24 July 2008.

Shah, I. H., and V. Chandra-Mouli. "Inequity and Unwanted Fertility in Developing Countries." *Bulletin of the World Health Organization*, Vol. 85, 2007, p. 86.

Sinding, S. W. "The Great Population Debates: How Relevant Are They for the 21st Century?" *American Journal of Public Health*, Vol. 90, 2000, pp. 1841–1845.

State University of New York. "Worst Environmental Problem? Overpopulation, ESF Faculty Says." Press Release, 21 April 2009.

"U.N. Adopts Plan to Slow Population." *Los Angeles Times*, 3 July 1999.

Wallace, B. "Debate Grows with Population." *Los Angeles Times*, 7 May 2008.

TOO MUCH CARBON: GLOBAL CLIMATE CHANGE

Summary of Threat

Nearly everyone has heard that the world is becoming warmer and that carbon emissions from human industries may be largely responsible. Higher temperatures, rising sea levels, and changes in the distribution of rainfall may transform the world we take for granted. Not everyone believes that this is happening; a decade ago, the author was not entirely convinced. But as of 2009, the evidence is overwhelming.

So What?

One effect of global warming, probably not the worst, will be a change in the distribution of certain infectious diseases. As the Earth grows warmer, malaria may disappear from some regions while expanding its range in others. If people in the newly affected areas lack immunity, the health impact could be significant. The forecast is clearer for some other pathogens and parasites, such as dengue fever, which has already spread into previously dengue-free areas at higher elevations than in the past (Chapter 3), and at least one plant fungus, phoma stem canker of canola (Chapter 5).

Other diseases that may become more prevalent in Europe or North America include chikungunya fever (Case Study 6-2), tickborne encephalitis, yellow fever, Chagas disease, hantavirus infections, and salmonellosis. Even the incidence of rabies may increase, as New World vampire bats move northward. Sadly, the lungworm parasite that infests the lungs of sheep has already expanded its range in Scotland, endangering the production of haggis. Another minor but scary player is the famous brain-eating amoeba (*Naegleria fowleri*), which lives in warm lakes.

Case Study 6-2: Chikungunya in Italy

In December 2007, the wire services picked up an unusual story. An outbreak of a dengue-like tropical disease called chikungunya—known as "chik" in countries where it is endemic—had occurred in the small Italian village of Castiglione di Cervia. More than 100 people became severely ill, with high fever, joint pain, and exhaustion. Chik, like dengue, has a low death rate; in 2005, a major chik epidemic infected 266,000 people on the island of Réunion, and about 200 died. But the 2007 Italian outbreak was the first known example in which a temperate region experienced a tropical disease outbreak as a clear result of global climate change. The tropical tiger mosquito, a chik vector, had colonized Italy about 10 years earlier, thanks to warmer winters. Then an Italian man visited India in 2007 and brought back chik, and the mosquitoes were waiting, along with the press of the world.

Can't Scientists Do Something about It?

As of 2009, most nations have signed the 1998 Kyoto Protocol (a treaty to limit greenhouse gas emissions), but the cost of implementation is an issue, and the United States has not yet ratified it. Meanwhile, a 2009 study by the U.S. National Oceanic and Atmospheric Administration (NOAA) concluded that reversing climate change will take over 1,000 years even if we can stop releasing greenhouse gases into the atmosphere. Measures to reduce emissions can slow the warming process and improve air quality, but the catch phrase "Stop global warming" is no longer an option if the NOAA model is correct. We may have waited too long, but that is no excuse for giving up. Perhaps the new mantra should be "Slow global warming."

The Numbers

A 2007 study in England showed that each child who is born represents a lifetime "social cost" of about £30,000 ($43,000 U.S.) in carbon emissions alone. By 2074, the UK population will increase by about 10 million, for a total social cost of over £300 billion ($430 billion U.S.).

For other relevant numbers, we refer the reader to Al Gore's 2006 documentary *An Inconvenient Truth*. No film or book can be 100 percent accurate, but his presentation is fair, and it gets the message across. See also Figure 6.2.

Discussion

In *Pygmalion* (1913), George Bernard Shaw observed that the safest conversational topics were the weather and everybody's health. It would appear that societal norms have changed, for these two topics now form the basis of a bitter debate about global climate change, in which each side claims that the other is ignoring scientific facts in favor of irrational fears.

Suffice it to say that everyone's worldview combines elements of both fact and belief; that the best available data appear to prove beyond any reasonable doubt that the Earth is in a warming trend; that said warming is partly the result of the greenhouse effect, in which carbon

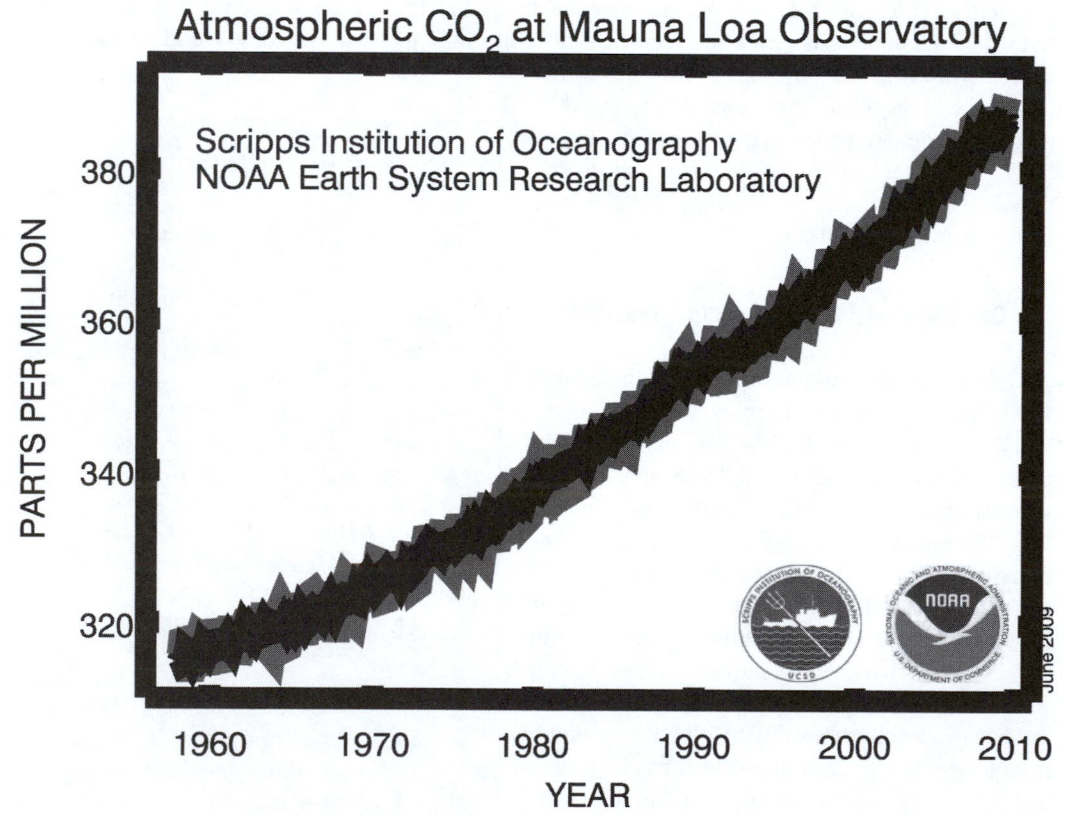

Figure 6.2 Monthly mean atmospheric carbon dioxide at Mauna Loa Observatory, Hawaii. These data, measured as the mole fraction in dry air, constitute the longest record of direct measurements of CO_2 in the atmosphere.

Source: U.S. National Oceanic and Atmospheric Administration.

emissions from human activities trap heat near the surface; and that the resulting long-term global climate change is likely to affect human societies in a number of ways, not all of them known or knowable.

In a previous book published in 2000, this author presented both sides of the global warming argument—at the publisher's request, in the interest of balance. But a decade later, little doubt remains. The world's climate is changing, and humans are at least partly responsible. Even if you live in a solar-powered cabin in the woods and drink from a lake, everything from the solar panels on the roof to the Sierra cup in your hand implies the existence of energy-guzzling technologies. Most of that energy comes from fossil fuel combustion, which creates greenhouse gases.

In 2008, the University of Illinois conducted a poll of 3,146 qualified earth scientists and found that 90 percent agreed that average global temperatures have risen in the last 200 years. Moreover, 82 percent agreed that human activity has been a significant factor in global warming. Given the respondent population, this is an overwhelming consensus. It would be hard to find any issue on which 100 percent of scientists agree, because science is not religion, and certainty is a big word.

Popular Culture

In a 1936 interview, the American climatologist Charles D. Reed (1875–1945) made the interesting observation that people in every era believe that the world is growing warmer.[1] He attributed this tendency to the fact that snow seems deeper to a child than to an adult. Yet Reed also acknowledged a measurable warming trend. The issue had already been controversial for at least 40 years, since 1896, when Swedish physicist Svante Arrhenius (1859–1927) showed that the accumulation of CO_2 in the atmosphere would trap heat near the Earth.

Scientists began to take global warming seriously in the 1950s, and comic books kept pace. *Strange Adventures* (DC Comics) ran stories on global warming in April 1955 ("The Day the Sun Exploded"), January 1956 ("The Earth-Drowners"), and September 1958 ("The Menace of Saturn's Rings").

Several motion pictures have explored global warming, including the 1995 quasi-blockbuster *Waterworld*. Critics complained that there is not enough water on Earth to submerge all the continents, but the message was clear enough: in a warmer world, ice will melt and sea levels will rise. In *The Arrival* (1996), an astronomer discovers that alien invaders who hate cold weather are responsible for global warming. Science fiction often employs artistic license to make a point; the 2004 movie *Day After Tomorrow* focuses on a real process rather than the aftermath, so it must compress the time frame of global climate change into a few days. It would be hard to get audiences to sit through a film in which a column of mercury rises slowly over a period of years or centuries.

In Michael Crichton's 2004 novel *State of Fear*, a scientist discovers that global warming is a hoax. As always, Crichton knew what the public wants to read about: evil doctors, dumb scientists, corrupt politicians, global conspiracies, and brave loners who know the truth. But the message is useful, because climatologists (like everyone else) may overstate their position on occasion.

The Future

This book can add little to the voluminous literature on global warming and the future of the Earth. The climate will grow warmer, on average, and possibly less predictable, and people will suffer and adapt as they have always done.

---

1. "Records Reveal Iowa Gradually Getting Warmer" (Iowa Daily Press Bureau, 6 January 1936).

It is not clear if malaria will become a worldwide problem, because vector control and window screens have eliminated malaria from many regions that are already warm enough to sustain it. At the right stage of its life cycle, the malaria parasite can even be killed by heat. Thus, its future distribution may depend not only on continued public health effort—a separate problem—but also on daily temperature fluctuations that are harder to predict than simple averages. We will live in interesting times.

Point/Counterpoint

P: Climate change is natural, whether humans cause it or not. Species will go extinct and resources will be depleted, but other species and resources will replace them. Without such changes, the great extinctions and biodiversity explosions of the past would never have happened. When our coastal cities are submerged, scuba divers will have fun exploring them. Building sea walls and new cities will create jobs and eliminate slums. Man adapts.

CP: Yes, change is natural and inevitable, and one day something else will dig up our fossils and try to figure out where the brain was located. This is all very interesting to think about. But disease and death are natural changes too, and yet we try to protect our children from needless suffering and give them the best world we can. Why have children, if we don't care what happens to future generations?

References and Recommended Reading

Bandyopadhyay, R., and P. A. Frederiksen. "Contemporary Global Movement of Emerging Plant Diseases." *Annals of the New York Academy of Sciences*, Vol. 894, 1999, pp. 28–36.

Bloom, J. "Is the World Ending, or What?" United Press International, 25 April 2002.

Borenstein, S. "Two Greenhouse Gases on the Rise Worry Scientists." Associated Press, 24 October 2008.

Brown, H. "Reducing the Impact of Climate Change." *Bulletin of the World Health Organization*, Vol. 85, 2007, pp. 824–825.

Campbell-Lendrum, D., et al. "Global Climate Change: Implications for International Public Health Policy." *Bulletin of the World Health Organization*, Vol. 85, 2007, pp. 235–237.

Carcavallo, R. U. "Climatic Factors Related to Chagas Disease Transmission." *Memórias do Instituto Oswaldo Cruz*, Rio de Janeiro, Vol. 94 (Suppl. 1), 1999, pp. 367–369.

Chung, J. "A Tropical Virus Moves North." *Los Angeles Times*, 29 December 2007.

"Climate Change May Alter Malaria Patterns." United Press International, 16 February 2009.

Connor, S. "The Methane Time Bomb." *The Independent*, 23 September 2008.

"Disease Outbreaks Blamed on Climate Change." Reuters, 12 June 2008.

Evans, N., et al. "Range and Severity of a Plant Disease Increased by Global Warming." *Journal of the Royal Society Interface*, Vol. 5, 2008, pp. 525–531.

Gilbert, M., "Climate Change and Avian Influenza." *Revue Scientifique et Technique*, Vol. 27, 2008, pp. 459–466.

"Global Warming: Return of Vampire Bats?" *Washington Post*, 6 November 1989.

"Global Warming Skeptics Target Students." United Press International, 5 May 2008.

Gore, Al. "An Inconvenient Truth: A Global Warning." Hollywood, CA: Paramount Pictures, 2006, 96 min. (DVD).

Gray, L. "Haggis at Risk from Global Warming." *The Telegraph*, 10 August 2008.

Gubler, D. J., et al. "Climate Variability and Change in the United States: Potential Impacts on Vector- and Rodent-borne Diseases." *Environmental Health Perspectives*, Vol. 109 (Suppl. 2), 2001, pp. 223–233.

Haines, A., et al. "Climate Change and Human Health: Impacts, Vulnerability, and Mitigation." *Lancet*, Vol. 367, 2006, pp. 2101–2109.

Harvell, C. D., et al. "Climate Warming and Disease Risks for Terrestrial and Marine Biota." *Science*, Vol. 296, 2002, pp. 2158–2162.

Hebert, H. J. "Heavy Editing Is Alleged in Climate Testimony." Associated Press, 24 October 2007.
IGAD Climate Prediction and Applications Centre. "Climate Change and Human Development in Africa: Assessing the Risks and Vulnerability of Climate Change in Kenya, Malawi and Ethiopia." Draft Report, United Nations Development Programme, 2007.
"Italian Village Hosts Tropical Disease." United Press International, 22 December 2007.
Khasnis, A. A., and M. D. Nettleman. "Global Warming and Infectious Disease." *Archives of Medical Research*, Vol. 36, 2005, pp. 689–696.
Lehman, J. A., et al. "Effect of Hurricane Katrina on Arboviral Disease Transmission." *Emerging Infectious Diseases*, Vol. 13, 2007, pp. 1273–1275.
Martin, V., et al. "The Impact of Climate Change on the Epidemiology and Control of Rift Valley Fever." *Revue Scientifique et Technique*, Vol. 27, 2008, pp. 413–426.
Michel, R., et al. "Risk for Epidemics after Natural Disasters." *Emerging Infectious Diseases*, Vol. 13, 2007, pp. 785–786.
Parry, M., et al. "Climate Change, Global Food Supply and Risk of Hunger." *Philosophical Transactions of the Royal Society of London, Series B, Biological Sciences*, Vol. 360, 2005, pp. 2125–2138.
Patz, J. A., et al. "The Effects of Changing Weather on Public Health." *Annual Review of Public Health*, Vol. 21, 2000, pp. 271–307.
Patz, J. A., and S. H. Olson. "Climate Change and Health: Global to Local Influences on Disease Risk." *Annals of Tropical Medicine and Parasitology*, Vol. 100, 2006, pp. 535–549.
Pinzon, J. E., et al. "Trigger Events: Enviroclimatic Coupling of Ebola Hemorrhagic Fever Outbreaks." *American Journal of Tropical Medicine and Hygiene*, Vol. 71, 2004, pp. 664–674.
Pogatchnik, S. "Belfast Environment Chief Bans Climate Change Ads." Associated Press, 9 February 2009.
"Poll: U.S. Not Panicked by Global Warming." United Press International, 21 April 2008.
Quarles, W. "Global Warming Means More Pests." *IPM Practitioner*, Vol. 29, 2007, pp. 1–8.
Reiter, P., et al. "Texas Lifestyle Limits Transmission of Dengue Virus." *Emerging Infectious Diseases*, Vol. 9, 2003, pp. 86–89.
Rose, J. B., et al. "Climate Variability and Change in the United States: Potential Impacts on Water- and Foodborne Diseases Caused by Microbiologic Agents." *Environmental Health Perspectives*, Vol. 109 (Suppl. 2), 2001, pp. 211–221.
Rosenthal, E. "As Earth Warms Up, Tropical Virus Moves to Italy." *New York Times*, 23 December 2007.
Rosenzweig, C., et al. "Attributing Physical and Biological Impacts to Anthropogenic Climate Change." *Nature*, Vol. 453, 2008, pp. 353–358.
"Salmon Disease Blamed on Warmer Climate." United Press International, 14 June 2008.
"Scientists Agree Human-Induced Global Warming Is Real, Survey Says." *ScienceDaily*, 21 January 2009.
Shulman, S., et al. "Smoke, Mirrors, and Hot Air: How ExxonMobil uses Big Tobacco's Tactics to Manufacture Uncertainty on Climate Science." Cambridge, MA: Union of Concerned Scientists, 2007, 68 pp.
"Six Die from Brain-Eating Amoeba." Associated Press, 28 September 2007.
Smith, E. "Despite Awareness of Global Warming, Americans Concerned More About Local Environment." News Release, University of Missouri, 26 March 2008.
Sokolov, A., et al. "Probabilistic Forecast for 21st Century Climate Based on Uncertainties in Emissions (without Policy) and Climate Parameters." *Journal of Climate*, 2009.
Solomon, S., et al. "Irreversible Climate Change due to Carbon Dioxide Emissions." *Proceedings of the National Academy of Sciences*, Vol. 106, 2009, pp. 1704–1709.
Sutherst, R. W. "Global Change and Human Vulnerability to Vector-borne Diseases." *Clinical Microbiology Reviews*, Vol. 17, 2004, pp. 136–173.
Traynor, K. "Warming Earth Could Face New Flu, Disease Threats." *American Journal of Health System Pharmacy*, Vol. 65, 2008, pp. 1112–1114.
U.S. Environmental Protection Agency. "Climate Change and Public Health." EPA 236-F-97-005, Office of Policy, Planning and Evaluation, October 1997.
"U.S. Power Plant Carbon Emissions Zoom in 2007." *ENS*, 18 March 2008.
Walsh, B. "Can Climate Change Make Us Sicker?" *Time*, 4 April 2008.
Watkins, K., et al. "Fighting Climate Change: Human Solidarity in a Divided World." New York: United Nations Development Programme, Human Development Report, 2007/2008.

Watson, J. T., et al. "Epidemics after Natural Disasters." *Emerging Infectious Diseases*, Vol. 13, 2007, pp. 1–5.

World Health Organization. "New Global Effort to Eliminate Chagas Disease: Partners Set Out Strategy Against the 'Kissing Bug' Disease." News Release, 3 July 2007.

NOT ENOUGH FOOD: FAMINE, PESTILENCE, DESTRUCTION, AND DEATH

Summary of Threat

Despite the successful Green Revolution of the 1960s and 1970s and more recent advances in agricultural technology, world hunger remains a widespread problem. In 2008 alone, an estimated 20 million people died from the effects of famine. Trends discussed in the two previous sections, human population growth and global warming, can only exacerbate this problem.

So What?

Aside from the cost in human suffering, hunger and malnutrition are underlying factors in many infectious disease epidemics, particularly in the tropics. An immune system weakened by an inadequate diet often falls prey to infections that would not threaten a healthy person. In a famine, some people starve to death, but many more die from disease and other indirect consequences of food shortage. Starving people may also resort to actions they would otherwise abhor, such as killing their neighbors and stealing their food, or blaming a specific ethnic group for the crisis. War, in turn, often promotes conditions (such as mass migrations and crop damage) that result in more famine and more disease.

Can't Scientists Do Something about It?

Scientists work constantly to detect plant and animal disease outbreaks, eliminate vectors, and develop resistant crops, but the task of fighting world hunger falls largely to economists and heads of state. At present, the world produces enough food for everyone, thanks in part to new crops and technologies developed during the Green Revolution. The problem is unequal distribution, and science can't do much about that. Also, prices are subject to the laws of supply and demand, and if the price of a staple food rises to the point where most local people are unable to buy it, growers or governments may export it instead, thus making the problem worse. In the not-so-distant future, when the global food supply actually falls short of demand, the role of science may eclipse these concerns.

The Numbers

According to a frequently quoted statistic, Europe has had at least 400 major famines in recorded history. China has fared even worse, with 1,828 major famines between 108 B.C. and A.D. 1911. Table 6.1 lists examples.

Several sources report that one child starves to death (or dies of causes related to malnutrition) every 5 or 6 seconds in the early twenty-first century. If true, that means about 6 million

Table 6.1 Some Major Famines in Human History

Date	Location	Estimated Deaths from Famine
1601–03	Russia	2 million
1630–31	India	2 million
1693–94	France	2 million
1695–97	Estonia	20% of the population
1702–04	India	2 million
1738–56	Timbuktu	Half the population
1770	India	15 million
1783	Iceland	20% of the population
1845–49	Ireland	1 million
1866–68	Finland	15% of the population
1876–78	India	5 million
1876–79	Northern China	13 million
1896–1902	India	19 million
1907	China	24 million
1921–22	Russia	5 million
1936	China	5 million
1941	China	3 million
1958–61	China	20 million
1984–85	Ethiopia	1 million
1996–98	North Korea	1.2 million
1998–2004	Congo	3.8 million

children die from starvation every year. But estimates of the total number of people (adults and children) who starve every year range from 5 million to 20 million or more. Part of the uncertainty results from the problem of defining causes "related" to malnutrition. If a child dies of measles because she was half-starved, does she also die of starvation? Is a death from new variant famine (Chapter 2) a death from HIV, or from hunger? Whatever the real numbers, they are too high.

Discussion

Throughout most of human history, wild fish and game populations were available as a hedge against famine. Survivalists still think they can live off the land after the Apocalypse. The problem is that there are far too many of us. For example, in 2008, the deer population of the United States reached a record level of 30 million —higher even than when the first Europeans got off the boat. But there were also about 300 million people in the United States in 2008. We can quibble about other wild resources, rabbits and berries and bark, but it is safe to conclude that the hunter-gatherer lifestyle is no longer practicable on a large scale.

Fish? According to a 2006 study by an international research team, most of the world's marine seafood populations are already in trouble and will collapse by 2050 as a result of overfishing, pollution, biodiversity loss, and—yes—climate change. The United Nations Food and Agriculture Organization (FAO) has warned for years that about 70 percent of fish species are in danger of collapse. "Collapse" does not mean that most edible seafoods will actually vanish, just that their numbers will decline by at least 90 percent, so that commercial fishing will no longer be viable (see Case Study 6-3, page 206).

Case Study 6-3: Aquaculture

In the 1950s, people wanted to farm the oceans as a solution to world hunger. By the 1980s, the focus had shifted to aquaculture, defined as the cultivation of aquatic organisms in tanks, ponds, or offshore enclosures. Aquaculture has great potential, but it also creates certain problems. For example, farm-raised fish need to eat something, and the most valuable species (such as salmon) are carnivores that normally eat other fish. So aquaculturists feed them—you guessed it—wild fish harvested from the oceans. By one estimate, every kilogram of farmed fish requires about 6 kilograms of wild fish. Aquaculture also generates large amounts of organic waste that often finds its way into the ocean or groundwater. In some parts of the world, particularly Southeast Asia, offshore aquaculture has destroyed mangrove forests. But there is no turning back. By 2004, aquaculture already contributed about one-third of total world fisheries production.

Large-scale famine seldom occurs in nations that are wealthy enough to buy pesticides, develop genetically modified disease-resistant crops, rent other people's bees, stockpile food for future use, invent robotic pollinators, and import whatever resources are in short supply. But anyone who watches the news must be aware that less prosperous societies, particularly in Africa, continue to suffer plagues of near-Biblical proportions. Foreign aid, when it comes, may be hampered by inefficient distribution or political problems.

Popular Culture

Bhabani Bhattacharya's 1947 novel *So Many Hungers* vividly depicts the Bengal Famine of 1943. Some movies and books about overpopulation (page 197) are also about hunger, since the two conditions often go together. For the literature of the Irish Famine, see Chapter 5.

In a gentler era, comic book publishers did their part. The May 1946 *Aquaman* (Adventure Comics) featured a story about world hunger, entitled "Four Fish to Fetch." The August 1958 issue of *Strange Adventures* (DC Comics) addressed the same theme in "The Boy Who Saved the Solar System."

A common folktale motif in many cultures is "The Magic Cauldron," usually a container that is always magically filled with food. In a popular *X-Files* episode, a fifteenth-century French peasant girl wishes for the three things everyone wants: a stout-hearted mule (transportation), a magic sack that is always full of turnips (food), and long life (long life). Transportation and food always seem to come first, but without the magic cauldron or sack or wok, the two will always be in conflict. Do we feed the mule, or eat it? Do we buy gasoline, hay, or milk?

Point/Counterpoint

P: The world actually has a surplus of food, but governments pay farmers not to grow crops. If they grow too much, the market prices fall. There is no food shortage, just problems with distribution. Humanitarian aid is self-defeating anyway—remember what happened in Somalia in 1993. But genetically modified crops and better farming methods will soon make these developing countries self-sustaining.

CP: Good idea in principle, but GM seeds are not available to most Third World farmers. Many can't afford them, and others don't know which ones to buy. They lose their farming skills and end up in debt. And if we just stop the food shipments, hundreds of millions more people will die. Someday that might be necessary, but not yet. Aside from humanity, there's expediency. If one government doesn't feed them, another will.

The Future

The widespread use of high-yield genetically modified (GM) crops and livestock appears to be inevitable. This approach has hit a few snags (Chapter 5), and a 2009 study concluded that genetic engineering has failed thus far to produce higher crop yields. Some GM organisms are disease-resistant, however, and wider use of this technology might at least reduce the need for pesticides and antibiotics.

Climate change may help farmers in some parts of the world (such as Canada) by extending the growing season, while further reducing the food supply in the tropics. But if food prices continue to rise, the Third World food price riots of 2007–2008 may provide the clearest glimpse of the future. During that crisis, developed nations pledged some $18 billion in food aid to poorer countries, but *Newsweek* reported in 2009 that most of the promised aid never materialized.

References and Recommended Reading

Avery, D. T. "Must We Suffer through Global Famine Again?" *Feedstuffs*, 19 May 2008.

Borenstein, S. "Overlooked in the Global Food Crisis: A Problem with Dirt." Associated Press, 8 May 2008.

Brown, L. R. "Why Ethanol Production Will Drive World Food Prices Even Higher in 2008." *Earth Policy Institute*, 24 January 2008.

Buerkle, T. "40 Countries Face Food Shortages Worldwide." News Release, United Nations Food and Agriculture Organization, 9 October 2006.

Charles, D. "Will a Warmer World Have Enough Food?" NPR.org, 29 October 2007.

Clover, C. 2004. *The End of the Line*. London: Ebury.

Clover, C. "Food Shortages: How Will We Feed the World?" *The Telegraph*, 22 April 2008.

Edwards, P., and I. Roberts. "Transport Policy Is Food Policy." *Lancet*, Vol. 371, 2008, p. 1661.

"Europe, Brazil Pledge Sustainable Biofuels Development." *Environment News Service*, 5 July 2007.

Foroohar, R. "Hungry Again." *Newsweek*, 30 January 2009.

"Global Food Shortage a 'Dangerous Threat.'" *Farmers Guardian*, 23 April 2004.

Gurian-Sherman, D. "Failure to Yield: Evaluating the Performance of Genetically Engineered Crops." Union of Concerned Scientists, April 2009.

Haile, M. "Weather Patterns, Food Security and Humanitarian Response in Sub-Saharan Africa." *Philosophical Transactions of the Royal Society of London, Series B, Biological Sciences*, Vol. 360, 2005, pp. 2169–2182.

Krisberg, K. "Global Food Shortages, Rising Prices Threaten Public Health: Advocates Call for Food Aid Restructuring." *The Nation's Health*, 1 June 2008.

Krugman, P. "Grains Gone Wild." *New York Times*, 7 April 2008.

Lawn, J. E., et al. "Countdown to 2015: Will the Millennium Development Goal for Child Survival Be Met?" *Archives of Disease in Childhood*, Vol. 92, 2007, pp. 551–556.

Menon, R. "Famine in Malawi: Causes and Consequences." United Nations Development Programme Occasional Paper, Human Development Report Office, 2007, 14 pp.

Monbiot, G. "Manufactured Famine." *The Guardian*, 26 August 2008.

Myers, R. A., and B. Worm. "Rapid Worldwide Depletion of Predatory Fish Communities." *Nature*, Vol. 423, 2003, pp. 280–283.

Perry, A. "Ethiopia: Pain Amid Plenty." *Time*, 6 August 2008.

Prentice, A. "Fires of Life: The Struggles of an Ancient Metabolism in a Modern World." *Nutrition Bulletin*, Vol. 26, 2001, pp. 13–27.

Prentice, A., et al. "Insights from the Developing World: Thrifty Genotypes and Thrifty Phenotypes." *Proceedings of the Nutrition Society*, Vol. 64, 2005, pp. 153–161.

Rukuni, M. "Africa: Addressing Growing Threats to Food Security." *Journal of Nutrition*, Vol. 132, 2002, pp. 3443S–3448S.

Sachs, J. "How to End the Global Food Shortage." *Time*, 24 April 2008.

Sachs, J. "Are Malthus's Predicted 1798 Food Shortages Coming True?" *Scientific American*, 25 August 2008.

Subasinghe, R., and D. Bartley. "Ensuring the Sustainability of Aquatic Production." Rome: Food and Agriculture Organization of the United Nations, FAO Aquaculture Newsletter No. 31, 2004.

United Nations World Food Programme. 2007. *World Hunger Series: Hunger and Health*. Rome: Earthscan, 212 pp.

Verdin, J., et al. "Climate Science and Famine Early Warning." *Philosophical Transactions of the Royal Society of London, Series B, Biological Sciences*, Vol. 360, 2005, pp. 2155–2168.

Vidal, J., and T. Radford. "One in Six Countries Facing Food Shortage." *The Guardian*, 30 June 2005.

von Braun, J. "The World Food Situation: New Driving Forces and Required Actions." Washington, D.C.: International Food Policy Research Institute Food Policy Report, December 2007, 27 pp.

Walt, V. "The World's Growing Food-Price Crisis." *Time*, 27 Feb 2008.

Worm, B., et al. "Impacts of Biodiversity Loss on Ocean Ecosystem Services." *Science*, Vol. 314, 2006, pp. 787–790.

TOO MUCH FOOD: METABOLIC SYNDROME AND TYPE 2 DIABETES

Summary of Threat

Obesity and its associated diseases are among the most urgent health problems facing developed nations today—an irony that is not lost on the Third World, which has the opposite problem. Doctors are not certain if the famous "metabolic syndrome" even exists, but its component conditions are all too real: obesity, glucose intolerance, high blood pressure, high LDL cholesterol, and hardening of the arteries.

So What?

Obese people are more likely than others to develop type 2 (adult onset) diabetes, which in turn is a major risk factor for tuberculosis, melioidosis, and other infectious diseases. Obesity itself is a risk factor for influenza and certain bacterial infections, because it tends to suppress the immune system. (A recent study showed that overweight people are less likely than others to die of pneumonia, but only because they are more likely to die from noninfectious conditions such as heart disease.)

Many overweight children become overweight, diabetic, chronically unhealthy adults. Yet studies show that parents often fail to realize that their children are fat. Many are in denial; in 2008, an obese woman made headlines by explaining that it was impossible for her to lose weight because fresh vegetables were too expensive. Besides compromising personal health, a mutually reinforcing culture of fatness imposes one more burden on the already overstrained healthcare system. According to a 2000 Surgeon General's report, the direct and indirect cost of obesity was $117 billion each year for the U.S. alone. More recent estimates are even higher.

Can't Scientists Do Something about It?

Scientists can study genetic and cultural factors that contribute to obesity, and publish the results. We can try to scare people by telling them what food is doing to their arteries and livers and kidneys. We can browbeat fast-food restaurant chains into reducing the trans fat content of their deep-fried glop, and we can advise school districts to serve salads. But individuals are ultimately responsible for their own food choices and exercise habits.

The Numbers

Diabetes drug spending in the United States rose from $6.7 billion in 2001 to $12.5 billion in 2007. As of 2009, nearly 20 million Americans have type 2 diabetes.

Denial may complicate the problem of obesity. In 2005–2006, the U.S. Centers for Disease Control and Prevention determined that 34 percent of American adults were obese when weighed, but only 26 percent admitted to being obese when interviewed by telephone.

Discussion

Obesity is largely a matter of eating too much and exercising too little. All living things that are capable of liking anything seem to like food, and they tend to overeat when a surplus presents itself. Humans in developed nations have more than a surplus; we are constantly bombarded with images of food that is fatter, sweeter, more brightly colored, and more readily available than anything in nature—what ethologists call a supernormal stimulus. Our ancestor *Homo erectus* would have loved a deep-dish pizza, which even today conjures images of a predator tearing into the belly of a still-warm quarry whose organs and blood and stomach contents are spilling onto the veldt. But greed can't be the whole story, because some people also seem to assimilate food more efficiently than others. Most people will never weigh 500 or 1,000 pounds, regardless of how much they eat.

Obesity, particularly morbid obesity, is often associated with type 2 diabetes (Case Study 6-4). But does obesity predispose to diabetes, or is it the other way around? Or does a third factor cause both? According to one theory, "thrifty genes" that favor survival under famine conditions also promote obesity and diabetes. According to another theory, these two conditions, plus several others—high blood pressure, insulin resistance in the liver, high cholesterol, and hardening of the arteries—constitute something called "metabolic syndrome." Some experts deny that there is any such thing; in 2005, the American Diabetes Association and the European Association for the Study of Diabetes issued a joint position statement to the effect that metabolic syndrome is nothing more than the sum of its parts. Even so, those parts are conditions best avoided.

A number of studies have implicated causal factors other than genetics or gluttony. Several viruses appear to trigger diabetes, including the arenavirus that causes lymphocytic choriomeningitis (Chapter 3). One type of airborne adenovirus appears to make fat cells multiply. Arsenic, diet soda, fast food, tobacco, the absence of ulcer-causing bacteria (*Helicobacter pylori*) in the stomach, the absence of intestinal parasites, and various pollutants of air, water, and soil have all been proposed as explanations for obesity or diabetes. But some 17 million American dogs are

Case Study 6-4: Gastric Banding and Diabetes

A 2008 study produced the latest in a mountain of reports that support the same conclusion: Most (not all) cases of type 2 diabetes appear to result from obesity, and losing weight is a good way to control diabetes and reduce the need for medication. One group of obese patients with diabetes had gastric banding surgery while a second group had conventional diabetes therapy with emphasis on lifestyle change. At follow-up, 73 percent of the surgical-banding group achieved remission of diabetes, as compared with 13 percent of the conventional-therapy group.

obese, too, yet few of them live on diet soda and fast food or use tobacco, and none of them are genetically related to their owners. (About 1 dog in 300 develops diabetes, as does 1 cat in 400.)

Popular Culture

Some preachers have used HIV or syphilis to illustrate the wages of sin, yet they rarely dismiss diabetics or quadruple bypass patients with the same sniff. Lust, gluttony, and sloth are all among the seven deadly sins, the last time we checked; yet hardly anyone seems to blame people for illnesses that result from overeating or lack of exercise. Benjamin Franklin's insightful conversation with his gouty big toe proves once again what a genius he was:

> FRANKLIN. Eh! Oh! Eh! What have I done to merit these cruel sufferings?
>
> GOUT. Many things; you have ate and drank too freely, and too much indulged those legs of yours in their indolence.[2]

Gout seems to be less common now than it was in Franklin's day, or at least less highly publicized than diabetes, which afflicts a similar population—essentially, overweight people who are susceptible by heredity, and who eat too much and exercise too little.

Diabetes has been a theme in at least 25 motion pictures, notably *Steel Magnolias* (1989), *The Godfather III* (1990), *Chocolat* (2000), and *It Runs in the Family* (2003). The 2004 movie *Super Size Me* is about fast food. *The Nutty Professor* (1996) and *Norbit* (2007) feature Eddie Murphy in a fat suit.

Traditional folk remedies for diabetes in North America and northern Europe include powdered mice, chickweed and comfrey tea, wormwood, geranium root, dandelion, boiled nettles, horsetail, Spanish moss, bugle weed, yarrow, sumac, huckleberry root, sage, and mullein leaf tea. Nonplant remedies included sulfur and molasses, goat's milk, buttermilk, honey, and vinegar. Extreme cures included drinking one's own urine or being bitten by a venomous snake.

In Irish folklore, walking on "hungry grass" was said to cause weakness and diabetes. Some sources identify it as creeping bentgrass (*Agrostis stolonifera*), which now grows on golf courses throughout the world without reported adverse health effects. Others claim that "hungry grass" referred to a patch of cursed grass near a body that was buried without benefit of clergy.

Point/Counterpoint

P: It's normal for the most successful, desirable people to be well fed. Having plenty of food has always been a sign of high social status. Look at Polynesian cultures. Look at the Venus of Willendorf. Look at the voluptuous women in Rubens' paintings. Western culture is hung up on a preadolescent ideal of beauty.

CP: Agreed! But we are talking about health, not beauty. Most (not all) overweight people are risking their health and their children's health. Do they really enjoy watching an obese mother and daughter on a television ad, discussing their favorite glucose meters or stomach-stapling procedures?

The Future

The medical and sociological issues related to obesity will not go away in the foreseeable future, but it is important to remember that overeating is not a problem for everyone. Many people

---

2. Benjamin Franklin, *Dialog Between Franklin and the Gout* (1780).

love food and accept being heavy as a normal part of life. One question that researchers need to answer is why many of these people develop diabetes and heart disease, while others stay healthy.

References and Recommended Reading

Albright, A. "What is Public Health Practice Telling Us about Diabetes?" *Journal of the American Dietetic Association*, Vol. 108 (Suppl. 4), 2008, pp. S12–18.

Altman, B., and A. Bernstein. 2008. *Disability and Health in the United States, 2001–2005*. Hyattsville, MD: National Center for Health Statistics, 89 pp.

Amar, S., et al. "Diet-Induced Obesity in Mice Causes Changes in Immune Responses and Bone Loss Manifested by Bacterial Challenge." *Proceedings of the National Academy of Sciences* (U.S.), 12 December 2008.

Benjamin, S. E., et al. "Obesity Prevention in Child Care: A Review of U.S. State Regulations." *BMC Public Health*, Vol. 8, 2008, p. 188.

Biddinger, S. B., et al. "Hepatic Insulin Resistance Is Sufficient to Produce Dyslipidemia and Susceptibility to Atherosclerosis." *Cell Metabolism*, Vol. 7, 2008, pp. 125–134.

Brownell, K. D., and D. Yach. "Lessons from a Small Country about the Global Obesity Crisis." *Globalization and Health*, Vol. 2, 2006, p. 11.

Chang, P.-C., et al. "Association Between Television Viewing and the Risk of Metabolic Syndrome in a Community-based Population." *BMC Public Health*, Vol. 8, 2008, p. 193.

Chowdhury, P., et al. "Surveillance of Certain Health Behaviors among States and Selected Local Areas—United States, 2005." *MMWR Surveillance Summaries*, Vol. 56, 2007, pp. 1–164.

Coronado, G. D., et al. "Attitudes and Beliefs among Mexican Americans about Type 2 Diabetes." *Journal of Health Care for the Poor and Underserved*, Vol. 15, 2004, pp. 576–588.

DeNoon, D. J. "Fat Fear Survey: Most Would Trade Family, Health, Wealth to Avoid Obesity." WebMD, 18 May 2006.

"Diabetes May Increase Risk of Developing TB." Reuters, 15 July 2008.

"Diabetes Rate Doubles in U.S. in Last 10 Years." Associated Press, 30 October 2008.

Dixon, J. B., et al. "Adjustable Gastric Banding and Conventional Therapy for Type 2 Diabetes: A Randomized Controlled Trial." *JAMA*, Vol. 299, 2008, pp. 316–323.

Drescher, K. M., and S. M. Tracy. "The CVB and Etiology of Type 1 Diabetes." *Current Topics in Microbiology and Immunology*, Vol. 323, 2008, pp. 259–274.

"Expert: Obesity is Epidemic of Our Time." United Press International, 13 November 2008.

Fee, M. "Racializing Narratives: Obesity, Diabetes, and the 'Aboriginal' Thrifty Genotype." *Social Science and Medicine*, Vol. 62, 2006, pp. 2988–2997.

"Hypereating Drives Many to Consume When Not Hungry." Associated Press, 21 April 2009.

Jaeckel, E., et al. "Viruses and Diabetes." *Annals of the New York Academy of Sciences*, Vol. 9, 2002, pp. 7–25.

Kahn, R., et al. "The Metabolic Syndrome: Time for a Critical Reappraisal." *Diabetes Care*, Vol. 28, 2005, pp. 2289–2304.

Ko, G. T., and J. C. Chan. "Burden of Obesity—Lessons Learnt from Hong Kong Chinese." *Obesity Reviews*, Vol. 9, Suppl. 1, 2008, pp. 35–40.

Kolata, G. "Overweight People Found Less Likely to Die from Some Diseases." *New York Times*, 7 November 2007.

Krasnoff, J. B., et al. "Health-related Fitness and Physical Activity in Patients with Nonalcoholic Fatty Liver Disease." *Hepatology*, Vol. 47, 2008, pp. 1158–1166.

Levine, S., and R. Stein. "Childhood Obesity Threat to Future Health, Longevity." *Washington Post*, 20 May 2008.

Meetoo D., et al. "An Epidemiological Overview of Diabetes Across the World." *British Journal of Nursing*, Vol. 16, 2007, pp. 1002–1007.

"Obesity and Immunity Linked." United Press International, 5 February 2008.

O'Connell, J. "Even a Thin Person Can Get Diabetes." MSNBC.com, 29 May 2008.

"Over the Long Haul, Most Americans Will be Fat, Long-term Study Suggests." Associated Press, 4 October 2005.

Park, A. "Study: Diabetes Linked to Cognitive Decline." *Time*, 5 January 2009.

Penn, L., et al. "Participants' Perspective on Maintaining Behaviour Change: A Qualitative Study within the European Diabetes Prevention Study." *BMC Public Health*, Vol. 8, 2008, p. 235.

Santos, A. C., et al. "Gender, Socio-economic Status and Metabolic Syndrome in Middle-aged and Old Adults." *BMC Public Health*, Vol. 8, 2008, p. 62.

Schulte, P. A., et al. "A Framework for the Concurrent Consideration of Occupational Hazards and Obesity." *Annals of Occupational Hygiene*, Vol. 52, 2008, pp. 555–566.

Schwartz, M. B., et al. "The Influence of One's Own Body Weight on Implicit and Explicit Anti-Fat Bias." *Obesity* (Silver Spring), Vol. 14, 2006, pp. 440–447.

Smith, A. G., et al. "Diet-Induced Obese Mice Have Increased Mortality and Altered Immune Responses When Infected with Influenza Virus." *Journal of Nutrition*, Vol. 137, 2007, pp. 1236–1243.

Speakman, J. R. "Thrifty Genes for Obesity and the Metabolic Syndrome—Time to Call Off the Search?" *Diabetes and Vascular Disease Research*, Vol. 3, 2006, pp. 7–11.

Uauy, R., and E. Diaz. "Consequences of Food Energy Excess and Positive Energy Balance." *Public Health and Nutrition*, Vol. 8, 2005, pp. 1077–1099.

Vasan, R. S., et al. "Estimated Risks for Developing Obesity in the Framingham Heart Study." *Annals of Internal Medicine*, Vol. 143, 2005, pp. 473–480.

"Western Diet Boosts Global Heart Attack Risk 30%." *HealthDay News*, 20 October 2008.

Willey, J. "Obesity Bug You Can Catch." *Daily Express*, 26 January 2009.

Wu, M. S., et al. "A Case-Control Study of Association of *Helicobacter pylori* Infection with Morbid Obesity in Taiwan." *Archives of Internal Medicine*, Vol. 165, 2005, pp. 1552–1555.

Yach, D., et al. "Epidemiologic and Economic Consequences of the Global Epidemics of Obesity and Diabetes." *Nature Medicine*, Vol. 12, 2006, pp. 62–66.

Yang, Z., and A. G. Hall. "The Financial Burden of Overweight and Obesity among Elderly Americans: the Dynamics of Weight, Longevity, and Health Care Costs." *Health Services Research*, Vol. 43, 2008, pp. 849–868.

TOO MANY SICK PEOPLE: THE HEALTHCARE CRISIS

Summary of Threat

Despite advances in medical science, the healthcare available on an everyday basis in the United States is often unaffordable (even with insurance), inaccessible, and substandard. Other countries have similar problems, but as of 2009, the United States claims the dubious honor of being the only developed nation that lacks some form of universal healthcare. The U.S. infant mortality rate is also among the highest in the developed world.

So What?

If people with AIDS, tuberculosis, and other contagious diseases cannot afford their medication, all of society suffers. If vaccine shortages interfere with childhood immunization programs, or if families cannot afford to pay their doctors, or if they avoid their doctors because of perceived inadequate care, deadly childhood diseases may return. If physicians overprescribe antibiotics as a quick fix, the problem of drug-resistant bacteria will become worse. And if hospitals neglect sanitation, the death toll from hospital-acquired infections will continue to rise.

Another aspect of the healthcare crisis is the allocation of research funding to popular programs (such as the war on bioterrorism) at the expense of real and present threats. In a 2007 Gallup poll to identify the most urgent U.S. health problems, the leading contenders were access and cost (Table 6.2).

Table 6.2 Most Urgent Health Problems, United States, 2007

Problem	Percentage of Respondents Saying This Was the Most Urgent
Access	30
Cost	26
Cancer	14
Obesity	10
AIDS	2
Diabetes	2
Heart disease	1
Finding cures	1
Flu	1
Drug/alcohol abuse	1
Smoking	<0.5
Bioterrorism	<0.5
Other	4
No opinion	9

Source: Gallup Poll, 11–14 November 2007.

Can't Scientists Do Something about It?

Scientists can invent better drugs, diagnostic tests, medical instruments, and sterilization procedures, but other people and agencies—such as governments, hospital administrators, health maintenance organizations, and insurance companies—ultimately determine how (or if) those advances will be used.

The Numbers

In 2004–2006, medical errors caused an average of about 80,000 preventable deaths per year in the United States (range, 44,000 to 98,000). Healthcare is a risky business, and some mistakes are inevitable, such as misdiagnosis or equipment failure. Other types of errors, such as mixing up patient records or prescribing the wrong dosage of a drug, should be largely preventable. The Agency for Healthcare Research and Quality estimates that medical errors cost the United States about $37.6 billion every year, including about $17 billion for preventable errors (Case Study 6-5).

In addition to medical errors per se, an estimated 2 million people develop infections in American hospitals every year, and about 100,000 of them die. It is not clear how many of these infections might be preventable if hospitals had better sterilization procedures, and if staff would refrain from reusing syringes.

Case Study 6-5: Sources of Stress

Patients are not the only ones who report dissatisfaction with hospital care. The doctors and nurses are not a happy bunch either, according to two 2008 reports. Hospital employees often work 24-hour shifts and worry that their exhaustion might kill patients. Thanks to understaffing or miscommunication during shift changes, heart attack victims are statistically more likely to die if they reach the hospital at night or on a weekend, and babies born at night are significantly more likely to die during the first month of life. To make matters worse, noise levels at a typical U.S. hospital far exceed maximum limits recommended by the World Health Organization.

According to a recent survey, 7 percent of respondents (or a household member) got married for the purpose of gaining access to health insurance. In addition, 23 percent took a new job, or kept a job they already had, primarily because of health insurance. As of 2009, an estimated 50 million Americans are uninsured, and even those with insurance often struggle to pay medical bills. Between 2004 and 2007, out-of-pocket healthcare costs for insured U.S. adults increased by 34 percent. In 2008, medical bills were a major factor in more than 60 percent of U.S. personal bankruptcies.

Discussion

In a 2007 speech at Harvard University, Microsoft chairman and philanthropist Bill Gates said: "Humanity's greatest advances are not in its discoveries—but in how those discoveries are applied to reduce inequity." To which we will add that the quest for equity should not depend on the generosity of individuals such as Mr. Gates, but should instead be a priority of every government.

During the 2008 presidential campaign, Barack Obama stated: "We need a health care system instead of a disease care system." Opponent John McCain's take on this issue was equally valid: "It is simply disgraceful that 43 million Americans cannot afford health care coverage." In 2009, it remains to be seen how the new president and other world leaders will address these issues.

Popular Culture

Movies that portray doctors or hospitals in a less than favorable light include *Dead Ringers* (1989), *The Boys from Brazil* (1978), *The Doctor* (1991), and *One Flew Over the Cuckoo's Nest* (1975). Novels in which the bad guys are doctors include Robin Cook's *Coma* (1977), *Godplayer* (1983), and *Shock* (2001); Michael Palmer's *Miracle Cure* (1998); and Neil Ravin's *Evidence* (1987). For the dark side of mental health care, try Sylvia Plath's *The Bell Jar* (1971) or Patrick McGrath's *Asylum* (1997).

Urban legends involving terrible hospitals are many (Figure 6.3). Some of the worst ones turn out to be true, such as stories of nurses who have sex with comatose or dead patients; or patients who die after waiting 24 hours in an emergency room lobby; or babies who are switched at birth, given an overdose of the wrong drug, or abducted by some fruitcake wandering through the ICU nursery. So when we hear a story about a doctor who amputates the wrong leg—which, of course, ends up in the hospital cafeteria meatloaf, where it is identified by the shade of the patient's toenail polish (and further investigation reveals that the patient had Ebola)—it's often hard to decide if we should take it seriously or not.

Other urban legends exaggerate real dangers. In one, a Gulf War veteran has an MRI without telling anyone that his body is full of shrapnel. Pieces of embedded metal emerge at the speed of bullets, destroying the machine and ripping holes in the patient—who clings to life until he sees the resulting $5 million hospital bill. In another, the patient is heavily tattooed with metallic inks, so the MRI rips out the metal and tears his skin off. In fact, embedded metal can cause local irritation during an MRI, and a few patients have found stray hairpins or paper clips lodged in their noses, but more serious problems are rare.

Point/Counterpoint

P: The U.S. healthcare system is the best in the world. Everyone has access to superb medical services from cradle to grave. COBRA ensures that people will not lose their health insurance even if they lose a job. We have the finest hospitals and medical schools with the latest equipment. People in Canada, the Netherlands, and other countries with national healthcare programs are always complaining about the low quality of care. They come here for treatment.

CP: Some (not all) low-income Americans qualify for Medicaid, but most doctors reject it because reimbursement rates are too low. In 2009, group health insurance premiums for a U.S. family of four cost on average $13,000 per year, not including copayments. COBRA costs even

Figure 6.3 This cartoon makes several statements about problems (real or perceived) in today's healthcare industry.

Source: Courtesy of CartoonStock (artist, Mike Baldwin).

more—often more than a family's income. Without insurance, the average cost of treatment for a simple broken leg is over $20,000. Even insured people can be driven to bankruptcy by the deductible and copayments for one major injury or illness.

The Future

For many American families, health insurance costs more than housing or food. But what do the premiums really buy? At least one-third of the cost of healthcare has nothing to do with medicine; it pays for insurance claims processing. Also, part of the reason doctors and hospitals charge so much is that they must pay their own malpractice insurance premiums. In other words, we pay for insurance to pay the hospital to bill insurance; and we pay our doctors so they can pay for more insurance. Eventually, someone will figure out how to exit this loop.

References and Recommended Reading

Abelson, Reed. "Deep Flaws Found in U.S. Health Care." *International Herald Tribune*, 17 July 2008.

"Autopsy: Blood Clots Caused Waiting Room Death." CNN, 11 July 2008.

Boyce, J. M. "Environmental Contamination Makes an Important Contribution to Hospital Infection." *Journal of Hospital Infection*, Vol. 65 (Suppl. 2), 2007, pp. 50–54.

Britt, R. R. "How Hospitals Can Turn Deadly." *LiveScience*, 1 October 2008.

"CDC: US Infant Mortality Drops to 29th." United Press International, 15 October 2008.

Chu, C., and P. A. Selwyn. "Current Health Disparities in HIV/AIDS." *AIDS Reader*, Vol. 18, 2008, pp. 144–146, 152–158, C3.

Commonwealth Fund Commission on a High Performance Health System. "Why Not the Best? Results from the National Scorecard on U.S. Health System Performance, July 2008." New York: The Commonwealth Fund, 64 pp.

Daschle, T. 2008. *Critical: What We Can Do about the Health-Care Crisis*. New York: Thomas Dunne Books, 226 pp.

"Death Rate 70 Percent Lower at Top Hospitals." *LiveScience*, 14 October 2008.

"Expert: Patient Safety Not Rocket Science." United Press International, 19 January 2009.

Fox, M. "Medical Bills Underlie 60 Percent of U.S. Bankruptcies: Study." Reuters, 4 June 2009.

Freed, G. L., et al. "Primary Care Physician Perspectives on Reimbursement for Childhood Immunizations." *Pediatrics*, Vol. 122, 2008, pp. 1319–1324.

Gabel, J. R., et al. "Trends in Underinsurance and the Affordability of Employer Coverage, 2004–2007." *Health Affairs*, Vol. 28, 2009, pp. 595–606.

Giuliano, M., and S. Vella. "Inequalities in Health: Access to Treatment for HIV/AIDS." *Annali dell'Istituto Superiore di Sanità*, Vol. 43, 2007, pp. 313–316.

Hertel-Fernandez, A. W., et al. "The Chilean Infant Mortality Decline: Improvement for Whom?" *Bulletin of the World Health Organization*, Vol. 85, 2007, pp. 798–804.

Hinman, A. R., et al. "Vaccine Shortages: History, Impact, and Prospects for the Future." *Annual Review of Public Health*, Vol. 27, 2006, pp. 235–259.

Kanato, M. "Drug Use and Health among Prison Inmates." *Current Opinion in Psychiatry*, Vol. 21, 2008, pp. 252–254.

"Many Psychiatric Patients Wait 24 Hours to Be Seen." *ED Management*, Vol. 20, 2008, pp. 105–106.

"Marrying to Access Spouse's Healthcare." United Press International, 29 April 2008.

Phillips, D. P., and C. C. Bredder. "Morbidity and Mortality from Medical Errors: An Increasingly Serious Public Health Problem." *Annual Review of Public Health*, Vol. 23, 2002, pp. 135–150.

Posse, M., et al. "Barriers to Access to Antiretroviral Treatment in Developing Countries: a Review." *Tropical Medicine and International Health*, Vol. 13, 2008, pp. 904–913.

"Rare Pneumonia Killed 22 Children." United Press International, 14 September 2008.

Rosenau, P. V., and C. J. Lako. "An Experiment with Regulated Competition and Individual Mandates for Universal Health Care: The New Dutch Health Insurance System. *Journal of Health Politics, Policy and Law*, Vol., 33, 2008, pp. 1031–1055.

Sharfstein, J. "Kids First?" *International Journal of Health Services*, Vol. 30, 2000, pp. 763–769.

Silber, J. H., et al. "Hospital Teaching Intensity, Patient Race, and Surgical Outcomes." *Archives of Surgery*, Vol. 144, 2009, pp. 113–120.

Smith, F. How Bad Does the Healthcare Crisis Have to Get? *Redbook*, 1 July 2007.

"Study: Medical Care, Spending Don't Match." United Press International, 18 July 2008.

Williams, N. "To Get Factory Jobs Back, U.S. Must Solve Health-Care Crisis." McClatchy-Tribune Regional News, 1 September 2008.

TOO MANY ANGRY PEOPLE: BIOTERRORISM

Summary of Threat

Bioterrorism is a real threat, but one that has not yet materialized, except for the 2001 anthrax mailings—and the responsible party (or one of the parties) in that case turned out to be an American scientist. Augmenting public health surveillance and biodefense-related research is a wise precaution, up to a point, but this shift of emphasis may have diverted resources from more immediate health issues.

So What?

As discussed in the previous section, the availability and affordability of healthcare have declined in recent years. There are many reasons, but the final common path is money. According to a team of researchers at the National Institutes of Health, the emphasis on diseases such as smallpox and anthrax since 2001 has taken government funding away from other important medical research and public health programs.[3]

Also, an atmosphere of paranoia may detract from more rational solutions to international tensions. When West Nile virus first appeared in the United States in 1999, some authorities immediately suspected bioterrorism,[4] although this virus already had a history of range expansions in other parts of the world. Similarly, in 2001, some journalists blamed Iraq for the anthrax mailings that were later attributed to U.S. scientist Bruce Ivins. A series of outbreaks of pneumonic tularemia near Martha's Vineyard, Massachusetts, in the summers of 2000–2004 also caused a certain amount of speculation.

Can't Scientists Do Something About It?

When the media reported the anthrax mailings in October 2001, the author was correcting page proofs for a previous book on biological hazards. Uncertain what to make of these recent events, she finally wrote: "Perhaps this was not terrorism in the usual sense, but the act of some over-zealous American researcher who wanted to scare the government into upgrading its public health system."[5] As it turned out, this assessment was not far off.

---

3. S. Altman et al., "An Open Letter to Elias Zerhouni" (*Science*, 4 March 2005).
4. R. Preston, "West Nile Mystery" (*The New Yorker*, 18 October 1999).
5. J. R. Callahan, *Biological Hazards* (Oryx Press, 2002), p. 278.

In other words: scientists can speculate until the cows come home, but identifying and addressing the causes of antisocial behavior is a problem for philosophers and heads of state.

The Numbers

Most large numbers involving biological weapons are based on projections rather than real-life events. There are a few disputed cases; for example, the 1978–1980 anthrax outbreak in Zimbabwe killed 182 people, but its origin is unknown. Tularemia infected more than 100,000 people near Stalingrad in 1942, but most analysts now believe that outbreak was natural. Disease outbreaks are a frequent consequence of war, with or without bioterrorism.

A 1993 study concluded that the release of 100 kilograms (about 220 pounds) of anthrax spores upwind from Washington, D.C., would kill between 130,000 and 3 million people. Other studies have concluded that a small vial of botulin toxin could kill all 7 billion people on Earth, given an effective delivery system.

The 2001 anthrax mailings infected 22 people, 5 of whom died. During the next 5 years, Congress responded to the incident by spending an estimated $28 billion on bioterrorism defense and related programs. It is impossible to estimate how many lives were saved or lost as a result of this shift in research priorities.

A 2007 Zogby poll asked 10,258 Americans which global health risk posed the greatest threat—AIDS or bioterrorism. And bioterrorism won by a landslide, particularly among those who ranked the U.S. healthcare system as excellent.

Discussion

Smallpox killed between 300,000 and 500,000 people worldwide in the twentieth century before its eradication in 1979 (Figure 6.4). Although the virus still exists in an unknown number of government freezers, there is no obvious reason why any foreign terrorist group or nation would reintroduce it to the world. The governments of wealthier nations would promptly immunize their people, and the disease would only add to the suffering of the populations that terrorists claim to represent.

The person(s) who released anthrax in 2001 probably knew it would not spread from person to person and would cause only a limited outbreak, thereby ensuring increased funding and public support for biodefense programs and anthrax vaccination without really endangering the United States or the world. It seems unlikely that anyone would be stupid enough to pull a similar stunt with smallpox.

In 2008, smallpox-related hysteria seemed to be subsiding, yet a bipartisan commission predicted a major terrorist attack using nuclear or biological weapons no later than 2013. Curiously, the commission apparently said little about chemical weapons, which many scientists regard as a greater threat.

Popular Culture

Hoaxes involving anthrax and other germs have been a feature of American life since at least 1923 (Case Study 6-6). German scientists worked on anthrax weapons as early as 1915, and soldiers returning from World War I knew about them. Author Aldous Huxley wrote about anthrax bombs in *Brave New World* (1932), and the British and Japanese both started testing real anthrax bombs by 1942.

In the 2007 motion picture *Death without Consent*, a young man learns that his father and brother died from the effects of a biological warfare experiment. *Anthrax* is a 2001 movie in which a sinister government research facility arouses the suspicion of Canadian cattle ranchers. When one dies of anthrax, a Royal Canadian Mounted Police officer investigates. Other motion

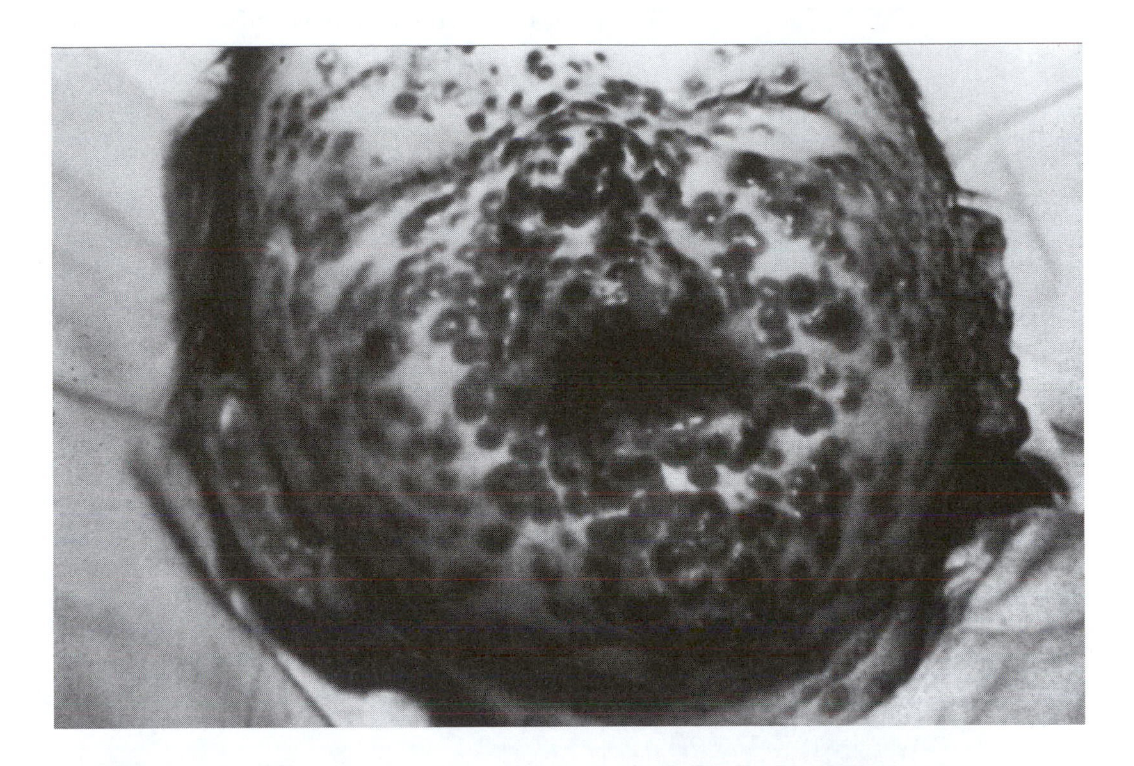

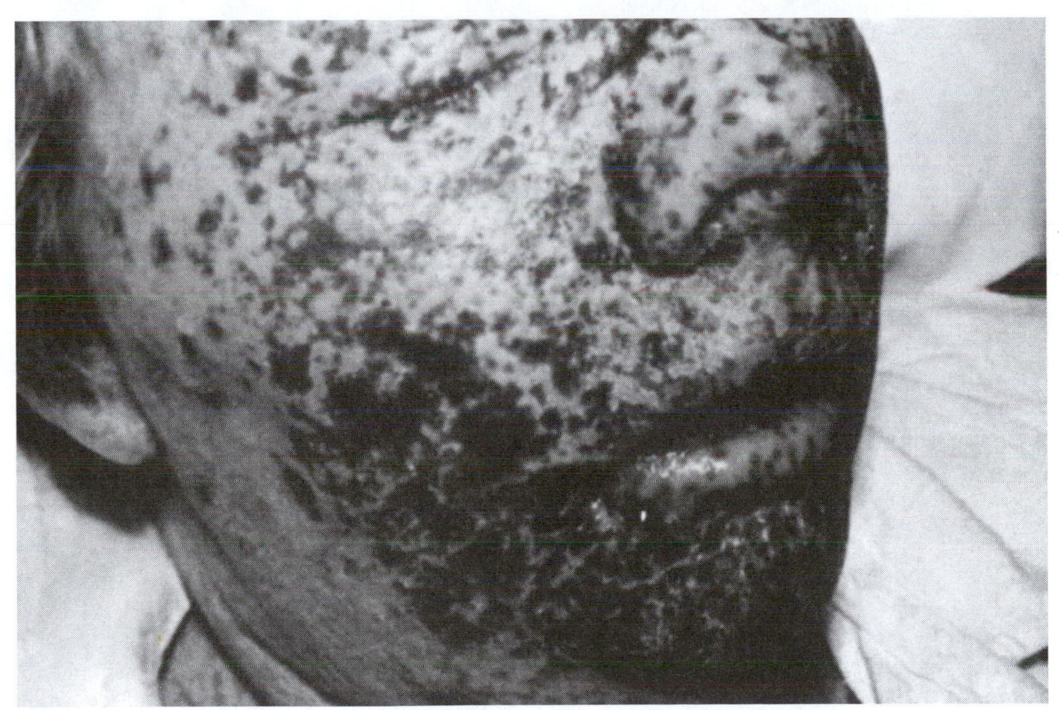

Figure 6.4 Smallpox was eradicated from the world in 1979, but samples of the virus still exist in government laboratories. Some experts fear that terrorists may release this dreaded disease. *Above:* A one-year-old infant with relatively mild smallpox. *Below:* A woman with fatal hemorrhagic smallpox:

Source: U.S. Centers for Disease Control and Prevention, Public Health Image Library.

Case Study 6-6: The Great Anthrax Plot of 1927

In November 1927, the police in Portland, Oregon, reported a strange incident. A man left a letter on a cafeteria table, and a waitress became suspicious. The police intercepted the letter and found that the sender was trying to obtain anthrax and leprosy germs from contacts in Siberia and South America. A search of the man's room turned up hundreds of letters that outlined a plan to bring anthrax and other diseases into the United States. One letter stated that anthrax germs were on their way to Portland via special messenger. Further investigation revealed that the same man had made a similar threat in 1923. But just when the local authorities had decided that he was crazy, an alleged federal investigator in Washington sent the Portland police a telegram, instructing them to hold the prisoner. The Associated Press reporter wrote that the suspect "may be a keystone man in an international plot."[1] Finally, it turned out that the "federal investigator" was a female welfare recipient who had read about the incident and sent the telegram as a hoax.

[1]"International Plot Linked to Germ Maniac," Associated Press, 9 November 1927.

pictures related to bioterrorism or biological warfare include *Twelve Monkeys* (1995), *Outbreak* (1995), *The Andromeda Strain* (1971), *Derailed* (2002), and the made-for-TV *Smallpox* (2005).

In Ketan Desai's 1999 novel *Germs of War*, a doctor at the Mayo Clinic hires a malcontent to develop a biological weapon, which he proceeds to test on the hospital staff. Robin Cook's 1999 novel *Vector* was based on a real-life 1979 incident in which a research facility in Russia accidentally released anthrax.

In October 2001, the Swedish auto manufacturer Volvo mailed small packets of a whitish powder to Volvo owners with reminders to have their vehicles serviced. The powder was a harmless vitamin supplement, but the timing was ghastly, since these packets reached their targets just as the 2001 anthrax letters began to dominate the news media in the wake of 9/11. Some recipients responded badly, and Volvo quickly ended its "Volvo for Life" campaign.

Point/Counterpoint

P: The war on terrorism is just the latest chapter in the eternal struggle between good and evil. Terrorists attack the West because they hate us for being free. They are not enemy combatants, and they are not entitled to due process. We must continue to study and stockpile biological weapons, because we know the terrorists have them and we must defend ourselves.

CP: People who commit violent acts for revenge or profit are criminals and should be treated as such. But some of the people we call terrorists may be fighting for causes that we have ignored. We don't know that terrorists intend to use biological weapons, but those weapons are just diseases, regardless of how you catch them. So yes, let's continue studying them anyway.

The Future

There is no rule that requires terrorists to deploy one weapon at a time. Some of the worst scenarios involve the simultaneous or successive release of different weapons, such as a chemical agent followed by a biological agent. Major confusion might result, since it is human nature to look for a single cause for multiple events or symptoms. One promising avenue of research is the development of pharmaceuticals that can protect exposed populations from more than one type of threat. In 2008, researchers announced a new experimental treatment that might protect against some chemical weapons, bacteria, viruses, and radiation.

References and Recommended Reading

Baillie, L. W. "Past, Imminent and Future Human Medical Countermeasures for Anthrax." *Journal of Applied Microbiology*, Vol. 101, 2006, pp. 594–606.

Bass, S. B., et al. "Mapping Perceptions Related to Acceptance of Smallpox Vaccination by Hospital Emergency Room Personnel." *Biosecurity and Bioterrorism*, Vol. 6, 2008, pp. 179–190.

Bates, B. "Bioterrorism and Vaccine Events Remain Threats." *Internal Medicine News*, 1 August 2008.

Becker, A. L. "Tularemia Cases on Martha's Vineyard Puzzle Experts." CIDRAP, 11 August 2004.

Belongia, E. A., and A. L. Naleway. "Smallpox Vaccine: The Good, the Bad, and the Ugly." *Clinical Medicine and Research*, Vol. 1, 2003, pp. 87–92.

Blaney, B. "Professor Thought Nothing Wrong with Shipping Bacteria." Associated Press, 21 November 2003.

Choi, C. "U.S. Crop Defense Network Tests Underway." United Press International, 11 September 2003.

Covarrubias, A. "Anthrax Hoaxes Becoming Bomb Scares of the '90s." Associated Press, 3 January 1999.

Crutchley, T. M., et al. "Agroterrorism: Where Are We in the Ongoing War on Terrorism?" *Journal of Food Protection*, Vol. 70, 2007, pp. 791–804.

Dembek, Z. F., et al. "Discernment between Deliberate and Natural Infectious Disease Outbreaks." *Epidemiology and Infection*, Vol. 135, 2007, pp. 353–371.

"Extortionist Threatens Tucson, Fails to Show Up for His Money." Associated Press, 17 September 1978.

"FBI Skeptics Seek Anthrax Probe Answers." United Press International, 7 September 2008.

Findley, S. "Clinton Sees Little Anthrax Threat to Civilians." *USA Today*, 17 December 1997.

Gordon, G. "Evidence Is Springing Up of $28 Billion Response to Bioterror Threat." Knight Ridder Washington Bureau, 8 August 2006.

Graham, B., et al. 2008. *World at Risk: The Report of the Commission on the Prevention of Weapons of Mass Destruction Proliferation and Terrorism*. New York: Vintage Books.

Hassler, K. "Agricultural Bioterrorism: Why It Is a Concern and What We Must Do." USAWC Strategy Research Project, 7 April 2003.

Hornick, R. "Tularemia Revisited." *New England Journal of Medicine*, Vol. 345, 2001, pp. 1637–1639.

Howell, W. N., and J. M. Maury. "Circling the Wagons: Community-Based Responses to Bioterrorism." Critical Incident Analysis Group Report, Spring 2002.

Jernigan, D. B., et al. "Investigation of Bioterrorism-Related Anthrax, United States, 2001: Epidemiologic Findings." *Emerging Infectious Diseases*, Vol. 8, 2002, pp. 1019–1028.

Khan, A. S., et al. "Precautions against Biological and Chemical Terrorism Directed at Food and Water Supplies." *Public Health Reports*, Vol. 116, 2001, pp. 3–14.

LaRussa, T. "Doctor Out of Work since Anthrax Probe." *Pittsburgh Tribune-Review*, 3 October 2005.

Laudisoit, A., et al. "Plague and the Human Flea, Tanzania." *Emerging Infectious Diseases*, Vol. 13, 2007, pp. 687–693.

L'vov, D. K., et al. "Smallpox is a Dormant Volcano." *Voprosy Virusologii*, July–August 2008. [Russian]

MacIntyre, C. R., et al. "Development of a Risk-Priority Score for Category A Bioterrorism Agents as an Aid for Public Health Policy." *Military Medicine*, Vol. 171, 2006, pp. 589–594.

Meinhardt, P. L. "Water and Bioterrorism: Preparing for the Potential Threat to U.S. Water Supplies and Public Health." *Annual Review of Public Health*, Vol. 26, 2005, pp. 213–237.

Pennington, H. "Smallpox and Bioterrorism." *Bulletin of the World Health Organization*, Vol. 81, 2003, pp. 762–767.

Ramshaw, E. "CDC Suspends A&M Research on Infectious Diseases." *Dallas Morning News*, 1 July 2007.

"Sabotage Considered in Virus Outbreak." United Press International, 14 September 2007.

Sayler, C. "Border Makes New Mexico Vulnerable." *New Mexican*, 25 March 2001.

"Scientists Protest Biodefense Priority." United Press International, 2 March 2005.

Sester, M., et al. "Vaccination of the Solid Organ Transplant Recipient." *Transplantation Reviews* (Orlando), Vol. 22, 2008, pp. 274–284.

Sinclair, R., et al. "Persistence of Category A Select Agents in the Environment." *Applied and Environmental Microbiology*, Vol. 74, 2008, pp. 555–563.

Smallman-Raynor, M. 2004. *War Epidemics: An Historical Geography of Infectious Diseases in Military Conflict and Civil Strife, 1850–2000*. New York: Oxford University Press, 805 pp.

Thomas, L. A. "Agricultural Bioterrorism Protection Act of 2002: Possession, Use, and Transfer of Biological Agents and Toxins." National Center for Import and Export, Laboratory Security Workshop, 7 September 2007.

"UPI Poll: Bioterrorism Seen as Top Threat." United Press International, 23 February 2007.

Vijay-Kumar, M., et al. "Flagellin Treatment Protects against Chemicals, Bacteria, Viruses, and Radiation." *Journal of Immunology*, Vol. 180, 2008, pp. 8280–8285.

Vitetta, E. S., et al. "A Pilot Clinical Trial of a Recombinant Ricin Vaccine in Normal Humans." *Proceedings of the National Academy of Sciences* (U.S.), Vol. 103, 2006, p. 2268–2273.

TOO MANY EXPERTS: THE BOGUS HEALTH INDUSTRY

Summary of Threat

Despite the availability of modern healthcare—or perhaps because of it—many people reject conventional treatment in favor of products and services collectively known as alternative medicine. It is becoming increasingly common for people to self-medicate with herbal products and to refuse childhood vaccines and other life-saving interventions. Health professionals must tread a fine line between public safety and individual liberty.

So What?

There was a time in history when people cooperated with the military draft, compulsory vaccination programs, and other unpopular mandates because they believed it was the right thing to do. Some died as a result, but the world often benefited in the long run. There was also a time when the public regarded inventions such as antibiotics, pasteurization, and water treatment as miracles of modern science. A million years ago, our ancestors possibly felt the same way about cooked meat. Now the pendulum has swung in the opposite direction.

Parents who oppose vaccination—often because of understandable concerns or ambiguous research findings—may start disease outbreaks. People who deny that HIV causes AIDS hindered public health efforts for many years. Others promote raw milk and juice, despite the disease risk (Table 6.3), or treat themselves and their families with megadoses of vitamins or herbal products that are ineffective or toxic. Still others harass water companies that practice fluoridation, clinics that distribute condoms, and authors who promote the germ theory of disease.

Can't Scientists Do Something About It?

Scientists can teach classes and write books, but people ultimately decide what they want to learn and which sources they trust. Charlotte the Spider was half right when she said that humans believe anything they see in print. Humans believe anything they see in print *if* it agrees with what they already believe. But dissent is healthy, and some "fringe" ideas turn out to be right (Case Study 6-7).

Table 6.3 Some Outbreaks Traced to Unpasteurized Dairy Products

Date	Location	Details
November 2006	Peterborough, Ontario, Canada	6 people infected with *Campylobacter* from raw milk
September 2006	Whatcom County, Washington	2 children infected with *E. coli* from raw milk
September 2006	Fresno County, California	4 children infected with *E. coli* O157:H7 from raw milk
June 2006	Waterloo, Ontario, Canada	15-year-old infected with *E. coli* from raw milk
Summer 2006	Waterloo, Ontario, Canada	2 children infected with *E. coli* and *Campylobacter* from raw milk
2006	Haldemand and Norfolk, Ontario, Canada	2 people became ill from unspecified bacteria in raw milk
December 2005	Yavapai County, Arizona	*Salmonella* found in raw milk from northern Arizona dairy; no reported illness
December 2005	Cowlitz County, Washington	6 children ages 5–14 infected with *E. coli* O157:H7 from raw milk
April 2005	Ontario, Canada (milk sold from the back of a truck)	4 people infected with *E. coli* O157:H7 from raw milk
2003	Washington State	11 children infected with *E. coli* from raw milk
December 2002	Illinois, Indiana, Ohio, and Tennessee	62 people sick, 2 children hospitalized due to *Salmonella* in raw milk from Ohio dairy
October 2002, February 2003	Edmonton, Alberta, Canada	13 people infected with *E. coli* O157:H7 from unpasteurized gouda cheese
December 2001	Wisconsin	75 people infected with *Campylobacter* from raw milk
August 2001	Vancouver Island, British Columbia, Canada	5 children infected with *E. coli* O157:H7 from raw goat's milk
2000	Austria	38 children infected with *Campylobacter* from raw milk
September 1998	Lancashire, UK	40 people infected with *Salmonella* from raw milk
1996–1998	Worcester County, Massachusetts	80+ people exposed to rabies by drinking raw milk from rabid cows
1997	Street vendors in California	31 people infected with *Salmonella* from raw milk and cheese
1992	Southern France	50+ people contracted Q fever from raw goat's milk
1992	Minnesota	50+ people infected with *Campylobacter* from raw milk
1990	Washington State	13 people infected with *Campylobacter* from raw milk
October 1985	California	23 people infected with *Campylobacter* from raw milk
March-April 1985	Northern Illinois	1,500+ people infected with *Salmonella* from inadequately pasteurized milk
1985	Chicago, Illinois (source)	16,000+ people in 6 states infected with *Salmonella* from raw milk that contaminated pasteurized milk

(Continued)

Table 6.3 (*Continued*)

Date	Location	Details
July 1984	Minnesota	23 people reported acute diarrhea, lasting 4+ weeks, after drinking raw milk.
June 1984	Vancouver Island, British Columbia, Canada	9 children infected with *Campylobacter* from raw milk
May 1984	Whittier, California	12 people infected with *Campylobacter* from raw milk
March 1984	Western Kentucky	16 people infected with *Salmonella* from raw milk
May 1983	Pennsylvania	57 people infected with *Campylobacter* from raw milk
1981–1983	California	239 people infected with *Salmonella* from raw milk
1981	Atlanta, Georgia	50+ people infected with *Campylobacter* from raw milk

Source: Food Safety Department, Kansas State University, 18 December 2007.

In 2004, a Texas woman wrote to the author of this book and complained that the 2002 biological hazards book did not mention the alleged danger of fluoridated drinking water. That reader threatened a boycott, and explained that she knew all about fluoridation because she had read about it in books. But which books? If people limit their reading to sources that echo what they already believe, how is it possible to learn anything? (Besides, even if fluoride in drinking water were hazardous, it would be a chemical hazard by the definition used here, not a biological one.)

Some mainstream physicians and hospitals have made an effort to accommodate alternative beliefs by including religious practitioners in treatment protocols, prescribing vitamin supplements or soothing music, or airing mutual grievances over a glass of raw cow colostrum. This approach appeals to some patients more than others.

Case Study 6-7: They Were Right

Until the early 1980s, most doctors believed that stomach ulcers resulted from lifestyle factors such as stress or rich food. Treatments ranged from years of bland diets and antacid drugs to freezing or removal of the stomach. Ulcers were also a risk factor for cancer. Then two Australian physicians, Drs. Barry Marshall and Robin Warren, proposed that stomach ulcers resulted from a common (and treatable) bacterial infection. When other doctors ridiculed their theory, one of the researchers infected himself with *Helicobacter pylori* and quickly developed ulcer symptoms. Later studies confirmed this result, and in 1994, the U.S. National Institutes of Health acknowledged that about 90 percent of ulcers result from infection. In 2005, Drs. Marshall and Warren received the Nobel Prize in Physiology or Medicine for their pioneering work.

The Numbers

As of 2008, at least 48 U.S. states allowed exemptions from childhood vaccination requirements on religious grounds, and 19 allowed exemptions based on other personal beliefs. The incidence of whooping cough (pertussis) in those 19 states was about 50 percent higher than in states with stricter vaccination requirements.

A 2004 study of 151,720 American children showed that unvaccinated children tended to have better-educated parents with higher incomes. This statement bears repeating, because we don't want anyone to miss it: people with *more* education and *more* money are *less* likely to vaccinate their children. The reason is unknown. Are these people too busy? Are they unable to afford preventive care that is free to lower-income people? Or does the content (vs. duration) of schooling make a difference?

A 2007 study of British consumers showed that more than 50 percent were so distrustful of vaccines that they would even avoid eating the meat of vaccinated chickens.

Discussion

Mistakes happen, and medical mistakes tend to be whoppers. The Cutter Laboratories disaster of 1954–1955 was one of the first examples that received extensive publicity: an early polio vaccine was rushed into human trials before it was safe, and several hundred children contracted polio as a result. Cutter made headlines again in 1983 for accidentally distributing HIV-contaminated blood products. But history wasn't finished with the polio vaccine. In 1999, the press revived an earlier finding that the Salk and Sabin vaccines given before 1962 contained a monkey tumor virus called SV-40. This virus now infects some 100 million people in the United States and Europe, and may increase their cancer risk.

In March 2008, autism advocacy groups proclaimed that Dr. Julie Gerberding, then head of CDC, had finally agreed with their claim that vaccines cause autism. This was inaccurate and unfair. What Dr. Gerberding really said—according to a published transcript—was that some children have an underlying mitochondrial disease that may produce some of the symptoms associated with autism when the children are stressed, by a fever or vaccination or any of a number of events.

As originally outlined, this section ended with an account of Morgellons disease, whose victims claim that multicolored fibers and other objects emerge from bumps on their skin. Most doctors write off Morgellons as delusional parasitosis or an allergic condition, but its advocates have a well-organized lobby. While reading about this topic in 2008, the author began to scratch—and stuck her finger on what seemed to be a thorn protruding from her skin. A series of similar objects emerged from the same spot over the next five days, prompting "Brundlefly" jokes from clever associates. When viewed under a microscope, these objects turned out to be glochids, small barbed spines that were probably acquired during a recent encounter with a cactus. Perhaps many cases of Morgellons result from misidentification of foreign materials that become embedded in the skin. In that case, both advocates and skeptics are right, but simply speaking different languages.

Popular Culture

Do people really need eight glasses of water every day? No, except under special circumstances, such as strenuous activity in hot weather.

Is canola oil poisonous? No, the grade of canola oil sold in grocery stores for human consumption is safe. (Industrial-grade canola oil is not.)

If you eat shrimp while you are taking vitamin C, will you die from arsenic poisoning? No, not unless you also take arsenic.

Do tampon manufacturers put poisonous chemicals on tampons to cause bleeding and cancer? No, it would be stupid to kill their customers.

Can strands of cut hair penetrate skin and start growing? Not exactly, but cut hairs can get into existing breaks in the skin and cause infections.

Does the average person's large intestine really contain 30 to 80 pounds of undigested red meat? No, it doesn't.

Can adhesive pads really suck poison out of the body through the soles of the feet? No, this is just another gimmick to convince people they are dirty.

Point/Counterpoint

P: People have the right to take control of their own lives. They can do that by seeking the advice of online herbal healers and by refusing to have their children injected with dangerous vaccines and drugs.

CP: People have the right to take control of their own lives. They can do that by improving their diet and exercise habits, and by taking the time to find a good doctor who will listen to their concerns and minimize risk.

The Future

Some observers feel that the recent proliferation of pseudoscience is just one symptom of a basic failure of evidence-based decision making. In 1964, the first international assessment of high school students showed that American teenagers ranked among the lowest-achieving mathematics students in the developed world. As of 2008, the United States had maintained that standard for 44 years. American students also rank near the bottom of the list on standardized tests of scientific literacy and problem solving skills. Perhaps the world is entering an era in which our leaders will place a higher priority on science education.

References and Recommended Reading

Armfield, J. M. "When Public Action Undermines Public Health: a Critical Examination of Antifluoridationist Literature." *Australia and New Zealand Health Policy*, Vol. 4, 2007, p. 25.

Bren, L. "Got Milk? Make Sure It's Pasteurized." *FDA Consumer Magazine*, September-October 2004.

Brody, J. E. "Potential for Harm in Dietary Supplements." *New York Times*, 8 April 2008.

Brown, J. C., and X. Jiang. "Prevalence of Antibiotic-Resistant Bacteria in Herbal Products." *Journal of Food Protection*, Vol. 71, 2008, pp. 1486–1490.

Chertoff, B. "Morgellons Disease Baffles Patients and Doctors." *Popular Mechanics*, June 2005.

"Court Says Vaccine Not to Blame for Autism." Associated Press, 12 February 2009.

Dangour, A. D., et al. "Nutritional Quality of Organic Foods: A Systematic Review." *American Journal of Clinical Nutrition*, 29 July 2009.

"Deadly Hib Infection Signals a Comeback." United Press International, 24 January 2009.

DeNoon, D. J. "Allergy Causes Most Mold Ills." CBSNews.com, 24 February 2005.

Doherty, B. "Parents Battle Medical Authorities for Control of their Children." *Reason Magazine*, February 2001.

Fishbein, D. B., and D. Raoult. "A Cluster of *Coxiella burnetii* Infections Associated with Exposure to Vaccinated Goats and their Unpasteurized Dairy Products." *American Journal of Tropical Medicine and Hygiene*, Vol. 47, 1992, pp. 35–40.

"Foolish Vaccine Exemptions." *New York Times*, 12 October 2006.

Gisselquist, D. "Denialism Undermines AIDS Prevention in Sub-Saharan Africa." *International Journal of STD and AIDS*, Vol. 19, 2008, pp. 649–655.

Harris, G. "Measles Cases Grow in Number, and Officials Blame Parents' Fear of Autism." *New York Times*, 22 August 2008.

Johnson, C. K. "Study Adds to Evidence of Vaccine Safety." Associated Press, 25 January 2009.

Kapp, C. "South African Court Bans Vitamin Trials for HIV/AIDS." *Lancet*, Vol. 372, 2008, p. 15.
Kimmel, S. R. "Vaccine Adverse Events: Separating Myth from Reality." *American Family Physician*, Vol. 66, 2002, pp. 2113–2120.
Kinney, T. "Supplement Maker Gets 25 Years in Fraud Case." Associated Press, 27 August 2008.
Lejeune, J. T., and P. J. Rajala-Schultz. "Food Safety: Unpasteurized Milk, a Continued Public Health Threat." *Clinical Infectious Diseases*, Vol. 48, 2009, pp. 93–100.
Neunder, M. "Quack Attacker Returns after Legal Threat." *Skeptical Inquirer*, 1 September 2007.
Newton, K. M., et al. "Treatment of Vasomotor Symptoms of Menopause with Black Cohosh, Multibotanicals, Soy, Hormone Therapy, or Placebo: A Randomized Trial." *Annals of Internal Medicine*, Vol. 145, 2006, pp. 869–879.
Offit, P. A. 2008. *Autism's False Prophets: Bad Science, Risky Medicine, and the Search for a Cure*. Irvington, NY: Columbia University Press.
Parry, J. "No Vaccine for the Scaremongers." *Bulletin of the World Health Organization*, Vol. 86, 2008, pp. 425–426.
Russell, S. "When Polio Vaccine Backfired." *San Francisco Chronicle*, 25 April 2005.
Savely, V. R., et al. "The Mystery of Morgellon's Disease." *American Journal of Clinical Dermatology*, Vol. 7, 2006, pp. 1–5.
Scudamore, J. M. "Consumer Attitudes to Vaccination of Food-Producing Animals." *Revue Scientifique et Technique*, Vol. 26, 2007, pp. 451–459.
Smith, P. J., et al. "Children Who Have Received No Vaccines: Who Are They and Where Do They Live?" *Pediatrics*, Vol. 114, 2004, pp. 187–195.
Taliani, G., et al. "Lumbar Pain in a Married Couple Who Likes Cheese: Brucella Strikes Again!" *Clinical and Experimental Rheumatology*, Vol. 22, 2004, pp. 477–480.
Thompson, D. 2008. *Counterknowledge: How We Surrendered to Conspiracy Theories, Quack Medicine, Bogus Science and Fake History*. New York: Norton, 176 pp.
Wakefield, A. J., et al. "Ileal-Lymphoid-Nodule Hyperplasia, Non-Specific Colitis, and Pervasive Developmental Disorder in Children." *Lancet*, Vol. 351, 1998, pp. 637–641.
Woolston, C. "Kinoki Foot Pads' Detox Claims Don't Stand Up to Science." *Los Angeles Times*, 22 September 2008.
Yeung, C. A. "A Systematic Review of the Efficacy and Safety of Fluoridation." *Evidence Based Dentistry*, Vol. 9, 2008, pp. 39–43.

TOO MANY DRUGS: SUBSTANCE ABUSE

Summary of Threat

Prescription drugs, over-the-counter drugs, alcohol, tobacco, and street drugs were all invented for good reasons. Many drugs save lives and relieve suffering. But when manufacturers promote potentially dangerous drugs for the relief of minor ailments, and when marijuana is the number-one cash crop in the United States, and when antidepressants are so widely used as to contaminate public water supplies, something is off.

So What?

Tobacco, alcohol, and many recreational drugs are biological threats in themselves, in the sense that they are derived from plants or fungi and potentially harmful. But overuse of some drugs can also make people more vulnerable to infectious diseases, including hepatitis C, HIV, and meningococcal meningitis. Recreational drugs can increase risk directly, by interfering with the immune system or damaging specific organs, or indirectly, by promoting behaviors such as needle-sharing or unsafe sex.

The most widely prescribed drugs are painkillers, which are potentially addictive and can mask underlying disease; antidepressants, which can predispose to pneumonia; and antibiotics, which promote the emergence of drug-resistant bacteria (Chapter 3).

Can't Scientists Do Something About It?

Scientists can study the effects of various elective drugs and the reasons why people want them. Sometimes the results can persuade millions of people to change their habits, as in the anti-smoking campaigns of the 1970s. But it's unlikely that anyone can eliminate the fundamental human urge to take drugs.

Many people who quit smoking reportedly turn to other habits that answer the same need, such as drinking or overeating. In other cases, people seem impossible to scare. In 2009, television commercials aggressively promoted a new anti-psoriasis drug, while acknowledging that it can cause side effects—including tuberculosis, cancer, and heart failure. Psoriasis can cause great discomfort, but who would risk these catastrophic illnesses in the hope of curing it? Many people, apparently.

The Numbers

The average number of drug prescriptions per elderly person in the United States increased from 20 in 1992 to about 40 in 2009. For the population as a whole in 2005, the average number of prescriptions per person was about 12 in Canada, the United States, and England.

When a person takes several drugs at the same time—often prescribed by different doctors—some drugs may interact, causing harmful effects that kill more than 200,000 people every year in the United States alone. Cascading drug therapy is another problem. A drug causes side effects, so the doctor adds a second drug to mask its effects. But that drug has side effects too, so the doctor adds a third drug, and so forth.

Americans use about 80 percent of the global supply of opioids (narcotic drugs similar to morphine) and 99 percent of all hydrocodone (Vicodin® and other trade names). The percentage of people who report using marijuana is higher in the United States and New Zealand (both 42 percent) than in any other country. As of 2008, the U.S. also leads the world in reported cocaine use (16 percent).

About one-third of all adults in the world use tobacco. In the U.S., about 10 percent of college graduates smoke, as compared to 35 percent of those with a high school education or less. This might mean that well-educated people have read more about cancer, or that they are more likely to work in smoke-free offices, or that they have other things to do. About 443,000 Americans die prematurely each year as a result of smoking or secondhand smoke. In a delightfully candid 2001 report, cigarette manufacturer Philip Morris concluded that smokers' premature deaths had saved the Czech Republic over $30 million per year in healthcare and pension costs.

According to a 2006 study, illegal marijuana production in the United States has a total cash value of $35.8 billion—greater than the combined value of corn ($23.3 billion) and wheat ($7.5 billion).

Discussion

Pharmacology and brewing are among the oldest professions. Archaeologists report that people in southeast Asia have used betel nuts for at least 10,000 years. In about 2,500 B.C., a wave of technologically advanced immigrants (whom we now call the Beaker People) brought the first alcoholic beverage to the British Isles, a fermented honey product called mead.

Chemical recreation is one of many ways to change the environment or our experience of it. We will skip the usual discussion of marijuana, cocaine, and heroin, because most readers probably know what they are. A less familiar example is the plant *Salvia divinorum* (Sally D or Holy Smoke). The Mazatec people of Oaxaca, Mexico have used this plant for centuries, but it was not widely known in the United States before about 1990. Reports indicate that it is not a "party drug" but a mild hallucinogen that tends to promote solitary meditation. Some people find the effects interesting, while others report no effects other than nausea.

Case Study 6-8: Restless Admen

Restless legs syndrome is an example of a common, minor discomfort that can become devastating if you focus on it hard enough and long enough. In 2005, the U.S. Food and Drug Administration approved the first drug designed to treat this condition, and an aggressive TV advertising campaign followed. By 2007, doctors had written 4.4 million prescriptions for this drug, at a reported cost of $491 million. In 2008, however, the FDA approved a less expensive generic version. Since this move would reduce sales of the brand-name product, the manufacturer stopped pushing it, and the restless legs "crisis" was over.

Just as hallucinogens promote unique experiences, other drugs offer the opportunity to feel and act like everyone else. People who think they are too fat, or too tired, or too shy, or too happy, or not happy enough, can find products to transform them quickly and without personal effort or insight. There are drugs to adjust even minor variances, such as short eyelashes, twitching legs (Case Study 6-8), poor study habits, a tendency to worry about bills, or a tendency to develop heartburn after overeating. Antidepressants are now the most frequently prescribed drugs in the United States, despite mounting evidence that these drugs may be no more effective than placebos. In 2005, U.S. doctors wrote about 118 million antidepressant prescriptions.

Popular Culture

Ira Levin's 1970 novel *This Perfect Day* depicts a future world in which every person receives regular injections of drugs that enforce tranquility and prevent disease. Everyone has a job and a home, and no one is unhappy by today's standards. But—read the book.

Examples of motion pictures about substance abuse include *The Lost Weekend* (1945), *The Man with the Golden Arm* (1955), *Days of Wine and Roses* (1962), *Altered States* (1980), *Dogs in Space* (1986), *Drugstore Cowboy* (1989), *The Doors* (1991), *Trainspotting* (1996), *Blow* (2001), and *Across the Universe* (2007). The pharmaceutical industry was a major player in the 1993 movie *The Fugitive*.

Is it true that red wine negates any cancer risk caused by smoking? No. Studies have shown that wine has certain health benefits (and risks), but does this mean that a person who smokes heavily should also start drinking heavily? That probably depends on overall personal goals.

Can the antidepressant drug clomipramine really cause people to have orgasms? Yes, but there are better ways to do it.

Do some people really take a recreational drug that is made from fermented sewage? Yes, apparently sewage is the source of a hallucinogenic inhalant in some parts of Africa. There appears to be no limit on what people will inhale or ingest to get high.

Point/Counterpoint

P: Drugs are good. Some of them take away pain and sorrow and anxiety, and some of them make us see pretty colors. Some of them fight infection. What's so bad about feeling good?

CP: Drugs are bad. They mask symptoms, deaden emotions, create antibiotic-resistant superbugs, and promote dependence on the pharmaceutical industry. What's so bad about feeling bad?

The Future

It is safe to predict that the war on drugs will never be "won," if winning means the eradication of all potentially dangerous recreational substances or overprescribed pharmaceuticals. At some point, however, any war may need restrategization. A useful first step might be to determine why the demand for drugs is greater in some countries than in others.

References and Recommended Reading

Abramson, J. 2004. *Overdosed America: The Broken Promise of American Medicine*. New York: HarperCollins, 352 pp.

Aleccia, J. "Without Ads, Restless Legs May Take a Hike." MSNBC.com, 14 May 2008.

"Caffeine Helps Muscles After Exercise." United Press International, 1 July 2008.

Chowdhury, P., et al. "Surveillance of Certain Health Behaviors among States and Selected Local Areas—United States, 2005." *MMWR Surveillance Summaries*, Vol. 56, 2007, pp. 1–164.

Cleck, J. N., and J. A. Blendy. "Making a Bad Thing Worse: Adverse Effects of Stress on Drug Addiction." *Journal of Clinical Investigation*, Vol. 118, 2008, pp. 454–461.

Coghlan, A. "Bipolar Children—Is the US Overdiagnosing?" *New Scientist*, 16 May 2007.

Cunningham, N. "Hallucinogenic Plants of Abuse." *Emergency Medicine Australasia*, Vol. 20, 2008, pp. 167–174.

Degenhardt, L., et al. "Toward a Global View of Alcohol, Tobacco, Cannabis, and Cocaine Use: Findings from the WHO World Mental Health Surveys." *PLoS Medicine*, Vol. 5, 2008, p. e141.

Dobson, R. "Being Unique Is a Disorder." *Sydney Morning Herald*, 4 September 2008.

"Eyelash-Boosting Drug Gets FDA Panel Approval." Associated Press, 5 December 2008.

Friedman, H., et al. "Addictive Drugs and their Relationship with Infectious Diseases." *FEMS Immunology and Medical Microbiology*, Vol. 47, 2006, pp. 330–342.

Gellad, Z. F., and K. W. Lyles. "Direct-to-Consumer Advertising of Pharmaceuticals." *American Journal of Medicine*, Vol. 120, 2007, pp. 475–480.

Golway, T. "A Cure for Everything: Just Pop a Few Pills." *New York Observer*, 27 May 2001.

Healy, J. "Miracle Drug Cures Everything." *Student BMJ*, Vol. 16, 2006, p. 174.

Henningfield, J. E., et al. "Tobacco Industry Litigation Position on Addiction: Continued Dependence on Past Views." *Tobacco Control*, December 2006.

Ioannidis, J. "Effectiveness of Antidepressants: An Evidence Myth Constructed from a Thousand Randomized Trials?" *Philosophy, Ethics, and Humanities in Medicine*, 27 May 2008.

Jureidini, J., and A. Tonkin. "Overuse of Antidepressant Drugs for the Treatment of Depression." *CNS Drugs*, Vol. 20, 2006, pp. 623–632.

Khiabani, H. Z., et al. "Cannabis Affects Driving Skills." *Tisskrift for den Norske Laegeforening*, Vol. 127, 2007, pp. 583–584. [Norwegian]

Kleinke, J. D. "Evidence-Based Medicine: Health Care's Next Holy War." *Journal of Managed Care Pharmacy*, Vol. 11 (Suppl. 4), 2005, pp. S3–6.

Lane, C. 2007. *Shyness: How Normal Behavior Became a Sickness*. New Haven, CT: Yale University Press.

Lee, S., and A. Mysyk. "The Medicalization of Compulsive Buying." *Social Science and Medicine*, Vol. 58, 2004, pp. 1709–1718.

Lynch, S. N. "An American Pastime: Smoking Pot." *Time*, 15 July 2008.
Manchikanti, L. "National Drug Control Policy and Prescription Drug Abuse: Facts and Fallacies." *Pain Physician*, Vol. 10, 2007, pp. 399–424.
Paccaud, F. "Implausible Diseases and Public Health." *European Journal of Public Health*, Vol. 17, 2007, p. 410.
Perrone, M. "Drugmakers' Push Boosts 'Murky' Ailment." Associated Press, 8 February 2009.
"The Perverse Prosperity of the Tobacco Industry." *Lancet*, Vol. 371, 2008, p. 276.
Pettus, A. "Psychiatry by Prescription." *Harvard Magazine*, July-August 2006, p. 38 ff.
Planta, M. D. "The Role of Poverty in Antimicrobial Resistance." *Journal of the American Board of Family Medicine*, Vol. 20, 2007, pp. 533–539.
Rising, K., et al. "Reporting Bias in Drug Trials Submitted to the Food and Drug Administration: Review of Publication and Presentation." *PLoS Medicine*, 25 November 2008.
Roth, B. L., et al. "Salvinorin A: A Potent Naturally Occurring Nonnitrogenous Kappa Opioid Selective Agonist." *Proceedings of the National Academy of Sciences* (U.S.), Vol. 99, 2002, pp. 11934–11939.
Saah, T. "The Evolutionary Origins and Significance of Drug Addiction." *Harm Reduction Journal*, 29 June 2005.
Shankar, P. R., and P. Subish. "Disease Mongering." *Singapore Medical Journal*, Vol. 48, 2007, pp. 275–280.
Slama, K. "Global Perspectives on Tobacco Control, Part I: The Global State of the Tobacco Epidemic." *International Journal of Tuberculosis and Lung Disease*, Vol. 12, 2008, pp. 3–7.
Smardon, R. "I'd Rather Not Take Prozac: Stigma and Commodification in Antidepressant Consumer Narratives." *Health* (London), Vol. 12, 2008, pp. 67–86.
"Smoking Kills 443,000 Prematurely in U.S." United Press International, 13 November 2008.
Spina, S. P., and A. Taddei. "Teenagers with Jimson Weed (*Datura stramonium*) Poisoning." *Canadian Journal of Emergency Medical Care*, Vol. 9, 2007, pp. 467–468.
Stephey, M. J. "Restless Legs Get Respect." *Time*, 19 July 2007.
Szalavitz, M. "Drugs in Portugal: Did Decriminalization Work?" *Time*, 26 April 2009.
Van Zee, A. "The Promotion and Marketing of OxyContin: Commercial Triumph, Public Health Tragedy." *American Journal of Public Health*, 17 September 2008.
Wanzor, L. "A Drug for Everything?" *New York Times*, 11 June 2002.
Zalesky, C. D. "Pharmaceutical Marketing Practices: Balancing Public Health and Law Enforcement Interests; Moving Beyond Regulation-through-Litigation." *Journal of Health Law*, Vol. 39, 2006, pp. 235–264.

TOO MUCH UV: STRATOSPHERIC OZONE DEPLETION REVISITED

Summary of Threat

The amount of ozone in the Earth's stratosphere has declined by about 4 percent every year since the late 1970s, largely as a result of chlorofluorocarbons (CFCs) and other chemicals that humans release into the atmosphere. Seasonal ozone "holes" have also appeared over the polar regions. Regulations that limit the use of these chemicals may be slowing or reversing the process.

So What?

The ozone layer reduces the amount of ultraviolet (UV) light that reaches the Earth. Besides promoting skin cancer, UV appears to increase the risk for certain infectious diseases by depressing the immune system. Animal experiments and epidemiological data show that UV may increase susceptibility to malaria, hepatitis B, herpes simplex, cutaneous leishmaniasis, and various bacterial skin infections. Increased UV also reduces yield of rice and other crops.

But UV protection also has a downside, because certain UV wavelengths cause the human body to synthesize vitamin D. Publicity about skin cancer has persuaded many people to avoid sunlight or use sunblock, and as a result, vitamin D deficiency is on the rise. A lack of vitamin D not only affects bone development, but also increases the risk of tuberculosis and other infectious diseases.

Can't Scientists Do Something About It?

They already have, and world leaders eventually listened. Researchers at the University of California published the first major warning about the role of CFCs in ozone depletion in 1974. In 1985, after the now-famous ozone hole appeared over Antarctica, 20 nations signed an agreement called the Vienna Convention, which established a framework for further regulation of ozone-depleting chemicals. Two years later, the Montreal Protocol was opened for signature. This treaty called for strict limits on the production and use of ozone-depleting chemicals, including a schedule for phasing out the use of CFCs. It became effective in 1989 and was amended in 1990 and 1992. As of 2009, 194 nations have signed (Case Study 6-9).

Despite some problems in enforcement of regulations and interpretation of data, the rate of ozone depletion appears to be slowing at last. A 2006 United Nations study predicted recovery by about 2065. However, if global warming interferes with recovery of the ozone layer, reversal of the damage may take a century or longer. The Antarctic ozone hole in 2006 matched the all-time record, and the 2008 ozone hole was the fifth largest on record.

Case Study 6-9: Mexico and the Ozone Layer

In 2005, the Mexican government announced that its largest chemical-manufacturing facility had stopped production of ozone-depleting chlorofluorocarbons (CFCs) 4 years ahead of the schedule established by the Montreal Protocol. The Quimibásicos plant in Monterrey, which had previously manufactured about 60 percent of all CFCs in Latin America, was able to reconfigure its equipment to produce a more environmentally friendly hydrogen-based refrigerant called HCFC-22 (Freon®22), thanks to a $30 million grant from the United Nations–sponsored Multilateral Fund for the Montreal Protocol. In addition, the Mexican government established several recycling centers to encourage residents to dispose of old refrigerators that contain CFCs. As of 2008, HCFC-22 is the world's most widely used refrigerant, but its byproducts have a slight impact on the ozone layer, so it is scheduled for phaseout by the year 2030.

The Numbers

In 2008, scientists developed a model to predict how the ozone layer would respond to a nuclear conflict involving 100 Hiroshima-sized bombs (totaling 1.5 megatons) detonated in the northern subtropics. The resulting fires would send an estimated 5 million metric tons of soot into the stratosphere, increasing chemical reactions that destroy ozone. Global ozone loss would average more than 20 percent, with losses of 50 to 70 percent at northern high latitudes, and 25 to 45 percent at middle latitudes. These losses would persist for 5 to 10 years, causing severe damage to human health and natural ecosystems. (When the ozone layer above Chile was reduced by 25 to 50 percent in 1986–1998, the real-life incidence of skin cancer increased by 66 percent.)

This scenario assumes a war that would use only 0.03 percent of the world's nuclear arsenal. A bigger war would, of course, be worse.

Discussion

Ozone depletion and climate change are not the same, but saturation media coverage of both topics has led to some confusion. Many of the same chemicals that deplete the ozone layer are also greenhouse gases that trap heat near the Earth. Thus, it might appear that the same measures should prevent both global warming and ozone depletion, but it isn't that simple. Most greenhouse gases, such as carbon dioxide and methane, are not classified as ozone-depleting substances under the Montreal Protocol. (The Kyoto Protocol is a separate international agreement that was intended to limit greenhouse gas emissions, as discussed earlier in the chapter.)

Short-wave UV, called UVB, causes sunburn and several forms of skin cancer. With enough exposure, UVB may also suppress the immune system. Longer-wave ultraviolet light, called UVA, penetrates the skin deeply and contributes to aging. It also increases the cancer-causing effect of UVB but is less likely to cause sunburn. Older sunscreen and sunblock products mainly served as protection from UVB, but since UVA is also harmful, it is important to read the label. An effective product should provide both UVA and UVB protection.

CFCs, reactive nitrogen, and other industrial chemicals are not the only cause of ozone depletion. The asteroid that struck the Earth 65 million years ago may have triggered a sequence of events that destroyed the ozone layer. More recently, in 1859, a large solar flare blasted the Earth and damaged the ozone layer, as evidenced by ozone-related nitrates preserved in Greenland ice cores. Observers on the ground saw the flare, and the resulting geomagnetic storm caused telegraph lines to work temporarily without being connected to batteries.

In 1994, the United Nations declared 16 September of each year as the International Day for the Preservation of the Ozone Layer. The date is appropriate, since the Antarctic ozone hole usually reaches its maximum in September.

Popular Culture

The 1991 motion picture *Highlander 2: The Quickening* depicts a nightmare future in which sunblock is obsolete. After the collapse of the ozone layer, bad immortals build an ultraviolet shield around the Earth, but good immortals tear it down again.

In the 1977 novel *Lucifer's Hammer,* an environmental activist mentions the ozone layer controversy, and the hero threatens to vomit if she doesn't shut up. The issue was widely misunderstood in that era; many people dismissed research findings as a left-wing conspiracy, while others claimed that the ozone hole would allow our oxygen to drain away into outer space.

The 2003 Spanish novel *La Muerte Rosa* ("The Pink Death"), by Chilean journalist Raul Sohr, is about an expedition to Patagonia to study ozone depletion and its effects on vegetation.

According to an urban legend, author John Howard Griffin (1920–1980) died of skin cancer after using ultraviolet lamps 15 hours per day to darken his skin. His biographers deny that he had cancer, although he suffered from chronic poor health, including malaria, heart disease, and osteomyelitis and died from complications of diabetes.

Point/Counterpoint

P: Ozone depletion was a government hoax. Nobody talks about it these days, except a few aging hippies and scientists who want grants. People stopped using spray cans, but I know somebody who got skin cancer last year anyway, and the ozone hole is still there. So it was all for nothing.

CP: Ozone depletion is real, and it has not gone away. The media just got tired of it. The ozone hole may be getting smaller, but we won't know for certain until more time has passed. It's lucky we stopped using CFCs. Skin cancer will always exist, because more than one thing causes it.

The Future

The future of the ozone layer is impossible to predict, but it is a safe bet that environmental laws will continue to interact in ways that their proponents do not anticipate. Ozone protection, for example, is designed to reduce the health effects associated with UV exposure. But a new, unrelated U.S. law will phase out incandescent light bulbs (as an energy-saving measure) between 2012 and 2014, requiring people to buy more fluorescent bulbs—which emit ultraviolet light. In theory, the shift to fluorescent light could exacerbate some autoimmune diseases, such as lupus, while reducing the incidence of vitamin D deficiency among indoor workers. It's hard to predict how this development will affect sunblock use, but it may increase demand for coffee, because a recent study shows that caffeine can reduce eye damage from UV exposure.

References and Recommended Reading

Aneja, V. P., et al. "Ammonia Assessment from Agriculture: U.S. Status and Needs." *Journal of Environmental Quality*, Vol. 37, 2008, pp. 515–520.

"China Shuts Five CFC Facilities to Stop Harm to Ozone Layer." Press Trust of India Ltd., 4 July 2007.

Cormier, Z. "The Hole Truth: A Success Story and Cautionary Tale." *This Magazine*, 1 November 2007.

de Gruijl, F. R. "UV-induced Immunosuppression in the Balance." *Photochemistry and Photobiology*, Vol. 84, 2008, pp. 2–9.

Diffey, B. "Climate Change, Ozone Depletion and the Impact on Ultraviolet Exposure of Human Skin." *Physics in Medicine and Biology*, Vol. 49, 2004, pp. R1–11.

Godar, D. E., et al. "Increased UVA Exposures and Decreased Cutaneous D(3) levels may be responsible for the increasing incidence of melanoma." *Medical Hypotheses*, 18 January 2009.

Hidema, J., and T. Kumagai. "Sensitivity of Rice to Ultraviolet-B Radiation." *Annals of Botany*, Vol. 97, 2006, pp. 933–942.

Langman, J. "Under the Hole in the Sky." *Newsweek International*, 3 December 2001.

McClam, E. "Scientists Add Up to 15 Years Recovery Time for Ozone Layer: UN Study." The Americas Intelligence Wire, 18 August 2006.

"Mexico Halts Production of Chemicals Harmful to Ozone Layer." *SourceMex Economic News & Analysis on Mexico*, 14 September 2005.

Mills, M. J., et al. "Massive Global Ozone Loss Predicted Following Regional Nuclear Conflict." *Proceedings of the National Academy of Sciences* (USA), Vol. 105, 2008, pp. 5307–5312.

Norval, M., et al. "The Effects on Human Health from Stratospheric Ozone Depletion and its Interactions with Climate Change." *Photochemical and Photobiological Sciences*, Vol. 6, 2007, pp. 232–251.

O'Hanlon, L. "Sun Burp Blasted Ozone Layer in 1859." *Discovery News*, 27 March 2007.

"Ozone-Friendly Replacement Gases Will Be Disastrous for Global Warming." *Space Daily*, 11 July 2008.

Rotman, D. "Remembering the Montreal Protocol." *Technology Review*, 1 January 2007.

Rowland, F. S. "Stratospheric Ozone Depletion." *Philosophical Transactions of the Royal Society of London, Series B, Biological Sciences*, Vol. 361, 2006, pp. 769–790.
Sanchez, C. F. "The Relationship between the Ozone Layer and Skin Cancer." *Revista Médica de Chile*, Vol. 134, 2006, pp. 1185–1190. [Spanish]
Shinkle, P. "Sara Lee to Pay Millions in Ozone-Layer Suit." Knight Ridder/Tribune Business News, 1 August 2003.
Sleijffers, A., et al. "Ultraviolet Light and Resistance to Infectious Diseases." *Journal of Immunotoxicology*, Vol. 1, 2004, pp. 3–14.
"Study: Ozone Restoration to Have Impact." United Press International, 25 June 2008.
Ungar, S. "Knowledge, Ignorance and the Popular Culture: Climate Change versus the Ozone Hole." *Public Understanding of Science*, Vol. 9, 2000, pp. 297–312.
van der Leun, J. C. "The Ozone Layer." *Photodermatology, Photoimmunology, and Photomedicine*, Vol. 20, 2004, pp. 159–162.
van der Leun, J. C., et al. "Climate Change and Human Skin Cancer." *Photochemical and Photobiological Sciences*, Vol. 7, 2008, pp. 730–733.
Varma, S. D., et al. "UV-B-induced Damage to the Lens In Vitro: Prevention by Caffeine." *Journal of Ocular Pharmacology and Therapeutics*, Vol. 24, 2008, pp. 439–444.
Velders, G. J., et al. "The Importance of the Montreal Protocol in Protecting Climate." *Proceedings of the National Academy of Sciences* (U.S.), Vol. 104, 2007, pp. 4814–4819.
von Hobe, M. "Revisiting Ozone Depletion". *Science*, Vol. 318, 2007, pp. 1878–1879.
Weatherhead, E. C., and S. B. Andersen. "The Search for Signs of Recovery of the Ozone Layer." *Nature*, Vol. 441, 2006, pp. 39–45.
Yamamoto, K., et al. "UV-B Irradiation Increases Susceptibility of Mice to Malarial Infection." *Infection and Immunity*, Vol. 68, 2000, pp. 2353–2355.

TOO MANY COOKS: ENVIRONMENTAL MANAGEMENT ISSUES

Summary of Threat

As the human population grows and the suburbs expand into former wilderness areas, wild animals and other natural resources increasingly require management in one form or another. At the same time, the popularity of charismatic megafauna has driven deer, bears, and other game species to record numbers. Recent consequences include habitat loss, Lyme disease outbreaks, and a $1 billion annual tab for vehicle collisions with deer.

So What?

There is an old saying to the effect that too many cooks spoil the broth. In other words, when many people and agencies all work independently toward similar goals but without adequate coordination, negative consequences are likely. Environmental management is a prime example, and the outcome is far more important than soup.

Wherever humans "manage" natural resources, diseases find new hosts. As discussed later, there is strong evidence that the deer population explosion in the northeastern United States has fueled the emergence of Lyme disease. Record numbers of wild geese in Canada may increase opportunities for avian influenza outbreaks. Wetland restoration has benefited West Nile virus, and wildlife translocation has expanded the range of rabies. The thousands of people injured in vehicle collisions with large animals every year contribute to the high cost of insurance and hospital overcrowding. Such problems are not unique to developed nations; Ebola and other hemorrhagic fever outbreaks seem related to human disturbance of tropical forests.

Can't Scientists Do Something About It?

Fundamental differences in philosophy often separate scientists, activists, hunters, and government resource agencies. Scientists typically want to protect and study ecosystems, while animal rights activists seek to protect individual animals, and hunters want to harvest their share of the Earth's resources. Government agencies are responsible for carrying out the will of the people, but under these conditions, compromise is often difficult. If deer or wild pigs overrun a community, is it better to shoot them, feed them, give them vasectomies, or allow them to be hit by cars? Unfortunately, this is not a joke.

The Numbers

In the United States in the early twenty-first century, there are about 1.8 million vehicle-deer collisions every year, resulting in the deaths of 150 to 200 people and over $1.1 billion in vehicle damage (Figure 6.5). Estimates of related medical and hospital costs range from $125 million to $150 million per year, not counting lost lives or collisions with other "managed" species, such as wild pigs, bears, and moose.

In 2008, there were approximately 30 million deer in the United States—at least as many as there were when the first Europeans arrived in North America. But only a fraction of the original habitat remains, much of it now fragmented by highways and suburbs. Deer overpopulation is worst in the northeast; California and Nevada actually report deer shortages. California had about 450,000 deer in 2008, 1 million fewer than in 1968, possibly due to protection of mountain lions (which eat deer). California also had about 150,000 deer hunters in 2008, of whom about 10 percent could reasonably expect to bag a deer. After Nevada's deer population fell from 240,000 in 1988 to 108,000 in 2008, wildlife officials decided to issue more mountain lion tags.

Goose statistics are also impressive. Thanks to limits on hunting, the North American population of greater snow goose increased from about 30,000 in 1965 to over 1 million in 2004 and is expected to exceed 2 million by 2010. These geese damage coastal marshes by stripping them of vegetation and also destroy winter crops in the mid-Atlantic area. Canada geese and other waterfowl have undergone a similar population explosion, and epidemiologists are concerned about the spread of diseases such as avian influenza.

Discussion

Who owns the Earth? We don't know, but the Department of Fish and Game owns California's wild pigs. For years, citizen groups have asked why state governments that issue hunting licenses and enforce wildlife regulations are not also responsible for the damage caused by their wildlife. The answer came in the spring of 2009, when the Monterey County Superior Court ordered the California Department of Transportation to pay $8.6 million to a motorcyclist who became disabled when his bike collided with one of many wild pigs on a highway. The argument was that the agency had sponsored an environmental restoration project that attracted large numbers of pigs, and then ignored the resulting road hazard.

(Factoid: An estimated 4 million non-native wild pigs roam the United States. They compete with native wildlife, destroy habitats, and transmit disease to domestic pigs. Wild pigs weigh up to 300 pounds, not counting outliers such as Hogzilla.)

Figure 6.5 Inept wildlife management policy has negative consequences for humans and animals alike. This photograph shows the aftermath of a 2005 vehicle collision with a deer, one of an estimated 1.8 million similar accidents that occur each year in the United States alone.

Source: Photo courtesy: Shannon Hicks/Newton Bee (CT).

Case Study 6-10: Deer Crossing

In 2007, the 287-acre campus of Goucher College in Baltimore, Maryland, had a deer population of 200 and a carrying capacity of 40. The deer were eating the foliage, spreading Lyme disease, and colliding with vehicles. Finally, the college administration hired bow hunters to reduce the deer population by about one-third. This action drew public outrage, but the alternatives were worse: inject the deer with birth-control drugs, release mountain lions or other predators to eat the deer, enlarge the campus, surround it with an 8-foot fence, or turn it into a deer preserve and tell the students to stay home. Relocating the deer may sound good, but in practice it's hard to find a suitable destination and get permission. (It is also expensive and hard on the deer.)

Environmental management is a complicated business, and problems are inevitable. In 1987, federal and state agencies established a program to augment wild turkey populations by breeding turkeys in pens and releasing them into the wild. As a result, infectious diseases spread rapidly among the crowded pen-raised birds and infected the wild populations. Wildlife rescue centers sometimes inadvertently cause similar problems when they expose wild animals to disease while nursing them back to health. The use of feeders and guzzlers, which concentrate game animals near food and water sources, also can increase disease transmission.

Another tricky management issue involves tropical fruit bats, believed to be the principal reservoir for some of the world's worst human diseases, such as SARS, Nipah virus, and Ebola and Marburg hemorrhagic fevers. It would be wrong to blame the bats, but what is the best solution? In the past, some countries tried to exterminate fruit bats—not because of these diseases, which had not been discovered yet, but to reduce competition for fruit. Right or wrong, these efforts ended because of collateral damage. Fruit bats pollinate some native plants; also, the hunters killed many insectivorous bats by mistake, and without these predators, mosquito populations exploded.

Another frustrating issue is the deer population explosion in the northeastern United States (Case Study 6-10) and its well-documented relationship to the rise of Lyme disease. Most people like to look at deer, and hunters also like to shoot them, so the idea of simply reducing the number of deer is unpopular. When asked about this issue in 2008, a Humane Society representative told the author: "Lyme disease has nothing to do with deer. It's caused by a little germ." Well, the last part is true—but like the rest of us, the little germ needs to live somewhere. The agent of Lyme disease happens to live in deer ticks (Figure 6.6), which spend part of their life cycle on deer. The ticks that transmit Lyme disease to humans are immature deer ticks that live temporarily on mice, but those ticks eventually grow up and move to deer.

Field studies have shown that reducing deer populations also reduces the incidence of Lyme disease. Many wildlife biologists suspect—but have not yet proven—that abnormally high deer population densities may also promote the spread of a deadly prion disease called chronic wasting disease (CWD), the deer equivalent of mad cow disease (Chapter 4). Hunters avoid herds infected with this disease because the meat is considered unfit for human consumption.

Popular Culture

In 2006, the following warning circulated on the Internet: "If someone comes up to your front door saying they are conducting a survey on deer ticks and asks you to take your clothes off and dance around with your arms up, do not do it. It is a scam." Duh.

Popular wisdom holds that a driver who is about to collide with a deer should hit the accelerator to raise the front end of the car and prevent the deer from going through the windshield.

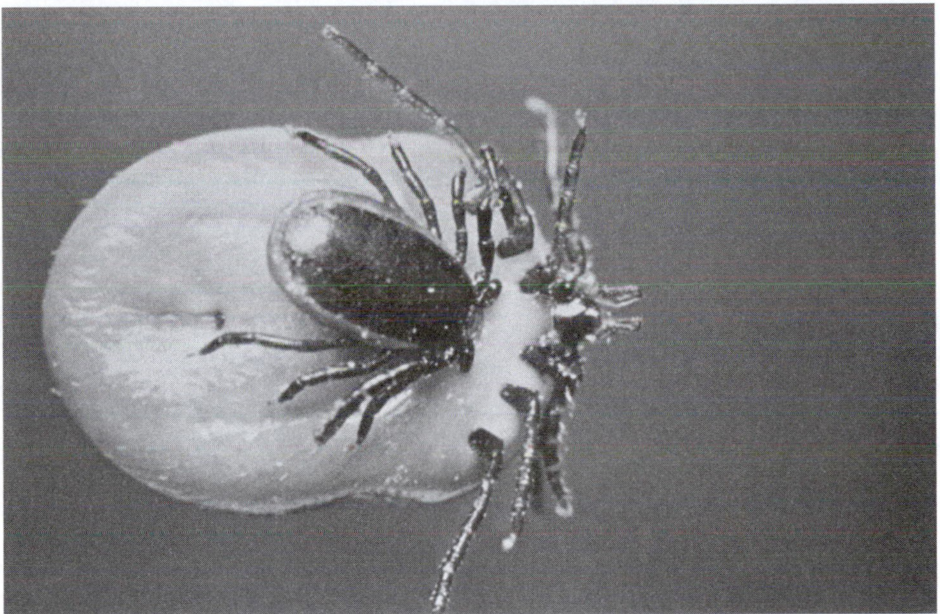

Figure 6.6 The recent Lyme disease epidemic has resulted partly from the deer population explosion in the northeastern United States, because the ticks that serve as vectors spend part of their life cycle on deer.

Above: Immature *Ixodes* tick.

Source: U.S. National Institute of Allergy and Infectious Diseases.

Below: Adult male and female *Ixodes ricinus* ticks mating.

Source: U.S. Centers for Disease Control and Prevention, Public Health Image Library.

As far as we can tell, this is wrong. Hit the brake instead, and try to miss the deer. Then write to your Member of Congress and raise hell about road conditions.

The 1995 motion picture *Outbreak* makes an interesting comment on the real-life association between hemorrhagic fever outbreaks and ecological perturbations: "He [a local healer] believes that the gods were awoken from their sleep by men cutting down the trees where no man should be. And the gods got angry."

In the 1994 motion picture *Ticks*, a marijuana farmer tries to enhance his crop by releasing chemicals with an unexpected side effect: local ticks grow to the size of dinner plates. They kill the farmer and then go after a group of troubled youths on an outing in the woods. But one of the kids is on steroids, and the tick that bites him ends up five feet long.

Point/Counterpoint

P: Resource agencies are not in the business of pure research. They answer to taxpayers, and millions of those taxpayers like to hunt. To keep those hunters happy, we need millions of deer and bears and wild pigs and turkeys, regardless of environmental or social impacts. Hunters have been around for a long time, and they aren't going away.

CP: Hunters have rights, but so do taxpayers with other agendas. Some want to preserve the natural environment in a baseline condition, whereas others see animals as beloved pets. Still others just want to drive to work without running into a deer. But hunting is old, and these concerns are new, so they might take a while to catch on.

The Future

The future of wildlife management is by no means hopeless. In the mid-1980s, Pinnacles National Monument in California had a large population of feral pigs that were trashing native wildlife habitats and eating enough acorns to halt regeneration of the forest. But Pinnacles is a unique place, and its advocates were willing to spend millions of dollars and ten years building a 30-mile pig fence around the perimeter of the park. By 2006, Pinnacles was officially pig-free.

References and Recommended Reading

Atzert, S., et al. "The Overpopulation of Two Native Species." Pleasantville, NJ: U.S. Fish and Wildlife Service, *Field Notes*, Spring 2004.

Bengis, R. G., et al. "The Role of Wildlife in Emerging and Re-emerging Diseases." *Revue Scientifique et Technique*, Vol. 23, 2004, pp. 497–511.

Bernstein-Wax, J. "Menacing Pigs Disappear from Hillsborough." *San Jose Mercury-News*, 12 October 2008.

Bohm, M., et al. "Wild Deer as a Source of Infection for Livestock and Humans in the UK." *Veterinary Journal*, Vol. 174, 2007, pp. 260–276.

Bradley, C. A., and S. Altizer. "Urbanization and the Ecology of Wildlife Diseases." *Trends in Ecology and Evolution*, Vol. 22, 2007, pp. 95–102.

Brewster, D. "Environmental Management for Vector Control: Is It Worth a Dam if It Worsens Malaria?" *British Medical Journal*, Vol. 319, 1999, p. 652.

Canning, A., and L. Ferran. "Targeting Deer in the Backyard." ABC News, 30 July 2008.

Charlton, K. G., et al. "Antibodies to Selected Disease Agents in Translocated Wild Turkeys in California." *Journal of Wildlife Diseases*, Vol. 36, 2000, pp. 161–164.

Chipman, R., et al. "Downside Risk of Wildlife Translocation." *Developments in Biologicals*, Vol. 131, 2008, pp. 223–232.

"Contractors Make a Killing Picking Up Deer Carcasses from Highways." Associated Press, 23 July 2000.
Daszak, P., et al. "Conservation Medicine and a New Agenda for Emerging Diseases." *Annals of the New York Academy of Sciences*, Vol. 1026, 2004, pp. 1–11.
"Deadly Animal-Vehicle Crashes on Rise; Sprawl Cited as Factor." Associated Press, 30 October 2008.
Devito, E. "Drastic Deer Damage Requires Drastic Deer Reduction." *New York Times*, 21 April 2008.
"Eradication of Feral Swine Priority." *Iowa Farmer Today*, 11 April 2007.
Esch, M. "It's Boom Time for Bears." Associated Press, 18 June 2009.
"Flu Blame Game." *New Scientist*, 21 February 2004.
Genne, D. "Bad Bats?" *Revue Médicale Suisse*, Vol. 3, 2007, pp. 2273–2274, 2276–2277.
Griffith, M. "Nevada Plans More Lion Hunts in Effort to Save Deer." Associated Press, 15 February 2009.
"Gunmen Ravage Colony of Rare Fruit Bats in Cyprus." Reuters, 29 November 2007.
Halpin, K., et al. "Emerging Viruses: Coming In on a Wrinkled Wing and a Prayer." *Clinical Infectious Diseases*, Vol. 44, 2007, pp. 711–717.
Karesh, W. B., et al. "Wildlife Trade and Global Disease Emergence." *Emerging Infectious Diseases*, Vol. 11, 2005, pp. 1000–1002.
"Many Freed Boars in Britain Still at Large." United Press International, 1 January 2006.
"Maryland Highway Workers Recycle Deer Carcasses." Associated Press, 27 July 2008.
McCabe, M. "No One's Rooting for These Pigs: Pinnacles Finishing 30-Mile Hog-Proof Fence." *San Francisco Chronicle*, 1 December 2001.
McNamara, R. "Deer, Deer, Everywhere." CBSNews.com, 17 February 2004.
Merianos, A. "Surveillance and Response to Disease Emergence." *Current Topics in Microbiology and Immunology*, Vol. 315, 2007, pp. 477–509.
Olival, K. J., and P. Daszak. "The Ecology of Emerging Neurotropic Viruses." *Journal of Neurovirology*, Vol. 11, 2005, pp. 441–446.
Olsen, B., et al. "Global Patterns of Influenza A Virus in Wild Birds." *Science*, Vol. 312, 2006, pp. 384–388.
O'Shaughnessy, P. T. "Parachuting Cats and Crushed Eggs: The Controversy Over the Use of DDT to Control Malaria." *American Journal of Public Health*, 17 September 2008.
Pasick, J., et al. "Susceptibility of Canada Geese (*Branta canadensis*) to Highly Pathogenic Avian Influenza Virus (H5N1)." *Emerging Infectious Diseases*, Vol. 13, 2007, pp. 1821–1827.
Piesman, J. "Strategies for Reducing the Risk of Lyme Borreliosis in North America." *International Journal of Medical Microbiology*, Vol. 296 (Suppl. 40), 2006, pp. 17–22.
"Psycho Killer Raccoons Terrorize Olympia." Associated Press, 22 August 2006.
Reed, K. D., et al. "Birds, Migration and Emerging Zoonoses: West Nile Virus, Lyme Disease, Influenza A and Enteropathogens." *Clinical Medicine and Research*, Vol. 1, 2003, pp. 5–12.
Revkin, A. C. "Out of Control, Deer Send Ecosystem into Chaos." *New York Times*, 12 November 2002.
Robbins, J. "As Cars Hit More Animals on Roads, Toll Rises." *New York Times*, 22 December 2007.
University of Arizona. "Can We Restore Wetlands and Leave the Mosquitoes Out?" *ScienceDaily*, 27 May 2004.
Willott, E. "Restoring Nature, without Mosquitoes?" *Restoration Ecology*, June 2004.
Woodward, C. "A View of Wildlife in Which People Are the Victims." Associated Press, 11 January 2002.
Zavaleta, J. O., and P. A. Rossignol. "Community-Level Analysis of Risk of Vector-Borne Disease." *Transactions of the Royal Society of Tropical Medicine and Hygiene*, Vol. 98, 2004, pp. 610–618.

CONCLUSION

Most of the topics in this chapter fall under the rubric of environmental problems. In recent opinion polls, Americans report a high level of concern about the environment, combined with a sense of futility and a wish for more effective government regulation. Those who take any action usually report minor lifestyle changes, such as recycling aluminum cans or washing laundry in cold water. As of 2009, more than half would accept higher utility bills to reduce global warming. Yet the percentage of survey respondents who believe the environmental movement probably or definitely "did more harm than good" more than doubled between 1992 and 2007, from 14 percent to 31 percent.[6] The environmental movement is diverse, and its most vocal components may not represent the will of the majority.

But if all these environmental and social issues are interfering with the war on infectious disease, is there anything we can do to improve the outlook instead of making it worse?

6. Gallup data on Pollingreport.com, downloaded 2008.

7

Fighting Back

> Thus the highest form of generalship is to balk the enemy's plans; the next best is to prevent the junction of the enemy's forces; the next in order is to attack the enemy's army in the field; and the worst policy of all is to besiege walled cities.
>
> —Sun Tzu, *The Art of War* (undated, circa 400 B.C.)

So the world is full of biological threats, and some human actions and inactions are making the situation worse. What can we do about it? Everyone has heard about "the war on disease," but there are many different ways to fight a war.

Most readers would probably agree that the world needs at least some of the following: disease-resistant crops, improved mosquito control, better vaccines and antibiotics, safer pesticides, cleaner air and water, better science education, some way to balance healthcare costs with personal income, economic models that don't depend on growth, and more effective partnerships between public health and religious institutions—and, of course, a global population with the knowledge and the will to look beyond the present moment. These objectives, and others, sort nicely into the four categories anticipated by Sun Tzu, a military strategist who did not necessarily exist, but whose advice (see top of page) has found favor with generations of soldiers and executives:

1. Balking the enemy's plans
2. Preventing the junction of the enemy's forces
3. Attacking the enemy's army in the field
4. Besieging walled cities

But there is at least one major difference between a military campaign and the war on disease. In the latter, neither side can win or surrender. Diseases, parasites, and predators have always limited the sizes of animal and plant populations and thus maintained the diversity and stability of ecosystems. These limiting factors were not the enemy until humans came along and invented the concept.

Now, with nearly 7 billion people to feed and protect, we must disregard the rules of engagement and fight dirty. Maintaining vast monocultures of humans or cows or wheat, without interference from the normal checks and balances of nature, requires us to defeat this equilibrium by the continuous input of energy. Defeating nature itself may prove impossible, and surrender would cost billions of lives. The result is an uneasy stalemate that requires a substantial arsenal and unending commitment.

PART 1: BALKING THE ENEMY'S PLANS

Although a microbe or pest cannot be said to have "plans," we can anticipate its future actions to some extent by observing and then blocking the conditions associated with its dispersal or multiplication. Thus, if we can learn enough about the enemy in advance, we may often prevent the battle before it starts. This form of generalship may be the most effective in a literal war, but in the realm of public health it is also the most expensive, the most likely to miss the mark, and the least popular with taxpayers. It is often easier to anticipate the plans of human adversaries than the mysterious forces of nature.

Effort at this level might focus on anything from the design of health science education materials to computer modeling of weather conditions that preceded the Four Corners hantavirus outbreak. Architects who design hospitals, historians who study the Plague of Athens, agronomists who find better ways to feed the Third World, and green industries that fight global warming all contribute to the goal of predicting and ultimately preventing infectious disease outbreaks—in this context, balking the enemy's plans.

Note, however, that there is a big difference between preventing natural outbreaks and blocking the efforts of bioterrorists. In the latter case, we are back to second-guessing human adversaries, an issue beyond the scope of this book.

Health Education

Summary of Topic

In a perfect world, people would voluntarily manage their own healthcare, nutrition, and reproduction without pressure from anyone. People seeking health information would know how to distinguish between science and quackery. Education is probably the only long-term solution to overpopulation, malnutrition, sexually transmitted diseases, and a half-dozen other public health problems.

How Will This Help?

Studies show that the children of more highly educated (usually more affluent) parents tend to have fewer chronic infections, such as herpes and hepatitis. The reason is unknown, but these parents might give their children better food and teach better health habits, live in cleaner neighborhoods, or avoid smoking indoors. They might also be more likely to have insurance that pays for preventive care.

But more education does not automatically mean better health education (Case Study 7-1). Studies have also shown that parents with more education and higher incomes are *less* likely than others to vaccinate their children (Chapter 6). Education properly teaches people to question authority, and without an effective health curriculum, this is one possible consequence.

Discussion

Surveys show that limited health literacy has reached epidemic levels in the United States. Without a fact-based education, people are more likely to act on impulse, particularly in the area of family planning.

Throughout history, people have responded to unstable economic conditions in one of two ways. They have many babies, consciously or unconsciously hoping that some will survive; or else they think about it, wait for safer conditions, and have fewer babies, so that most are likely to survive. Ethologists and social scientists have compared these strategies to *r* and *K* selection in other species (Chapter 6). Nowadays, public assistance provides the safety net that often shifts the balance in favor of so-called *r* selection, but some countries have found that educating girls is an effective way to promote family planning and reduce the birthrate.

Case Study 7-1: Freedom of the Press

We cannot have a well-informed public without a well-informed press. Example: In 2008, a news release claimed that HIV "is constantly mutating in an effort to gain drug resistance."[1] If taken literally, this statement is nonsense. Is HIV, a virus with no brain, really making an "effort" to do anything? No, it is simply in the nature of this virus to mutate often, and some of the variations that arise at random enable it to resist specific drugs. A high mutation rate works to its advantage, but that has nothing to do with effort or intent. Scientists sometimes make informal statements like this as a kind of shorthand, assuming that other scientists know what they mean, and the reporter may have quoted a speaker out of context. But the statement as written not only confused some readers, but also provided fodder to the mill of every group that opposes the teaching of modern science. This is just one of hundreds of garbled statements that turn up in science press releases every year, and their cumulative impact on health literacy is unknown. For another example, see the preface (page xi).

[1]"Scientists Watch as HIV Matures" (UPI, 8 October 2008).

Popular Culture

After the Irish Famine of 1846–1849, the average age of marriage for Irish women increased from 20 to 30. Nobody made them wait; no law was passed. But on an island with limited resources and a recent disastrous population crash, this was exactly the right thing to do, because it resulted in smaller families. How did this change come about?

Some authors have produced explanations based on reproductive strategies of birds, but Famine survivors and their descendants tell a different tale. Those who survived in Ireland, and those who emigrated after watching their neighbors starve to death, tended to think twice before starting families of their own. Individuals made more or less conscious decisions to achieve financial stability first. Some survivors went farther and rejected everything Irish, from the Gaelic language to the custom of having many children. Young Irish women had grown up in a culture where men made all the decisions, and that gave the women a sense of security—until the Famine came, and the men could do nothing to stop it. Many women emigrated and took jobs, and those who were successful often postponed marriage or stayed single.

The Future

Education in general, and science and health education in particular, seem to be among the first things to go when a state or nation falls on hard economic times. More retired scientists might volunteer as teachers if given incentives beyond the rewards of altruism. (Tax breaks, for example.)

References and Recommended Reading

Amanna, I., and M. K. Slifka. "Public Fear of Vaccination: Separating Fact from Fiction." *Viral Immunology*, Vol. 18, 2005, pp. 307–315.

Connolly, C. “Some Abstinence Programs Misleading.” *Washington Post*, 2 December 2004.
Dowd, J. B., et al. “Early Origins of Health Disparities: Burden of Infection, Health, and Socioeconomic Status in U.S. Children.” *Social Science and Medicine*, 17 January 2009.
Henderson, L., et al. “Perceptions of Childhood Immunization in a Minority Community: Qualitative Study.” *Journal of the Royal Medical Society*, Vol. 101, 2008, pp. 244–251.
Kent, J. P. “On the Decline of Marriage in Rural Ireland 1851–1911: The Role of Ecological Constraints and/or Developing Philopatry.” *Population and Environment*, Vol. 23, 2002, pp. 525–540.
Kulczycki, A. “Ethics, Ideology, and Reproductive Health Policy in the United States.” *Studies in Family Planning*, Vol. 38, 2007, pp. 333–351.
Kurz-Milcke, E., et al. “Transparency in Risk Communication: Graphical and Analog Tools.” *Annals of the New York Academy of Sciences*, Vol. 1128, 2008, pp. 18–28.
Parker, R. M., et al. “Preparing for an Epidemic of Limited Health Literacy: Weathering the Perfect Storm.” *Journal of General Internal Medicine*, Vol. 23, 2008, pp. 1273–1276.
Pitrelli, N., and G. Sturloni. “Infectious Diseases and Governance of Global Risks through Public Communication and Participation.” *Annali dell’Istituto Superiore di Sanità*, Vol. 43, 2007, pp. 336–343.
Renner, B., et al. “Preventive Health Behavior and Adaptive Accuracy of Risk Perceptions.” *Risk Analysis*, Vol. 28, 2008, pp. 741–748.
Trenholm, C., et al. “Impacts of Abstinence Education on Teen Sexual Activity, Risk of Pregnancy, and Risk of Sexually Transmitted Diseases.” *Journal of Policy Analysis and Management*, Vol. 27, 2008, pp. 255–276.
Wood, S. “Opening Data to the World: Why Health Numbers Matter.” *Bulletin of the World Health Organization*, Vol. 85, 2007, p. 736.
Zhongdan, C., et al. “The 100% Condom Use Program: A Demonstration in Wuhan, China.” *Evaluation and Program Planning*, Vol. 31, 2008, pp. 10–21.

Better Food

Summary of Topic

“Better food” in this context does not mean gourmet cooking. It means a diverse, affordable diet that meets or exceeds human nutritional requirements and that does not promote obesity or other health problems. Long-term nutritional planning also includes genetic engineering and selective breeding to develop better crops, and seed banking to increase the stability of the food supply.

> **Case Study 7-2: The Doomsday Vault**
>
> Every now and then, someone executes a plan that truly surpasses the wildest dreams of science fiction. The Svalbard Global Seed Vault (Figure 7.1) is one of the best examples in recent memory. Constructed in 2008 on a Norwegian island north of the Arctic Circle, this underground repository was designed to protect humanity’s key food crops by storing seeds at low temperatures that will preserve them for centuries. If an epidemic, an asteroid collision, or other disaster destroys crops, this archived gene pool will enable farmers to start over—assuming, of course, that anyone survives the disaster itself, and has some means of transportation to Norway. During its first year of operation, the vault received more than 400,000 seed samples.

How Will This Help?

Good nutrition increases resistance to many infectious diseases, and bad nutrition has the opposite effect. Vitamin D deficiency is a risk factor for tuberculosis and influenza. Protein-energy malnutrition (PEM) in Third World children increases risk for HIV, tuberculosis, and malaria. Unfortunately, nutritional interventions in developing countries sometimes backfire. For example, giving anemic children an iron supplement can make them more susceptible to malaria, and introducing western foods can promote western diseases, such as atherosclerosis and diabetes.

Figure 7.1 Svalbard Global Seed Vault. Constructed in 2008 on a Norwegian island north of the Arctic Circle, this underground seed repository was designed to protect humanity's food crops by storing seed samples.

Source: Global Crop Diversity Trust.

Examples in previous chapters have made it clear that food production based on monoculture invites widespread famine. Improved agricultural practices and seed banking (Figure 7.1, Case Study 7-2) may alleviate this problem.

Discussion

In one of history's greatest ironies, western healthcare workers are trying to persuade more African women to breastfeed their children, something they did without instruction for thousands of years. The problem apparently started when European colonists discouraged public breastfeeding and conveyed the message that bottle-feeding was more modern. Then AIDS appeared in Africa, and with it the risk of HIV transmission from mother to child. But since clean water and refrigeration are luxuries in many parts of Africa, bottle-fed infants often develop severe diarrhea—not from the formula itself, but from bacterial contamination. As a result, even HIV-positive mothers are now urged to breastfeed.

Breast milk is clean and requires no preparation, and it contains most of the nutrients that a human infant needs. Long-term benefits include reduced risk for asthma, obesity, and some forms of childhood cancer. Nothing is perfect; a 2001 study showed that prolonged breastfeeding is also associated with reduced arterial function in adulthood, and exclusive breastfeeding with inadequate sunlight exposure can cause vitamin D deficiency and rickets. Breastfed infants may also develop deficiencies of iron and vitamins A and K, and if the mother is HIV-positive, the estimated risk of the child contracting HIV is about 1 percent per month of breastfeeding. But almost anything is better than risking a fatal case of diarrhea from contaminated formula.

Popular Culture

In 2008, a California couple became briefly famous after surviving for one month on a food budget of one dollar per day. It must have been a slow news day, because this is not difficult. Some of the cheapest foods, such as beans and rice, are also among the most nutritious (Figure 7.2). Beans with cornbread, beans with tortillas, and homemade ham bean soup are other delicious combinations. Yet this couple reported that the cheapest foods available were junk foods, such as candy

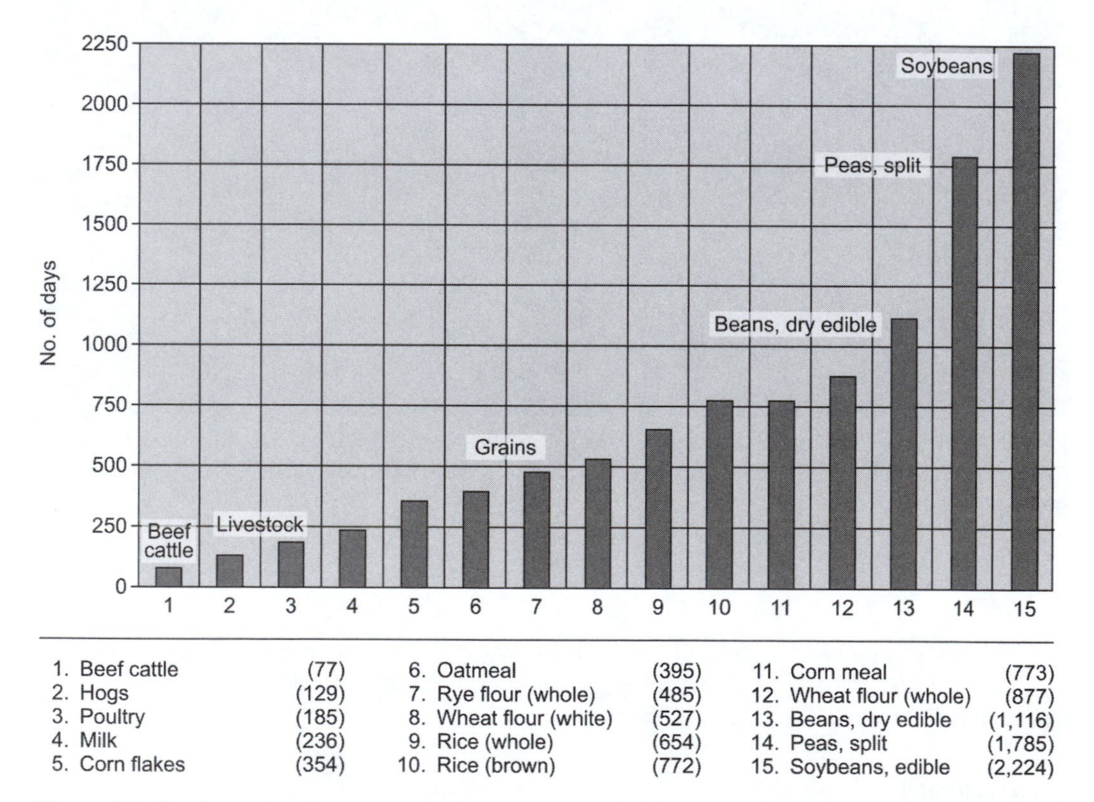

Figure 7.2 Graph showing how many days of adequate protein intake a male adult can obtain from 1 acre of selected crops or livestock. For example, a man can live for four months on one acre of beef cattle, or six years on one acre of soybeans.

Source: Adapted from L. H. Bean, "Closing the World's Nutritional Gap with Animal or Vegetable Protein" (*FAO Bulletin* 6, 1966).

and chips. The author also lives in California, but apparently shops in different stores—or else these people limited their choices to foods that could be eaten immediately without cooking. For many, the ability to prepare simple, cheap foods from scratch is a lost art.

A widespread urban legend claims that fresh fruits and vegetables are dangerous, because fruits contain cyanide and vegetables are contaminated with *E. coli*. There is no need to worry about cyanide from fruit, unless you eat huge quantities of apple seeds or the kernels inside peach pits. As for vegetables, it's a good idea to wash them (and cook them if appropriate), but any food—including fast food—can be contaminated.

In 2002, Subway Restaurant spokesman Jared Fogle appeared on a *South Park* episode in which he mentioned that "aides" helped him lose weight. He was referring to a personal trainer and dietician, but in the usual spirit of *South Park*, the people thought he meant AIDS and asked him to leave town.

The Future

The disparities between populations with too much food and those with too little will probably continue to grow. One trend that can benefit both is the rebirth of the victory garden—a home or neighborhood vegetable garden that is intended to save money, improve health, and promote civic responsibility. The original victory gardens became popular during the two World Wars as an opportunity for citizens to contribute to the war effort by reducing food shortages. The present war on diseases of excess and poverty is no less urgent.

References and Recommended Reading

Ambrus, J. L. "Nutrition and Infectious Diseases in Developing Countries and Problems of Acquired Immunodeficiency Syndrome." *Experimental Biology and Medicine*, Vol. 229, 2004, pp. 464–472.

Anderson, R., and C. Morgan. "A Life Saver Called 'Plumpynut.'" CBSNews.com, 22 June 2008.

Bhandari, N., et al. "Mainstreaming Nutrition into Maternal and Child Health Programmes: Scaling Up of Exclusive Breastfeeding." *Maternal and Child Nutrition*, Vol. 4, Suppl. 1, 2008, pp. 5–23.

"Breastmilk Can Save 1 Million Lives Yearly." United Press International, 5 August 2004.

Cavatassi, R., et al. "The Role of Crop Genetic Diversity in Coping with Agricultural Production Shocks: Insights from Eastern Ethiopia." UN Food and Agricultural Organization, ESA Working Paper No. 06–17, December 2006.

Dewey, K. G. "Increasing Iron Intake of Children through Complementary Foods." *Food and Nutrition Bulletin*, Vol. 28, Suppl. 4, 2007, pp. S595–609.

Drakesmith, H., and A. Prentice. "Viral Infection and Iron Metabolism." *Nature Reviews Microbiology*, Vol. 6, 2008, pp. 541–552.

Graham, J. P., et al. "Growth Promoting Antibiotics in Food Animal Production: An Economic Analysis." *Public Health Reports*, January–February 2007, pp. 79–87.

Greer, F. R. "Are Breastfed Infants Vitamin K Deficient?" *Advances in Experimental Medicine and Biology*, Vol. 501, 2001, pp. 391–395.

Greer, F. R. "Do Breastfed Infants Need Supplemental Vitamins?" *Pediatric Clinics of North America*, Vol. 48, 2001, pp. 415–423.

"IAEA Says Plant Breeding Technique Can Help Beat World Hunger." Press Release, International Atomic Energy Agency, 2 December 2008.

Lee, W. T., and J. Jiang. "The Resurgence of the Importance of Vitamin D in Bone Health." *Asia Pacific Journal of Clinical Nutrition*, Vol. 17, Suppl. 1, 2008, pp. 138–142.

Leeson, C. P. M., et al. "Duration of Breast Feeding and Arterial Distensibility in Early Adult Life: Population-Based Study." *BMJ*, Vol. 322, 2001, pp. 643–647.

Mellgren, D. "Norway to House Seeds in Doomsday Vault." Associated Press, 18 June 2006.

Normile, D. "Reinventing Rice to Feed the World." *Science*, Vol. 321, 2008, pp. 330–333.

"Question Formula-Feeding by AIDS Moms." *Science Online*, 23 July 2007.

Roth, D. E., et al. "Acute Lower Respiratory Infections in Childhood: Opportunities for Reducing the Global Burden through Nutritional Interventions." *Bulletin of the World Health Organization*, Vol. 86, 2008, pp. 356–364.
Schaible, U. E., and S. H. Kaufmann. "Malnutrition and Infection: Complex Mechanisms and Global Impacts." *PLoS Medicine*, Vol. 4, 2007, pp. 806–812.
Scrimgeour, A. G., and H. C. Lukaski. "Zinc and Diarrheal Disease: Current Status and Future Perspectives." *Current Opinion in Clinical Nutrition and Metabolic Care*, Vol. 11, 2008, pp. 711–717.
"Undernutrition Can Be Deadly for Kids." United Press International, 17 June 2004.

A Higher Power

Summary of Topic

Throughout history, people have bashed organized religion (usually someone else's) as a source of various social evils, including disease epidemics. Since others have addressed this topic, we will instead examine ways in which the world's great religions have made positive contributions to public health. Consideration of this topic may reveal opportunities for further improvement.

How Will This Help?

Balking the enemy's plans may require a level of public commitment beyond the reach of law or education. To make sacrifices in the name of a common goal, people must believe in what they are doing. Many scientists find sufficient motivation in the pursuit of knowledge and may find it hard to understand less tangible motives. For example, hundreds of Catholic nuns and priests volunteer for potentially dangerous medical studies every year. Other religious groups have modified some of their traditional practices for the common good, as discussed below.

Discussion

All religions have certain things in common, including the goal of improving their members' lives. Achieving this goal often requires making exceptions to doctrine. The following examples appear in random sequence. (See also Case Study 7-3.)

Case Study 7-3: The Holy City

Any large gathering can increase the risk of disease transmission, and the annual *Hajj* (Muslim pilgrimage to Mecca) is no exception. In 1987, for example, Muslim pilgrims taking part in this ritual suffered outbreaks of meningococcal disease. Crowding also increases the risk of tuberculosis. Other disease risks are specific to this pilgrimage; men traditionally have their heads shaved, and some barbers reuse razor blades, a practice that can spread hepatitis B, hepatitis C, and HIV. To reduce some of these risks, Saudi authorities now require certificates of vaccination, and some Muslim clerics have recently called for modification of the laws governing the pilgrimage.

- Spiritual water—holy water fonts, baptisteries, and containers of water left outdoors for Catholic saints or Afro-Cuban deities—provides breeding places for mosquitoes in Cuba. To reduce the risk of dengue transmission, people have reportedly modified this custom by covering the containers.
- A tribe in Kenya had a major problem with tapeworm until investigators found the reason. During a water shortage, the people trained their dogs to lick the children's faces clean, and the dogs transferred tapeworm eggs to the children's mouths. Nearby Muslim communities did not share this problem, because Muslims regard dogs as unclean. Their neighbors benefited by following their lead.

- The Amish, who traditionally do not vaccinate their children, have allowed exceptions to protect their neighbors from epidemics. (The Amish do not question the efficacy of vaccines, any more than they claim that automobiles or telephones don't work. They simply avoid modern technologies.)
- African healers often use the same razor blade to draw blood from successive patients. Some healers have acknowledged that this practice spreads disease, and at least one recently resumed the older practice of drawing blood with porcupine quills—presumably, a new quill for each patient.
- In Asia, many Taoists have discontinued the ceremony of releasing prayer birds for fear of contributing to the spread of avian influenza.
- In some Christian churches, the Lord's Supper requires large numbers of people to drink from the same cup. Many congregations now dip bread in the cup instead or serve the beverage in small individual containers.
- Some Catholic cardinals support the use of condoms if a husband or wife is HIV-positive, although Pope Benedict XVI has ruled otherwise. Surveys show that many Catholics already use condoms, even if neither partner is infected.
- A 2008 study examined the effects of traditional Yom Kippur fasting on participants with diabetes and concluded that fasting is safe if a treatment plan is prepared in advance.
- In 2009, Muslim clerics in Nigeria announced that they no longer oppose vaccination and will join the fight against polio. With their help, the goal of worldwide polio eradication may at last be possible.

Popular Culture

In the sixteenth century, Martin Luther wrote that "melancholy" —by which he apparently meant a spiritual crisis, or what we now call depression—is the devil's work, but that "sicknesses and plagues," unless present from birth, are the wages of personal sin. Yet he supported the treatment of disease, irrespective of its source.

A modern urban legend claims that airlines avoid assigning a Christian pilot and copilot to the same plane. The usual explanation is that, should the Rapture occur while the plane is in flight, both the pilot and copilot would be taken up, and the plane would crash. (The author called three airlines to inquire about this policy, but all three hung up.)

Perhaps the most famous example of religion accommodating public health appears in Luke 14:1–6, when Jesus defends his decision to cure a sick man on the Sabbath.

The Future

Faith will continue to be an important aspect of healthcare for many people, and public health programs will benefit by enlisting the cooperation of religious leaders. Although much progress has been made, a few areas remain problematical, such as family planning and stem cell research.

References and Recommended Reading

Ahmed, Q. A., et al. "Health Risks at the Hajj." *Lancet*, Vol. 367, 2006, pp. 1008–1015.

"Archbishop Asks His Clergy to Test a Vaccine for AIDS." *New York Times*, 12 March 1990.

Atrah, H. I., et al. "Blood Exchanged in Ritual Ceremonies as a Possible Route for Infection with Hepatitis C Virus." *Journal of Clinical Pathology*, Vol. 47, 1994, p. 87.

Cates, W. "Contraception and Prevention of HIV Infection." *JAMA*, Vol. 296, 2006, p. 2802.

Ciftçi, E., et al. "Traditions, Anthrax, and Children." *Pediatric Dermatology*, Vol. 19, 2002, pp. 36–38.

Clark, P. A., et al. "Mandatory Neonatal Male Circumcision in Sub-Saharan Africa: Medical and Ethical Analysis." *Medical Science Monitor*, Vol. 13, 2007, pp. 205–213.

"Could Choosing Condoms Mean Choosing Life?" *USA Today*, 17 March 2009.

Drain, P. K., et al. "Male Circumcision, Religion, and Infectious Diseases: an Ecologic Analysis of 118 Developing Countries." *BMC Infectious Disease*, Vol. 6, 2006, p. 172.

Goldman, R. "A Healthy Habit: Nuns Used in Medical Research." ABC News, 12 June 2007.

Grajower, M. M. "Management of Diabetes Mellitus on Yom Kippur and Other Jewish Fast Days. *Endocrinology Practice*, Vol. 14, 2008, pp. 305–311.

Hasnain, M., et al. "Influence of Religiosity on HIV Risk Behaviors in Active Injection Drug Users." *AIDS Care*, Vol. 17, 2005, pp. 892–901.

Henry, J., and L. Donnelly. "Female Muslim Medics 'Disobey Hygiene Rules.'" *The Telegraph*, 4 February 2008.

Janeke, J. B., and G. de Bruin. "Holy Communion—Chalice or Challicles?" *South African Medical Journal*, Vol. 95, 2005, p. 544.

Kantor, L. M., et al. "Abstinence-Only Policies and Programs: An Overview." *Sexuality Research and Social Policy*, Vol. 5, 2008, pp. 6–17.

Kopelman, L. M. "If HIV/AIDS Is Punishment, Who Is Bad?" *Journal of Medicine and Philosophy*, Vol. 27, 2002, pp. 231–243.

Manangan, L. P., et al. "Risk of Infectious Disease Transmission from a Common Communion Cup." *American Journal of Infection Control*, Vol. 26, 1998, pp. 538–539.

"Muslim Hajj Pilgrimage: Minimizing the Risks of Infectious Disease." *Eurosurveillance*, Vol. 10, 2005, p. E050120.5.

"Outbreak of Measles is Spreading Among Amish." United Press International, 2 June 1988.

"Poll: Catholics Divided on Sex Teachings." United Press International, 27 July 2008.

"The Pope and Science." *Lancet*, Vol. 371, 2008, p. 276.

Sinn, D. "Bird Ritual Elevates Fears." Associated Press, 14 January 2007.

Srikanthan, A., and R. L. Reid. "Religious and Cultural Influences on Contraception." *Journal of Obstetrics and Gynaecology Canada*, Vol. 30, 2008, pp. 129–137.

Subayi-Cuppen, T., and K. O. Okanla. "Voodoo and HIV/AIDS." World Bank website, May 2005 (downloaded October 2008).

"Traditional Healers, New Partners against HIV/AIDS." UN Integrated Regional Information Networks, 25 February 2003.

Van Howe, R. S., and J. S. Svoboda. "Neonatal Circumcision is Neither Medically Necessary nor Ethically Permissible: A Response to Clark et al." *Medical Science Monitor*, Vol. 14, 2008, pp. LE7–13.

Wacharapluesadee, S., et al. "Drinking Bat Blood May Be Hazardous to Your Health." *Clinical Infectious Diseases*, Vol. 43, 2006, p. 269.

Winfield, S. "Cardinal Says Condoms Are 'Lesser Evil.'" Associated Press, 22 April 2006.

Basic Research

Summary of Topic

The usual purpose of basic research is to increase our knowledge of the universe. Scientists conduct research in many areas, such as biodiversity and number theory, with no immediate practical application in mind. But besides enriching human experience, basic research often yields unexpected long-term rewards in the form of practical applications.

How Will This Help?

We could fill another book with examples of basic research that have benefited mankind by leading to breakthroughs in applied science or technology (Case Study 7-4). Researchers studying viruses in lemurs have made fundamental discoveries that have increased our understanding of HIV. Analysis of data collected during past epidemics has helped in the development of models to control the spread of other diseases. Historians have analyzed government responses to past epidemics to identify effective (and ineffective) public health measures. Events as ancient as the

Plague of Athens in 430 B.C. may hold clues to the nature of hemorrhagic fever outbreaks.

Discussion

The media often poke fun at research that appears to serve no useful purpose, but sometimes the scientists get the last laugh. A famous example was the 1969 discovery of an unusual bacterium that lives in hot water at Yellowstone National Park. In 1976, researchers learned how this bacterium is able to survive: its enzymes, unlike those of most organisms, remain stable at high temperatures. This finding was of interest mainly to academic scientists until 1983, when Dr. Kary Mullis invented the polymerase chain reaction (PCR)—an industrial process that enables geneticists to copy and manipulate DNA. But to make PCR feasible, the researchers needed bacterial enzymes that could withstand heat, and the geyser bacteria had exactly what they needed. By the time Dr. Mullis won the Nobel Prize in 1993, the biotechnology revolution was well underway. PCR is now an indispensable tool for DNA analysis.

Case Study 7-4: Eyes on the Prize

In 2007, biologists studying the development of tadpoles discovered that a certain enzyme, when injected into an embryo, triggered the formation of ocular tissues—in other words, the precursors of eyes. Although this research was still in its early stages when this book was written, the implications are huge. Perhaps someday it will be possible to use stem cells to grow functional human eyes, thanks to an unexpected consequence of basic research on frogs. For the millions of people blinded by trachoma and other infectious diseases every year, the prospect of eye replacement is, indeed, a light at the end of a dark tunnel.

Popular Culture

Much of what we call popular culture depends on the unspoken assumption that basic research is valuable. Why else do people try to prove that the Loch Ness monster is real, or that Elvis is alive, or that a second sniper was on the grassy knoll? What difference does it make, if not to satisfy a fundamental urge to know the truth?

There are many urban legends about serendipitous discoveries. According to one, the inventor of the artificial sweetener aspartame was trying to make a new ant poison in 1965 and forgot to wash. He then put his hand in his mouth and noticed that the chemical residue tasted sweet. (In fact, the inventor of aspartame was trying to invent an artificial sweetener, and that was exactly what he got. The person who discovered saccharin in 1879 was the one who forgot to wash his hands.)

Any research project can sound ridiculous if taken out of context. A few years ago, standup comics had a field day with a $102,000 government grant to find out if tequila or gin was more effective in getting a sunfish drunk. But sunfish are commonly used for testing the effects of many toxic chemicals, including alcohol. (Whether this practice is right or wrong is a separate issue.)

The Future

After 2003, U.S. government funding for basic research declined sharply due to budget deficits and defense spending. At the same time, biomedical research support shifted to projects with a direct link to bioterrorism or other near-term threats. A 2008 American Association for the Advancement of Science (AAAS) study predicts that basic research funding will continue to decline in the near term.

References and Recommended Reading

Altman, S., et al. "An Open Letter to Elias Zerhouni." *Science*, Vol. 307, 2005, pp. 1409–1410.

Brock, T. D. "The Value of Basic Research: Discovery of *Thermus aquaticus* and other Extreme Thermophiles." *Genetics*, Vol. 146, 1997, pp. 1207–1210.

Esler, P. "Funding Basic Research Brings Unexpected Benefits." *Nature*, 18 October 2007.
"Genetic Code of Malaria Parasite Sequenced." United Press International, 9 October 2008.
Jeffrey, W. "Importance of Basic Research to United States' Competitiveness." Testimony before the U.S. Senate, 29 March 2006.
Olson, P. E., et al. "The Thucydides Syndrome: Ebola Déjà Vu?" *Emerging Infectious Diseases*, Vol. 2, 1996, pp. 155–156.
Pappas, S. "Stanford Scientists' Discovery of Virus in Lemur Could Shed Light on AIDS." Stanford University, News Release, 1 December 2008.
Robbins, J. "The Search for Private Profit in the Nation's Public Parks." *New York Times*, 28 November 2006.
Scarrow, G. D. "The Athenian Plague: A Possible Diagnosis." *Ancient History Bulletin*, Vol. 2, 1988, pp. 4–8.
Stoskopf, M. K. "Observation and Cogitation: How Serendipity Provides the Building Blocks of Scientific Discovery." *ILAR Journal*, Vol. 46, 2005, pp. 332–337.
Stutz, B. "Megadeath in Mexico." *Discover*, 21 February 2006.
Swaminathan, N. "Accidental Discovery Could Lead to Creation of Human Eyes in a Lab." *Scientific American*, 24 October 2007.

Water, Toilets, and Garbage

Summary of Topic

A more euphemistic title for this section would be inappropriate. Water and sewage treatment, waste management, and related services that the developed world takes for granted are among the most important achievements of any civilization. These simple precautions can greatly reduce disease transmission and increase productivity in Third World nations.

How Will This Help?

As of 2008, about 2 billion people worldwide do not have access to safe drinking water or sanitation. By 2015, the United Nations hopes to reduce the percentage of people lacking these services by one-half, as a Millennium Development Goal adopted by the General Assembly. Contaminated drinking water is a factor in an estimated 80 percent of fatal childhood illnesses in developing nations. Water contamination, in turn, often results from the absence of sanitary waste disposal and water treatment facilities.

By some estimates, isolating humans from their own waste products would eliminate about 85 percent of all helminth (worm) infestations in the Third World. A 2003 study showed that infectious diarrhea was more frequent in U.S. communities with a higher density of septic tanks per unit area. In addition to the loss of life, waterborne illnesses reduce workforce productivity and contribute to a nation's economic problems.

Discussion

For nearly two decades, some authors have claimed that the South American cholera epidemic of 1991 resulted from Peru's decision to stop chlorinating its water, while others have claimed that the story was nothing more than a politically motivated urban legend. As it turns out, the truth is somewhere in between. The cholera epidemic was real enough; about 1 million people contracted the disease, and at least 13,000 died. But was unchlorinated water responsible? Yes, according to studies published in the journals *Nature* and *Lancet*. The unresolved question is whether the Peruvian government decided to stop chlorinating, or whether the water treatment system simply deteriorated or was inadequate in the first place.

People in developed nations often ask why Third World residents and others who lack access to clean water cannot simply boil their water before drinking it. Sometimes they do, but boiling requires fuel and opportunity, commodities that are often in short supply (Case Study 7-5). Better solutions include deep well drilling, closed water storage containers, water purification tablets, and portable water filters.

Some outsiders visiting rural Africa have noticed that people dislike the taste (or lack of taste) associated with pure water. Surface water contaminated with suspended particulates and plant material is not necessarily unsafe, and it probably tastes like a thin soup of familiar flavors. Even western consumers who want everything sterilized complain that distilled water tastes flat. By contrast, well water from a moss-covered bucket is delicious, and most—not all—of its occupants are edible and do not cause illness. Thus, the ideal treatment system would allow drinking water to retain some local character.

Case Study 7-5: The Bataan Death March

It's a hot day in the Philippines in 1942, and you are one of 72,000 men participating in a forced 60-mile march to a prison camp after the fall of Bataan. For five days, you will have no food and no clean water, and if you say anything, the guards will probably kill you. Thousands of your comrades are dying on all sides; about 18,000 will not make it as far as the camp. So you do something that you would never do in Los Angeles, or wherever you come from. You dip your hands into a mud puddle beside the road and drink quickly before the guard notices. When you are this thirsty, any water seems delicious. The local people drink from the same puddles, and many of them are sick—but they, at least, have some resistance to local diseases. Weeks later, having survived the death march, you will die of dysentery instead.

The availability of clean water reduces but does not eliminate the need for waste management (WM), a range of services that includes sewage treatment, waste disposal, and landfill management. Without WM, sewage pathogens often contaminate farm crops, groundwater, and coastal waters. The goal of sewage treatment is to make the resulting effluent clean enough to dump somewhere, or even clean enough to return to the consumer as drinking water. Waste disposal means removing solid waste from households and businesses, recycling or composting part of it, and burying the rest in landfills. In many countries, households store organic waste indoors for days at a time, a practice associated with health problems that include allergies, skin rashes, and diarrhea.

Popular Culture

The abundance of urban legends about toilets that explode or otherwise malfunction suggests the cultural significance and anxiety associated with this invention. But the real importance of plumbing burst on the American consciousness after Hurricane Katrina, when officials packed some 30,000 displaced people into the storm-damaged New Orleans Superdome, soon overwhelming its bathroom facilities and creating a public relations mess unequaled in recent history. Popular culture has variously sought to exaggerate or downplay this story, but eyewitnesses will never forget.

In any disaster, clean water may quickly become a precious commodity. On September 11, 2001, when an ambulance driver responding to the World Trade Center disaster approached a local restaurant and asked for water to treat survivors, employees reportedly sold them three cases of bottled water for $130.

The Future

Water and sanitation are not strictly Third World problems. Every year, more rural areas of the United States report unsafe well water due to groundwater pollution from dairy farms, industrial waste, or aging infrastructure. Many communities receive regular notices of high nitrate levels in tap water. A large percentage of consumers no longer feel safe drinking unfiltered tap water, and this trend will probably continue. Bottled water use appears to be leveling off in favor of faucet filters.

References and Recommended Reading

Anderson, C. "Cholera Epidemic Traced to Risk Miscalculation." *Nature*, 28 November 1991, p. 255.

Bijapur, N. "Singapore's Success with Water Treatment." *Deccan Herald*, 6 November 2007.

Boadi, K. O., and M. Kuitunen. "Environmental and Health Impacts of Household Solid Waste Handling and Disposal Practices in Third World Cities." *Journal of Environmental Health*, Vol. 68, 2005, pp. 32–36.

Borchardt, M. A., et al. "Septic System Density and Infectious Diarrhea in a Defined Population of Children." *Environmental Health Perspectives*, Vol. 111, 2003, pp. 742–748.

Clasen, T., et al. "Interventions to Improve Water Quality for Preventing Diarrhoea: Systematic Review and Meta-analysis." *BMJ*, Vol. 334, 2007, pp. 755–756.

Clift, E. "A Colossal Failure of Leadership." *Newsweek*, 2 September 2005.

Dutzik, T., and A. Gomberg. "Public Health at Risk: the Dangers Posed by Sewage Pollution in Ohio's Lake Erie Basin." *Ohio PIRG Education Fund*, Spring 2006.

Gleick, P. H. "Global Freshwater Resources: Soft-Path Solutions for the 21st Century." *Science*, Vol. 302, 2003, pp. 1524–1528.

Godfree, A., and J. Farrell. "Processes for Managing Pathogens." *Journal of Environmental Quality*, Vol. 34, 2005, pp. 105–113.

Herr, C. E., et al. "Health Effects Associated with Indoor Storage of Organic Waste." *International Archives of Occupational and Environmental Health*, Vol. 77, 2004, pp. 90–96.

Mintz, E., et al. "Not Just a Drop in the Bucket: Expanding Access to Point-of-Use Water Treatment Systems." *American Journal of Public Health*, Vol. 91, 2001, pp. 1565–1570.

Quick, R. E., et al. "Narrow-Mouthed Water Storage Vessels and In Situ Chlorination in a Bolivian Community: A Simple Method to Improve Drinking Water Quality." *American Journal of Tropical Medicine and Hygiene*, Vol. 54, 1996, pp. 511–516.

Oldham, D., et al. "What Is the Best Portable Method of Purifying Water to Prevent Infectious Disease?" *Journal of Family Practice*, Vol. 57, 2008, pp. 46–48.

Reynolds, K. A., et al. "Risk of Waterborne Illness via Drinking Water in the United States." *Reviews of Environmental Contamination and Toxicology*, Vol. 192, 2008, pp. 117–158.

Royte, E. "A Tall, Cool Drink of . . . Sewage?" *New York Times*, 10 August 2008.

Sanderson, H. "One-Third of China's Yellow River Polluted." Associated Press, 25 November 2008.

Sobsey, M. D., and S. Bartram. "Water Quality and Health in the New Millennium: The Role of the World Health Organization Guidelines for Drinking-Water Quality." *Forum of Nutrition*, Vol. 56, 2003, pp. 396–405.

"Study: Toilets, Safe Water Can End Poverty." United Press International, 20 October 2008.

Tickner, J., and T. Gouveia-Vigeant. "The 1991 Cholera Epidemic in Peru: Not a Case of Precaution Gone Awry." *Risk Analysis*, Vol. 25, 2005, pp. 495–502.

Walsh, B. "Toilet Tales: Inside the World of Waste." *Time*, 18 November 2008.

PART 2: PREVENTING THE JUNCTION OF THE ENEMY'S FORCES

Bring out the trebuchet when the enemy's lances first appear above the crest of the hill, or when the first sentry falls. Build a wall and stop the invader in its tracks. Arm the people, clean up the camp, and confuse the enemy. This is the next line of defense, short of hand-to-hand combat: public health surveillance, immunization against diseases that we think are heading our way—with a few inevitable false alarms—and obvious barriers such as window screens, mosquito nets,

Figure 7.3 A simple but effective precaution. Graphics such as this may help reduce the spread of influenza and other diseases among school children.

Source: Dubuque Community School District, Dubuque, Iowa.

and sneeze-in-your-sleeve campaigns (Figure 7.3), as well as less obvious measures such as sterile insect release.

If successful, these actions prevent the enemy from joining battle without causing major risk or inconvenience. Why spray pesticides if nets alone will work? Why treat disease if we can prevent it instead? Pesticides and drugs, like chemical weapons and small arms fire, so often cause collateral damage.

Lookouts: Surveillance and Screening

Summary of Topic

Public health surveillance is the collection, analysis, and interpretation of data needed for disease prevention and control. In the United States, the Centers for Disease Control and Prevention (CDC) collects and publishes data on reportable diseases, while the Animal and Plant Health Inspection Service (APHIS) reports comparable data for animal and plant diseases.

Case Study 7-6: Monkeypox in America

In 2003, the CDC announced that at least 37 people in three Midwestern states had contracted monkeypox—an exotic African virus, closely related to smallpox—from pet prairie dogs that, in turn, had contracted the disease from African rodents at a pet shop. (Monkeypox is not specific to monkeys, but can also infect rodents, rabbits, and humans.) Local doctors reported the disease to CDC, as required by law, and CDC made a thorough investigation. As a result, this catastrophic-sounding outbreak was a nonstarter. In Africa, the human case fatality rate for monkeypox is between 1 and 10 percent, but many of the victims are malnourished or have other diseases. The death rate should be lower in the United States, where people are better fed and usually have access to hospital care. There were no known human fatalities in the 2003 outbreak, and CDC banned further imports of African rodents.

Screening, defined as testing to determine presence or absence of a disease, is an important aspect of surveillance.

How Will This Help?

Public health agencies need to identify and monitor health-related events in order to plan effective countermeasures, just as military commanders need intelligence on enemy weapons, numbers, and movements. Examples of "health-related events" include infectious diseases (Case Study 7-6), injuries, toxic exposures, or high-risk behaviors. To coin a phrase, we can't fix it without knowing what is broken.

Lookouts can also watch for plant diseases. Before Asian soybean rust (Chapter 5) reached North America, the United States prepared for its arrival by establishing sentinel plots and developing a strategic action plan. The fungus arrived anyway, but at least we had time to develop better fungicides and to learn from the research findings of other nations where soybean rust was already established.

Discussion

Definitions vary, but surveillance is usually the first step in a four-part process designed to anticipate, identify, and deal with infectious disease outbreaks or other public health issues:

1. Surveillance: What is the problem?
2. Risk factor identification: What is the cause?
3. Intervention evaluation: What works?
4. Implementation: How do we do it?

The CDC defines public health surveillance as "the ongoing, systematic collection, analysis, and interpretation of data (e.g., regarding agent/hazard, risk factor, exposure, health event) essential to the planning, implementation, and evaluation of public health practice, closely integrated with the timely dissemination of these data to those responsible for prevention and control."

The global effort to combat infectious disease depends on screening campaigns that identify infected persons early enough for treatment to be effective. Inevitably, screening raises issues regarding the rights of the individual and the community. People seldom object to testing for infections such as strep throat, but may feel insulted when offered routine screening for sexually transmitted diseases such as HIV. Screening enables infected people to seek treatment or avoid high-risk behavior, but many regard testing as a violation of privacy, or would simply prefer not to know.

As a result, until 2006, the CDC recommended HIV testing only for persons at high risk. The latest guidelines, however, recommend HIV testing for everyone between the ages of 13 and 64. In 2004, Canada passed the Mandatory Testing and Disclosure Act (MTDA), which requires the source of an infectious disease (usually HIV or hepatitis) to provide a bodily substance for testing.

The Influenza Surveillance Network, established by WHO in 1952, consists of institutions in nearly 100 countries that collect specimens and isolate influenza viruses for shipment to one of five centers for advanced antigenic and genetic analysis. The results enable WHO to recommend the composition of influenza vaccine each year and to alert the world to any strain that might become pandemic.

Another global surveillance effort is the GeoSentinel Surveillance Network, which the CDC created in 1996 in partnership with the International Society for Travel Medicine. Its purpose is to collect sentinel surveillance data from travelers and estimate travel-related disease risk.

Many people may be unaware of the important role of poultry in guarding public health by serving as an early warning system. Caged chickens deployed at strategic locations receive periodic blood tests to determine whether they have been exposed to diseases such as West Nile virus or avian influenza.

Popular Culture

New strains of the influenza virus appear every year, and public health surveillance enables scientists to find out which flu strains are circulating in order to prepare effective vaccines. Unfortunately, "surveillance" has more than one meaning. In 2008, a couple of websites advanced the theory that doctors use flu shots as an excuse to inject small tracking devices into unwitting subjects, or even to infect them with diseases in preparation for harvesting antiserum. This is nonsense.

Just as scientists monitor real disease outbreaks, sociologists and folklorists monitor fictitious ones. Bioterrorism hoaxes, strange nonexistent diseases, and exaggerated versions of real ones are staples of urban legend. Until recently, the CDC maintained a Health Related Hoaxes and Rumors website to assuage public fears, but there were so many hoaxes that it was shut down.

The Future

For public health surveillance to work, people must cooperate, but cooperation often involves sacrificing some degree of privacy or personal freedom. For example, most people would readily admit catching monkeypox from a pet prairie dog, but fewer are willing to divulge the source of a sexually transmitted disease. Solving this problem remains a challenge for public health workers.

References and Recommended Reading

Capua, I., and D. J. Alexander. "Ecology, Epidemiology and Human Health Implications of Avian Influenza Viruses: Why Do We Need to Share Genetic Data?" *Zoonoses and Public Health*, Vol. 55, 2008, pp. 2–15.

Chevalier, V., et al. "Use of Sentinel Chickens to Study the Transmission Dynamics of West Nile Virus in a Sahelian Ecosystem." *Epidemiology and Infection*, Vol. 136, 2008, pp. 525–528.

Chowdhury, P., et al. "Surveillance of Certain Health Behaviors Among States and Selected Local Areas—United States, 2005." *MMWR Surveillance Summaries*, Vol. 56, 2007, pp. 1–164.

Domenech, J., et al. "Regional and International Approaches on Prevention and Control of Animal Transboundary and Emerging Diseases." *Annals of the New York Academy of Sciences*, Vol. 1081, 2006, pp. 90–107.

Gostin, L. O. "HIV Screening in Health Care Settings: Public Health and Civil Liberties in Conflict?" *JAMA*, Vol. 296, 2006, pp. 2023–2025.

Guarner, J., et al. "Monkeypox Transmission and Pathogenesis in Prairie Dogs." *Emerging Infectious Diseases*, March 2004.

Koenen, F., et al. "Real-Time Laboratory Exercises to Test Contingency Plans for Classical Swine Fever: Experiences from Two National Laboratories." *Revue Scientifique et Technique*, Vol. 26, 2007, pp. 629–638.

Malhotra, R., et al. "Should There Be Mandatory Testing for HIV Prior to Marriage in India?" *Indian Journal of Medical Ethics*, Vol. 5, 2008, pp. 70–74.

Marcus, P. I., et al. "Super-Sentinel Chickens and Detection of Low-Pathogenicity Influenza Virus." *Emerging Infectious Diseases*, Vol. 13, 2007, pp. 1608–1610.

Pitrelli, N. and G. Sturloni. "Infectious Diseases and Governance of Global Risks through Public Communication and Participation." *Annali dell'Istituto Superiore di Sanità*, Vol. 43, 2007, pp. 336–343.

Poglayen, G. 2006. "The Challenges for Surveillance and Control of Zoonotic Diseases in Urban Areas." Annali dell'Istituto Superiore di Sanità Vol. 42, No. 4, pp. 433–436.

"Senate Passes BioShield Bill." Associated Press, 19 May 2004.

Keller, R. "Sentinel Plots Are Key." *AgProfessional*, January 2008.

Tourre, Y. M., et al. "Early Warning Planning for Mosquito-Borne Epidemics." *ArcNews Online*, Spring 2008.

Touze, J. E., et al. "New Concepts in Epidemiological Surveillance in French Army." *Bulletin de l'Académie Nationale de Médecine*, Vol. 188, 2004, pp. 1143–1151.

Zetola, N. M. "Association between Rates of HIV Testing and Elimination of Written Consents in San Francisco." *JAMA*, Vol. 297, 2007, p. 1061.

Arming the People: Vaccination

Summary of Topic

Every year, the CDC and equivalent agencies in other countries publish schedules of recommended vaccinations for children and adults, based on the latest surveillance data, occupational risk, and other factors. Parents who do not want their children vaccinated can easily obtain waivers if they are willing to accept the consequences.

How Will This Help?

As "weapons of mass protection," vaccines have enabled doctors to eliminate or greatly reduce the incidence of many serious diseases. For example, in the United States in 1987, there were 41 reported cases of *Haemophilus influenzae* B (Hib) meningitis per 100,000 children under the age of 5. After introduction of the Hib vaccine in 1990, the incidence fell to only 0.11 per 100,000 in 2007. In economic terms, public health officials consider a vaccine cost-effective if the cost is less than $100,000 U.S. per year of life gained, and most vaccines qualify.

Sometimes it also makes sense to vaccinate domesticated animals, and even wildlife. Mandatory rabies vaccination of dogs has greatly reduced the incidence of rabies in developed countries. Individually injecting wild animals is seldom feasible, but baits containing an oral rabies vaccine are available for animals such as coyotes and raccoons.

Discussion

Tables 7.1 and 7.2 summarize CDC recommended vaccine schedules for children and adults as of 2009. A 2008 study showed that 77 percent of U.S. children had all recommended vaccinations. Less than 1 percent of children between the ages of 19 and 35 months had received no vaccines.

The day is long past when most people viewed doctors as authority figures. Thanks to the Internet, consumers have access to more medical information (and misinformation) than ever

Table 7.1 Child Immunization Schedule Recommended by the U.S. Centers for Disease Control and Prevention (2009).

Vaccine	Doses	Ages	Notes
Hepatitis B	3 or 4	0 (soon after birth) 1–2 months 24 weeks or older	Recommended before hospital discharge. Catch-up doses at age 7–18 if needed.
Rotavirus	2 or 3	6–14 weeks 4 months 6–8 months (optional)	Do not begin after 15 weeks.
Diphtheria, Tetanus, Pertussis (DTaP)	6+	6 weeks–2 months 4 months 6 months 12–18 months 4–6 years 11–12 years	Diphtheria and tetanus toxoids and acellular pertussis vaccine (DTaP) has replaced DPT. Catch-up dose at age 13–18 if needed.
Hemophilus influenzae type b (Hib)	3 or 4	6 weeks–2 months 4 months 6 months (optional) 12–15 months	Combined DTaP/Hib vaccine should not be given at 2, 4, or 6 months.
Pneumococcal	4 or 5	2 months 4 months 6 months 2+ years (optional)	PPSV at age 2–18 years for children with high-risk conditions.
Inactivated Poliovirus (IPV)	4	2 months 4 months 6–18 months 4–6 years	IPV has replaced oral polio vaccine since 2000. Catch-up doses at age 7–18 if needed.
Influenza (trivalent inactivated vaccine or live attenuated influenza vaccine)	Yearly	6 months–18 years	Minimum age 6 months for TIV, 2 years for LAIV.
Measles, Mumps, Rubella (MMR)	2	12–15 months 4–6 years (or earlier)	Doses 28+ days apart. Catch-up doses at age 7–18 if needed.
Varicella	2	12–15 months 4–6 years (or earlier)	Doses 3+ months apart. Catch-up doses at age 7–18 if needed.
Hepatitis A	2	12–23 months	Doses 6+ months apart. Older children at risk should also be vaccinated.
Meningococcal	1	2–10 years (optional) 11–12 years	Recommended for younger children at risk.
Human Papillomavirus (HPV)	3	9–12 years (females only)	Allow 2 months between doses 1 and 2, 6 months between doses 2 and 3. Catch-up dose at 13–18 if needed.

Table 7.2 Adult Immunization Schedule Recommended by the U.S. Centers for Disease Control and Prevention (2009).

Vaccine	Doses	Ages	Notes
Tetanus, Diphtheria, and Acellular Pertussis (Td/Tdap)	1	19–65+ years	Tdap for adults who have not received a dose; otherwise Td booster every 10 years. Pregnant women: Td in trimester 2–3 or postpartum.
Human Papillomavirus (HPV)	3	9–26 years (females only)	Catch-up dose if needed (see Table 7.3).
Varicella	2	19–65+ years	Allow 4–8 weeks between doses. Not given during pregnancy.
Herpes Zoster	1	60–65+ years	Recommended for most adults 60+, with some exceptions (HIV, etc.).
Measles, Mumps, Rubella (MMR)	1 or 2	19–49 years	Adults born <1957 assumed immune to measles. Second dose if risk factors present (workplace exposure, travel, etc.). Not given if pregnant or immunocompromised.
Influenza	Yearly	50+ years	Also at ages 19–49 if risk factors are present (diabetes, HIV, etc.)
Pneumococcal (Polysaccharide)	1 or 2	65+ years	Also at ages 19–65 if risk factors are present (diabetes, lung disease, etc.)
Hepatitis A	2	19–65+ years	Only if risk factors are present (liver disease, workplace exposure, etc.).
Hepatitis B	3	19–65+ years	Only if risk factors are present (liver disease, workplace exposure, etc.).
Meningococcal	1+	19–65+ years	Recommended for first-year students in dormitories, military recruits, healthcare personnel, others at risk.

before. As a result, many people read exaggerated warnings about the side effects of vaccination and refuse to participate. No vaccine is perfect, and a few people suffer adverse effects every year. In most (but not all) cases, these adverse effects are minor and temporary.

Ironically, many other people apparently skip vaccines for the opposite reason—not because they are worried, but because the doctor failed to remind them, or the patient failed to listen. In one study, only 25 percent of African-American female adolescents had the HPV vaccine; most were willing to have it, but their caregivers did not remember the doctor mentioning it. Still other people avoid vaccination because they are too busy, or for unknown reasons (Case Study 7-7). Yet a 2007 study showed that many U.S. residents have booster vaccinations sooner than necessary. It seems that most adults are either under- or overvaccinated.

Some new vaccines inevitably cause problems, and the resulting publicity tends to alienate consumers. For example:

Case Study 7-7: Physician, Heal Thyself

Consumer health education highlights the importance of vaccines in preventing disease outbreaks, yet recent studies have shown that only about 40 percent of U.S. healthcare workers get their flu shots. (By comparison, a 2008 RAND survey showed that 47 percent of the general U.S. population intended to have flu shots, and about 33 percent actually had them.) In England, the situation is apparently even worse; a 2009 report indicates that only about 14 percent of UK healthcare workers had flu shots. The reasons for this low participation rate are probably the same for all three groups: denial ("I never get sick"), crowded schedules, general lack of confidence in the healthcare industry, or fear of mishaps, such as contaminated needles or defective vaccines. What is the solution? The author doesn't know. She has never had a flu shot, either.

- The manufacturer of a Lyme disease vaccine that received FDA approval in 2000 took it off the market in 2004, apparently due to complaints about side effects.
- A rotavirus vaccine introduced in 1998 was discontinued in 1999 after causing severe side effects in about 20 children (of an estimated 1 million who received it).
- Some U.S. military personnel deployed to Iraq have refused anthrax vaccination, because it often has side effects and serves no clearly defined purpose.

According to the theory of herd immunity, the individual incentive to be vaccinated disappears at high coverage levels. In other words, if most people in a community have already been vaccinated, we are unlikely to catch the disease, since there is nobody to catch it from. At that point, we may decide not to be vaccinated (or to have our children vaccinated), thus allowing others to assume the risk instead. But this strategy can backfire, because unless a disease has actually been eradicated, it can reappear unexpectedly.

Popular Culture

Periodic vaccine shortages may result from production problems or fear of litigation, but the rumor mill holds that these shortages—and shortages of other consumer products, such as gasoline or popular toys—are staged for the purpose of increasing demand. People who would otherwise postpone their shots indefinitely might suddenly want them if they are told they can't have them.

The widespread belief that vaccines cause autism is partly the fault of the medical profession (Chapter 3). A 1998 journal paper reported a tentative link between autism and the measles-mumps-rubella (MMR) vaccine, but the authors retracted their findings when later studies exonerated the vaccine. In 2009, the vice president of a major autism advocacy group resigned because of the organization's continued anti-vaccination stance. It takes courage to change a long-held personal belief in response to new information.

The Future

Some promising developments include experimental DNA vaccines, which are cost-effective and avoid the risk of infection, and combined vaccines, which save time and reduce the required number of shots. Scientists have also developed edible vaccines for several diseases, including a hepatitis B vaccine in transgenic potatoes and an experimental malaria vaccine in transgenic tomatoes. But will most people want to eat these transgenic foods?

References and Recommended Reading

"Americans May Be Over-Vaccinating." United Press International, 8 November 2007.

Andre, F. E., et al. "Vaccination Greatly Reduces Disease, Disability, Death and Inequity Worldwide." *Bulletin of the World Health Organization*, Vol. 86, 2008, pp. 140–146.

Awasthi, S. "Next Generation of Human Vaccines: What Does the Future Hold?" *Human Vaccines*, Vol. 4, 2008, pp. 344–346.

Barnard, J. "Don't Vaccinate? Doctors Paying to Hear Views." Associated Press, 9 January 2009.

Bauch, C. T., and D. J. D. Earn. "Vaccination and the Theory of Games." *Proceedings of the National Academy of Sciences* (USA), Vol. 101, 2004, pp. 13391–13394.

"Biologists Seek Safer Whooping Cough Shots." Associated Press, 7 November 2006.

"Black Teens Open to Getting HPV Vaccine." United Press International, 6 February 2009.

Bruyand, M., et al. "Yellow Fever Vaccination in Non-Immunocompetent Patients." *Medecin et Maladies Infectieuses*, 19 August 2008.

Chen, X., et al. "Differences in Perception of Dysentery and Enteric Fever and Willingness to Receive Vaccines among Rural Residents in China." *Vaccine*, Vol. 24, 2006, pp. 561–571.

Greene, L. C. "Polio Vaccine 50 Years Old." *San Bernardino County Sun*, 11 April 2005.

Jojola, S. M., et al. "Oral Rabies Vaccine (ORV) Bait Uptake by Captive Striped Skunks." *Journal of Wildlife Diseases*, Vol. 43, 2007, pp. 97–106.

Kalb, C. "This Question Has Been Asked and Answered." *Newsweek*, 28 January 2009.

Kharabsheh, S., et al. "Mass Psychogenic Illness Following Tetanus-Diphtheria Toxoid Vaccination in Jordan." *Bulletin of the World Health Organization*, Vol. 79, 2001, pp. 764–770.

Lombard, M., et al. "A Brief History of Vaccines and Vaccination." *Revue Scientifique et Technique*, Vol. 26, 2007, pp. 29–48.

Nara, P. L., et al. "Perspectives on Advancing Preventative Medicine through Vaccinology at the Comparative Veterinary, Human and Conservation Medicine Interface: Not Missing the Opportunities." *Vaccine*, Vol. 26, 2008, pp. 6200–6211.

"Nearly Half of Adults Say No Flu Shot." United Press International, 12 December 2008.

Oxford, J. S. "The End of the Beginning: Vaccines for the Next 25 Years." *Vaccine*, Vol. 26, 2008, pp. 6179–6182.

"Polio Shots Launched Era of Vaccines." Associated Press, 11 April 2005.

Roberts, J. "Children Fall Ill after Measles Jabs." The Independent, 1 October 1995.

Roush, S. W., and T. V. Murphy. "Historical Comparisons of Morbidity and Mortality for Vaccine-Preventable Diseases in the United States." *JAMA*, Vol. 298, 2007, pp. 2155–2163.

Sachs, J. S. "Seven Vaccines You Need Right Now." CNN.com, 7 May 2008.

"Unvaccinated NHS Workers Blamed for Flu." United Press International, 28 February 2009.

"Vaccine Rate for U.S. Children Is High." United Press International, 4 September 2008.

Welte, M. S. "Flu Shots a Tough Sell to Health Care Workers." Associated Press, 19 October 2008.

Wortley, P. M., et al. "Predictors of Smallpox Vaccination among Healthcare Workers and Other First Responders." *American Journal of Preventive Medicine*, Vol. 32, 2007, pp. 538–541.

Holding the Line: Convenient Barriers

Summary of Topic

Window screens, bed nets, some insect repellents, and hand sanitizer gels are examples of convenient, harmless measures that can keep many disease vectors and pathogens away from people and domestic animals. People tend to avoid less convenient barriers and measures, such as face masks, moon suits, and quarantine, until the enemy is at the gates (as discussed later in the chapter).

How Will This Help?

Simple barriers that keep disease agents and vectors away from hosts are among our most powerful defenses. Doctors and nurses routinely use hand sanitizer gels while on the job. Wildlife biologists, funeral home workers, and many other occupational groups use disposable gloves. Even office equipment may have antimicrobial coatings (Case Study 7-8).

Case Study 7-8: Really Clean Pens

Responding to public demand for infection barriers and clean surfaces, the PM Company sells a line of antimicrobial ballpoint pens for use in banks and other public places. The secret is AgION(r), a silver- and zinc-containing zeolite matrix that retards the growth of bacteria and fungi when used as a coating—not only on stainless steel pens, but also on other frequently contaminated objects such as keyboards, cell phones, water filters, or shoes. This material provides continuous protection by gradually releasing silver ions from the treated surface. Unfortunately, a 2009 study suggests that nanosilver particles in some consumer products may harm human health and the environment, by killing good bacteria as well as bad ones.

A less obvious barrier is the common window screen. According to some historians, the elimination of malaria from the United States in the 1930s owed more to window screens and the invention of radio (and, later, television) than to all the swamp drainage channels of the New Deal. When people spend their evenings inside screened houses, mosquitoes are less likely to bite them. In Third World nations, studies have confirmed that the installation of screens and bed netting reduces the incidence of malaria and dengue.

Discussion

Various cultures have invented and reinvented window screens. The ancient Egyptians used linen curtains to keep flies out of buildings. Medieval European "screens" were decorative grills, capable of stopping things the size of birds, but not insects. Wire fabric was invented in about 1757, and in 1868 an American sieve manufacturer patented a screened window to exclude cinders from rail cars. Most references to window screens before the 1860s apparently meant blinds or shutters. As early as 1869, advertisements offered adjustable window screens to keep out insects, but these products did not really take off until the early 1900s. Kentucky manufacturer Henry Higgin (1844–1926) invented some of the first modern metal-framed screens, and his company filed several related patents between 1915 and 1930. Even before people knew that mosquitoes transmitted disease, the bites must have been annoying, and it seems incredible that this simple device took so long to catch on.

Not every barrier is a physical wall or semipermeable membrane. In some cases, distance alone can be a barrier. A 2007 study showed that room size—the volume of space per person—was the most important factor determining whether healthcare workers became infected while caring for multidrug-resistant tuberculosis patients.

A barrier can also be a zone of habitat that is unsuitable for a disease vector or pathogen. As an example, the recent arrival of West Nile virus in North America has focused attention on the need for a dry zone surrounding homes or public areas. This strategy requires the removal or covering of

Figure 7.4 Water that collects in outdoor containers provides breeding sites for mosquitoes that transmit disease. This problem has increased during the recent economic recession due to numerous abandoned or repossessed properties.

Source: Skye Compton.

standing water where mosquitoes can breed, such as flowerpot saucers, rain barrels, or other open containers (Figure 7.4). For outdoor pets, automatic watering devices are available to replace the traditional bowl of stagnant water. Studies have turned up some unexpected sources of mosquito habitat, including septic tanks and even the water trays of self-defrosting refrigerators.

Are insect repellents "convenient" barriers? They are effective, but should people use them every day, or save them as a last resort? This is a borderline case, because some repellents are potentially harmful or unpleasant. The CDC recommends products that contain the chemical DEET for use on skin or clothing, but the required precautions are many. DEET should not be applied to broken or irritated skin, or under clothing, or in excessive quantities, or in enclosed spaces, or near the eyes or mouth; and parents should not spray DEET on children's bedding, or allow children to handle it. If used incorrectly, DEET can cause rashes or even seizures. Small wonder that some people prefer to take their chances with West Nile—but many outdoor workers use DEET and report no problems with it. (Researchers have found that the strong smell of this chemical confuses mosquitoes, so the example also belongs to the category of deception.)

Popular Culture

President Barack Obama, in his book *The Audacity of Hope*, describes a 2005 incident in which then-President Bush sterilized his hands with antiseptic gel immediately after a handshake. A few years earlier, this habit might have seemed strange or rude, but the world is changing.

There are many urban legends about bad things (spiders, perverts) entering a house through a torn window screen. There are many others about unlikely-sounding products that double as mosquito repellents, including specific brands of dishwashing liquid, fabric softener sheets, bath oil, or mouthwash. Older traditions include the use of a plant called beautyberry (*Callicarpa americana*) to repel mosquitoes. A 2006 USDA study confirmed that the leaves contain chemicals that work nearly as well as DEET.

Starting in about 1940, the film industry contributed to the war on disease by producing a series of motion pictures that taught the benefits of window screens. For example, the 1944 Walt Disney animated short entitled *The Winged Scourge* depicted the Seven Dwarfs (from the 1937 movie *Snow White and the Seven Dwarfs*) installing window screens and spraying insecticides in huts.

The Future

As of 2009, air-filtering masks are widely perceived as overkill, even during a flu epidemic. Masks are uncomfortable, they conceal facial features that humans use to communicate, and the wearer may be perceived as dangerous or paranoid. Masks even abridge the God-given right to wear a beard, which may prevent a tight seal. Yet we predict that face masks may soon become standard apparel, at least for airline and subway passengers.

References and Recommended Reading

Barrera, R., et al. "Unusual Productivity of *Aedes aegypti* in Septic Tanks and Its Implications for Dengue Control." *Medical and Veterinary Entomology*, Vol. 22, 2008, pp. 62–69.

Corbel, V., et al. "Evidence for Inhibition of Cholineserases in Insect and Mammalian Nervous Systems by the Insect Repellent DEET." *BMC Biology*, Vol. 7, 2009, p. 47.

Cowan, M. M., et al. "Antimicrobial Efficacy of a Silver-Zeolite Matrix Coating on Stainless Steel." *Journal of Industrial Microbiology and Biotechnology*, Vol. 30, 2003, pp. 102–106.

Fedunkiw, M. "Malaria Films: Motion Pictures as a Public Health Tool." *American Journal of Public Health*, Vol. 93, 2003, pp. 1046–1057.

Franchi, A., et al. "Room Size Is the Major Determinant for Tuberculin Conversion in Health Care Workers Exposed to a Multidrug-Resistant Tuberculosis Patient." *International Archives of Occupational and Environmental Health*, Vol. 80, 2007, pp. 533–538.

Ghosh, S. K., et al. "A Community-Based Health Education Programme for Bio-Environmental Control of Malaria through Folk Theatre (*Kalajatha*) in Rural India. *Malaria Journal*, 15 December 2006.

Kumar, A., et al. "Silver-Nanoparticle-Embedded Antimicrobial Paints Based on Vegetable Oil." *Natural Materials*, Vol. 7, 2008, pp. 236–241.

Ooi, E. E., et al. "Dengue Prevention and 35 Years of Vector Control in Singapore." *Emerging Infectious Diseases*, Vol. 12, 2006, pp. 887–893.

Simmons, G., et al. "A Legionnaires' Disease Outbreak: A Water Blaster and Roof-Collected Rainwater Systems." *Water Research*, Vol. 42, 2008, pp. 1449–1458.

U.S. Centers for Disease Control and Prevention. "Updated Information Regarding Mosquito Repellents." 8 May 2008.

Van den Berg, H., et al. "Reducing Vector-borne Disease by Empowering Farmers in Integrated Vector Management." *Bulletin of the World Health Organization*, Vol. 85, 2007, pp. 561–565.

Vazquez, M., et al. "Effectiveness of Personal Protective Measures to Prevent Lyme Disease." *Emerging Infectious Diseases*, Vol. 14, 2007, pp. 210–216.

Weese, J. S. "Barrier Precautions, Isolation Protocols, and Personal Hygiene in Veterinary Hospitals." *Veterinary Clinics of North America Equine Practice*, Vol. 20, 2004, pp. 543–559.

White, C. "The Effect of Hand Hygiene on Illness Rate among Students in University Residence Halls." *American Journal of Infection Control*, Vol. 31, 2003, pp. 364–370.

Zhang, A., et al. "Isolongifolenone: A Novel Sesquiterpene Repellent of Ticks and Mosquitoes." *Journal of Medical Entomology*, Vol. 46, 2009, pp. 100–106.

The Fifth Column: Ringers and Decoys

Summary

It is rarely possible to "trick" bacteria and viruses directly, since their capacity for independent action is limited, but higher organisms such as insects often are amenable to this strategy. Insect lures and traps, the use of chemicals such as pheromones that affect behavior, and the release of sterile male insects are among the options in this category.

How Will This Help?

Methods such as lures, traps, pheromones, and sterile insect release may never entirely replace pesticides, because pests typically breed in extremely large numbers. These methods are, however, an important component of integrated pest management (IPM), because they can reduce the cost and human health risks associated with pesticide use. Unfortunately, some of the most effective insect lures—such as the famous Shell No-Pest® Strips, first used in 1963 and pulled off the market in 1979—contained chemicals that turned out to be potentially neurotoxic to humans, thus giving all chemical traps an undeserved bad reputation.

Discussion

Armies in a real war often construct fake tanks and artillery pieces to trick the opposing forces into disclosing their capabilities while wasting their bombs and manpower on dummy targets. Other forms of deception include the use of concealed landmines and "fifth columnists" who spread disinformation while helping the invaders. Analogous strategies are commonly used in the war on disease.

For example, when scientists release large numbers of sterile male insects of a given pest or parasite species, the females who mate with these unwitting subversives waste their entire reproductive investment. This method was first used in 1951 on a parasitic insect called screwworm (*Cochliomyia hominivora*), and by 1966 the species was all but eradicated from the United States. Other weapons in the deception category are as simple as electronic bug-zappers, roach motels, and flypaper.

There are also chemical weapons that contribute to the fog of war, such as pheromones that control the behavior of insects and other arthropods. Scientists have discovered "assembly" pheromones that disrupt mating behavior in various species, including citrus leafminers (*Phyllocnistis citrella*) and ixodid ticks.

Popular Culture

The archetypal folktale in this category must be the story of the Pied Piper of Hamelin, a mysterious stranger who arrives one day and offers to lead the rats out of the German town of Hamelin by playing a flute. But in real life, lures don't work unless the target pest or vector mistakes the lure for something it wants.

Another example of this theme is Joel Chandler Harris's 1881 trickster tale "The Wonderful Tar-Baby Story," in which the hapless Brer Rabbit assaults a statue made of tar, trying to force it to acknowledge his presence, and becomes stuck. Brer Fox designed this lure after studying his adversary's weaknesses, which included an unwillingness to be ignored. But like any real-life pest, Brer Rabbit still had a trick or two up his sleeve.

The famous story of the Three Little Pigs illustrates yet another principle of disinformation: Make the enemy believe that your troops will be exposed on a specific road at a specific time, and then move them by another road or at another time instead (see also Case Study 7-9).

Case Study 7-9: Out of Synch

For indoor bed nets to be effective against malaria vectors, the vectors must be inclined to bite at times when most people are in bed. One problem with this strategy is that the vectors can adapt (at least in theory) by changing their schedules, but people rarely have the same luxury. In other words, natural selection can produce mosquitoes that do most of their biting in the daytime, but it would be hard for a farmer to respond by working his fields at night and sleeping indoors during the day. Thus, the vectors may be more adaptable than we are.

The Future

The types of biological controls discussed in this section will continue to be valuable IPM tools, particularly as the long-term environmental and health effects of pesticide use become apparent. Genetic engineering may soon enable scientists to deploy more sophisticated biological controls, such as mosquitoes that are incapable of serving as hosts for malaria.

References and Recommended Reading

Alphey, L., et al. "Insect Population Suppression Using Engineered Insects." *Advances in Experimental Medicine and Biology*, Vol. 627, 2008, pp. 93–103.

Benedict, M. Q., and A. S. Robinson. "The First Releases of Transgenic Mosquitoes: An Argument for the Sterile Insect Technique." *Trends in Parasitology*, Vol. 19, 2003, pp. 349–355.

Borges, L. M., et al. "Efficacy of 2,6-Dichlorophenol Lure to Control *Dermacentor nitens*. *Veterinary Parasitology*, Vol. 147, 2007, pp. 155–160.

Handler, A. M., et al. "Development and Utilization of Transgenic New World Screwworm, *Cochliomyia hominivorax*." *Medical and Veterinary Entomology*, Vol. 23 (Suppl. 1), 2009, pp. 98–105.

Navarro-Llopis, V., et al. "Evaluation of Traps and Lures for Mass Trapping of Mediterranean Fruit Fly in Citrus Groves." *Journal of Economic Entomology*, Vol. 101, 2008, pp. 126–131.

Sonenshine, D. E. "Pheromones and Other Semiochemicals of Ticks and Their Use in Tick Control." *Parasitology*, Vol. 129 (Suppl), 2004, pp. S405–425.

Sonenshine, D. E. "Tick Pheromones and their Use in Tick Control." *Annual Review of Entomology*, Vol. 51, 2006, pp. 557–580.

Stelinski, L. L., et al. "Mating Disruption of Citrus Leafminer Mediated by a Noncompetitive Mechanism at a Remarkably Low Pheromone Release Rate." *Journal of Chemical Ecology*, Vol. 34, 2008, pp. 1107–1113.

A Clean Camp: Home, School, and Workplace

Summary of Topic

Not every part of a living space needs to be clean, and there is some evidence that a totally clean environment may be bad for the immune system. The most important rule is to keep specific things from coming in contact with one another, such as dead animals with live ones, waste with food, or disease vectors with people. In place of the standard warning "You don't know where it's been," substitute inference: where has it been?

How Will This Help?

Surveys have shown that people worry about highly publicized but unlikely disease hazards, such as brain-eating amoebas, tropical hemorrhagic fevers, or the accidental or deliberate release of biological weapons. In fact, most infectious disease transmission results from ordinary contacts with other people and pets, or from food, water, or objects that have become contaminated by accident. Many of these infections are preventable without major lifestyle changes.

Discussion

There are two basic types of indoor spaces: those you can mostly control, such as a private home or an office that you manage; and those you can't control, such as a business where you shop or work, a dormitory bathroom, or a prison. In the first situation, proactive measures are best, such as setting rules about taking out the garbage. In the second, reactive measures such as handwashing or shower shoes may be all you have.

Regardless of the setting, the most important rule is to keep certain things from coming in contact with each other. Many preventable diarrhea outbreaks start with scenarios that sound like no-brainers. Pet turtles and frogs should not live on the kitchen counter where food is prepared, the baby's diaper pail should not be near the refrigerator, and nothing intended for human consumption should be under a bird cage or near a toilet. The kitchen sink drain is the dirtiest place in most homes—if food falls in there, wash or boil it before eating it. And don't forget the attic! Bats, flying squirrels, mice, and pigeons all contribute hazards of their own.

Communal areas may be challenging, such as gym showers and shared laundry rooms. Most people already know about athlete's foot and meningitis, but there is a popular misconception that washing machines make everything clean. On the contrary, washing clothes in cool water and hanging them to dry is a good way to share other people's lice, ticks, and many bacteria. Some cities have ordinances about water temperatures in commercial laundries, but building owners often save money by turning down the thermostat. Tenants can respond by turning it up again or washing clothes in the sink.

Not every home has a white picket fence and window screens. Homeless shelters, cardboard boxes, doorways, and freeway overpasses can also be homes. Associated risks include not only interpersonal violence and malnutrition, but also mosquitoborne and louseborne diseases. Some communities have programs that provide homeless people with small, portable shelters, but any long-term solution must start by determining why so many people are marginalized in the first place.

In prisons, severe overcrowding and sexual tension create ideal breeding grounds for HIV and hepatitis. Some states provide condoms for inmates, but it is rarely possible to enforce their use. Inmates have limited opportunities to protect their health, except by washing their hands, postponing tattoos, and not sharing cigarettes or personal items such as razors.

In developed nations, most employers make protective equipment available to workers who are exposed to biohazards. But accidents are inevitable, and workers sometimes take shortcuts. Hospital staff and first responders sometimes reuse syringes, handle needles carelessly, or improvise solutions to unexpected biohazards (Case Study 7-10). Waste management and sewer workers are at risk for hepatitis, leptospirosis, and tetanus; poultry and swine workers are exposed to influenza and salmonellosis. Postal workers have always risked hazardous exposure, but the 2001 anthrax mailings elevated these concerns to a new level. Bartenders and waitresses catch meningitis by handling glasses contaminated with human saliva. Triathlon runners have contracted leptospirosis by splashing through contaminated streams. Sex workers can catch any communicable disease their clients happen to have.

Schools are workplaces for teachers, but they are crucibles of social evolution for students. Preschool children slobber all over each other (and the class turtle), eat food off the floor, and make a mess in the bathroom, but their immune systems need the exercise. Parents need to accept the fact that children will eventually bring home roseola, pinkeye, and other childhood diseases—even icky-sounding things like head lice and pinworms. Adolescents are at risk for mononucleosis, meningitis, and anything an adult might catch. Recreational activities such as beer pong, mud wrestling, fursuiting, passing joints, and sharing needles all transfer bacteria and viruses.

Internet shopping reduces the potential for disease transmission, but it will never entirely replace the traditional marketplace. Is a salad bar more dangerous than a hookah bar? No one seems to know, but many people already avoid pay phones and drinking fountains. Shopping carts are contaminated with saliva, mucus, urine, and fecal matter; public armrests are worse. With practice, it's possible to use a public bathroom without coming in contact with any surface (other than the shoe/floor interface). But step away from the toilet quickly if it's the self-flushing variety, which can eject a cloud of nasty airborne droplets. Signs may be useful in persuading people to wash their hands before leaving public restrooms.

The Future

At least in theory, telecommuting would be a fine way to avoid the expense of workplace biosafety programs. Yet it is not clear whether telecommuting will ever really catch on for most occupations. Some companies use it already, but many others have found the results unsatisfactory, usually citing insurance or supervision problems.

References and Recommended Reading

Badiaga, S., et al. "Preventing and Controlling Emerging and Reemerging Transmissible Diseases in the Homeless." *Emerging Infectious Diseases*, Vol. 14, 2008, pp. 1353–1359.

Case Study 7-10: Legacy of a Low Bidder

In 1998, a city in the southwestern United States opened the doors of a much-needed tertiary care center. New buildings often require a few months of shakedown, but this hospital's inaugural mishaps were destined to last more than 10 years.

First, patients noticed that changes in wind direction redirected the odor of sewer gas from the vent pipes into their rooms. The contractor fixed that problem, but a new mystery soon impacted offices in the basement near the computer room. What appeared to be raw sewage would periodically run down the walls and saturate carpets, making work areas unusable. The source was elusive, because the contractor had gone out of business, leaving no blueprints of the final plumbing layout. The infection control staff surrounded the contaminated area with taped plastic curtains and warning signs, while a repair crew identified the cause: the contractor had substituted a cheaper grade of pipe that could not withstand chemicals that the cafeteria (directly above) used for cleaning. Employees in adjacent offices learned to recognize a gurgling sound as a warning that sewage was about to erupt from a toilet.

The repair crew eventually dug a large hole in the basement floor and left it open for the next year and a half while they tried to repair the main sewer pipe, which was also crumbling. Every so often, the hole filled with wastewater, and the basement smelled like an open pit toilet. Whenever it got too bad, the computer staff would lift the false floor and pour a pitcher of water into an open drain, thus filling the trap and preventing the escape of sewer gas. This was the status quo until 2009, when staff arrived one morning to an overpowering sense of déjà vu. In several of the previously cleaned-up work areas, a material described by our man on the scene as "stinking black goo" had dripped from overhead pipes that ruptured during the night. Thus, the problem entered its eleventh year, and another local government learned the hard way that poorly designed buildings are hard to keep clean.

Boyce, J. M. "Environmental Contamination Makes an Important Contribution to Hospital Infection." *Journal of Hospital Infection*, Vol. 65 (Suppl. 2), 2007, pp. 50–54.

Cliver, D. O. "Cutting Boards in Salmonella Cross-Contamination." *Journal of AOAC International*, Vol. 89, 2006, pp. 538–542.

Curtis, V., et al. "Hygiene in the Home: Relating Bugs and Behaviour." *Social Science and Medicine*, Vol. 57, 2003, pp. 657–672.

Evans, M. R., et al. "An Outbreak of Viral Gastroenteritis Following Environmental Contamination at a Concert Hall." *Epidemiology and Infection*, Vol. 129, 2002, pp. 355–360.

Finn, R., et al. "Cluster of Serogroup C Meningococcal Disease Associated with Attendance at a Party." *Southern Medical Journal*, Vol. 94, 2001, pp. 1192–1194.

Goldhammer, K. A., et al. "Prospective Study of Bacterial and Viral Contamination of Exercise Equipment." *Clinical Journal of Sports Medicine*, Vol. 16, 2006, pp. 34–38.

Harrison, L. H., et al. "Risk Factors for Meningococcal Disease in Students in Grades 9–12." *Pediatric Infectious Disease Journal*, Vol. 27, 2008, pp. 193–199.

Kagan, L. J., et al. "The Role of the Home Environment in the Transmission of Infectious Diseases." *Journal of Community Health*, Vol. 27, 2002, pp. 247–267.

Kassem, I. I., et al. "Public Computer Surfaces Are Reservoirs for Methicillin-Resistant Staphylococci." *ISME Journal*, 31 May 2007.

Larson, E. L., et al. "Predictors of Infectious Disease Symptoms in Inner City Households." *Nursing Research*, Vol. 53, 2004, pp. 190–197.

Lee, M. B., and J. D. Greig. "A Review of Enteric Outbreaks in Child Care Centers: Effective Infection Control Recommendations." *Journal of Environmental Health*, Vol. 71, 2008, pp. 24–32.

Meyer, T. E., et al. "West Nile Virus Infection among the Homeless, Houston, Texas." *Emerging Infectious Diseases*, Vol. 13, 2007, pp. 1500–1503.

Niveau, G. "Prevention of Infectious Disease Transmission in Correctional Settings: a Review." *Public Health*, Vol. 120, 2006, pp. 33–41.

Parry, S. M., et al. "A Case-Control Study of Domestic Kitchen Microbiology and Sporadic *Salmonella* Infection." *Epidemiology and Infection*, Vol. 133, 2005, pp. 829–835.

Reynolds, K. A., et al. "Occurrence of Bacteria and Biochemical Markers on Public Surfaces." *International Journal of Environmental Health Research*, Vol. 15, 2005, pp. 225–234.

Saenz, R. A., et al. "Confined Animal Feeding Operations as Amplifiers of Influenza." *Vector Borne and Zoonotic Diseases*, Vol. 6, 2006, pp. 338–346.

Scott, E. "Food Safety and Foodborne Disease in 21st Century Homes." *Canadian Journal of Infectious Diseases*, Vol. 14, 2003, pp. 277–280.

Scott, E., et al. "A Pilot Study to Isolate *Staphylococcus aureus* and Methicillin-Resistant *S. aureus* from Environmental Surfaces in the Home." *American Journal of Infection Control*, Vol. 36, 2008, pp. 458–460.

Tunç, K., and U. Olgun. "Microbiology of Public Telephones." *Journal of Infection*, Vol. 53, 2006, pp. 140–143.

Vonesch, N., et al. "Emerging Infectious Diseases among Swine Workers." *Giornale Italiano di Medicina del Lavoro ed Ergonomia*, Vol. 29, 2007, pp. 401–404. Italian.

Weiss, M. M., et al. "Disrupting the Transmission of Influenza A: Face Masks and Ultraviolet Light as Control Measures." *American Journal of Public Health*, Vol. 97 (Suppl. 1), pp. S32–37.

Winther, B., et al. "Environmental Contamination with Rhinovirus and Transfer to Fingers of Healthy Individuals by Daily Life Activity." *Journal of Medical Virology*, Vol. 79, 2007, pp. 1606–1610.

PART 3: ATTACKING THE ENEMY'S ARMY IN THE FIELD

If prevention or exclusion fails, the next form of generalship focuses on the use of antibiotics, antiviral drugs, antisera, pesticides, and other measures to kill (or inactivate) imminent threats and the vectors that transport them. At this stage, we deploy chemical weapons, bring out the masks, remind people to get their booster shots, and close the movie theaters.

This approach is sometimes difficult with an emerging disease or pest, whose forces are likely to become fully entrenched before we can mount an effective defense. At this point, strategy 3 often collapses into strategy 4, the least effective of all. In a literal war, we have a fairly good idea how to kill our fellow human beings, if it comes to that; but every nonhuman biohazard is unique, and designing weapons may require months or years of study. But in the case of an old adversary such as tuberculosis, which retreated and regrouped and then launched a new offensive, we can often draw upon the lessons of the past.

Killing the Enemy: Snipers and WMDs

Summary of Topic

Some diseases and agricultural pests inevitably sneak past the health inspectors and sacrificial chickens, and then it's all-out war. In the case of human diseases, the weapons of choice must kill the infectious agent without injuring the host. With livestock and plants, we cannot always afford to be so fussy, as the histories of foot-and-mouth disease and citrus canker demonstrate all too well (Chapters 4 and 5).

How Will This Help?

Antimicrobial, antiviral, and antiparasitic drugs can save many lives and shorten disease outbreaks, just as agricultural pesticides can save individual farms or fields. Sometimes these measures can drive back the enemy, at least temporarily. An example is the ongoing battle to eradicate the Mediterranean fruit fly, or Medfly (*Ceratitis capitata*), which first reached California in 1980 but has been held back (more or less) by years of aggressive pesticide spraying campaigns combined with trapping and the release of sterile males. In other cases, victory can happen by accident, as in the case of the Rocky Mountain locust (Chapter 5), which farmers eradicated in the course of plowing their fields. That solution was analogous to driving a jeep over the sleeping enemy in the dark. But sometimes the enemy eludes our best efforts (Case Study 7-11).

Discussion

It seems that every powerful weapon has some drawbacks. Chapter 3 described the problem of drug-resistant bacteria that have evolved after years of exposure to the antibiotics that were once hailed as invincible miracle drugs. Pharmaceutical companies continually develop new antibiotics to replace the old ones, but these drugs are effective only until new resistant bacterial strains appear. Unfortunately, some antibiotics have major side effects. After the 2001 anthrax mailings, for example, doctors prescribed the antibiotic ciprofloxacin to prevent

Case Study 7-11: Surprise Attack

Despite public health surveillance programs, diseases and pests sometimes slip through the net and attack without warning. In some cases, a disease is already well known in one part of the world but makes an unexpected leap to another. A prime example is the 1999 arrival of West Nile virus in the United States, where it quickly spread from coast to coast and became established in all 48 conterminous states by 2005. As of 2009, researchers have not yet developed a human vaccine for this disease. In other cases, a "new" disease seems to drop from the sky. SARS first appeared in Asia in 2003, causing more than 800 deaths and worldwide panic before mysteriously vanishing again. By 2004, scientists had developed a SARS vaccine that worked on mice, but the outbreak was already over. Will the vaccine work if SARS returns? Or will the enemy change tactics? As of 2009, we just don't know.

exposed postal workers from contracting pulmonary anthrax; but some of them reportedly found the headaches and other side effects of ciprofloxacin so intolerable that they stopped taking it. (Not only new antibiotics have side effects; one of the oldest ones, streptomycin, sometimes caused deafness.)

Antimalarial drugs also are notorious for their side effects, and many travelers to the tropics risk contracting malaria rather than take these drugs (Chapter 2). Another chemical weapon with an unwelcome kick is a controversial pesticide called endosulfan, which kills a number of crop pests but may cause severe health and environmental problems. In response, some countries have banned its use.

Popular Culture

According to a short-lived rumor, the highly publicized "toxic fumes" death of a southern California woman in 1994 resulted from exposure to the organophosphate pesticide malathion, which had recently been sprayed in the area as part of a campaign to eradicate the Mediterranean fruit fly. Newspapers first reported (but later retracted) a doctor's statement that people exposed to the unidentified fumes tested positive for organophosphates. Studies have shown that malathion can become a more toxic chemical called malaoxon when sprayed on environmental surfaces, and it might undergo further transformation on a person's skin, but apparently investigators eliminated that possibility in the 1994 case.

Weapons have always been valuable trade goods. The 1949 motion picture *The Third Man* is a film adaptation of the Graham Greene classic about the moral dilemma of a black market drug dealer in postwar Vienna, who sells a watered-down version of a valuable new drug called penicillin.

The Future

Again, the world needs better drugs and pesticides. High on our wish list are drugs that attack pathogens in ways that reduce the probability of resistance, improved drug delivery systems, and pesticides that kill crop pests without harming people or the environment.

References and Recommended Reading

Cherkasov, A., et al. "Use of Artificial Intelligence in the Design of Small Peptide Antibiotics Effective against a Broad Spectrum of Highly Antibiotic-Resistant Superbugs." *ACS Chemical Biology*, 4 December 2008.

Coates, A. R., and Y Hu. "Novel Approaches to Developing New Antibiotics for Bacterial Infections." *British Journal of Pharmacology*, Vol. 152, 2007, pp. 1147–1154.

Fountain, H. "Researchers Find Bacteria That Devour Antibiotics." *New York Times*, 8 April 2008.

Gaspar, M. M., et al. "Developments on Drug Delivery Systems for the Treatment of Mycobacterial Infections." *Current Topics in Medicinal Chemistry*, Vol. 8, 2008, pp. 579–591.

Jagusztyn-Krynicka, E. K., and A. Wyszyńska. "The Decline of Antibiotic Era—New Approaches for Antibacterial Drug Discovery." *Polish Journal of Microbiology*, Vol. 57, 2008, pp. 91–98.

Linares, J. F., et al. "Antibiotics as Intermicrobial Signaling Agents Instead of Weapons." *Proceedings of the National Academy of Sciences* (U.S.), Vol. 103, 2006, pp. 19484–19489.

Marty, M. A., et al. "Assessment of Exposure to Malathion and Malaoxon due to Aerial Application over Urban Areas of Southern California." *Journal of Exposure Analysis and Environmental Epidemiology*, Vol. 4, 1994, pp. 65–81.

Miles, M. R., et al. "Soybean Rust: Is the U.S. Soybean Crop at Risk?" American Phytopathological Society, June 2003.

Naqvi, S. M., and C. Vaishnavi. "Bioaccumulative Potential and Toxicity of Endosulfan Insecticide to Non-Target Animals." *Comparative Biochemistry and Physiology* C, Vol. 105, 1993, pp. 347–361.

Patrick, D. M. "Antibiotic Use and Population Ecology: How You Can Reduce Your 'Resistance Footprint.'" *Canadian Medical Association Journal*, Vol. 180, 2009, pp. 416–421.
"Powerful Antibiotic Battles MRSA." United Press International, 22 October 2008.
"Scientists Develop SARS Vaccine." United Press International, 19 July 2006.
Talbot, G. H., et al. "Bad Bugs Need Drugs: An Update on the Development Pipeline from the Antimicrobial Availability Task Force of the Infectious Diseases Society of America." *Clinical Infectious Diseases*, Vol. 42, 2006, pp. 657–668.

Disabling the Enemy's Transportation: Inconvenient Barriers

Summary of Topic

Just as soldiers have jeeps or planes, the agents of disease have their own means of transportation, such as mosquitoes, ticks, infected people or cows, body fluids, or the wind. If minor control efforts fail, the next level includes such measures as pesticides (this time for vectors rather than pests), quarantine, protective masks and clothing, male and female condoms, and the slaughter of exposed livestock.

How Will This Help?

Controlling the movement of vectors or hosts may be enough to control the disease itself. Most people no longer have body lice, which were once important vectors for epidemic typhus and trench fever. Bubonic plague tends not to spread through a human population unless its members are infested with fleas. Another public health victory that targeted transportation (hosts) rather than the disease itself was the 2001 conquest of a foot-and-mouth disease outbreak in England. Despite understandable public protest, the health authorities slaughtered an estimated 4 million cattle and brought the invasion to a screeching halt.

Discussion

We started with the easy examples as usual. Nobody really wanted fleas or lice in the first place, and many people can rationalize the slaughter of millions of cows to stop a disease outbreak. But would we kill a flock of beautiful wild swans that *might* harbor bird flu? Maybe. What about a group of infected people? This strategy of blocking transportation has obvious limits.

Since killing sick people is out of the question, the usual solution is to isolate them instead. A century ago, locked facilities were full of people suffering from contagious diseases such as tuberculosis, typhoid, and Hansen's disease (leprosy). This practice has nearly disappeared from developed countries due to human rights issues, but a good scare might bring it back. A few people with diseases such as multidrug-resistant TB have been briefly incarcerated in recent years (Chapter 2).

As another example of situation ethics, consider the Asian tiger mosquito (*Aedes albopictus*), which invaded the United States in the 1980s. It is a potential vector of dengue fever, and its arrival caused some concern, although not enough to risk bombarding it with pesticides, which often cause environmental damage or illness. But if dengue begins to spread in the United States, drastic measures may become easier to justify.

In some cases, blocking transportation of a pathogen or pest may be physically impossible. Tropical storms and hurricanes enabled citrus canker and soybean rust to reach North America, and it's rarely possible to stop the wind, although barriers may help in local situations. Hurricane prevention is pure science fiction at present.

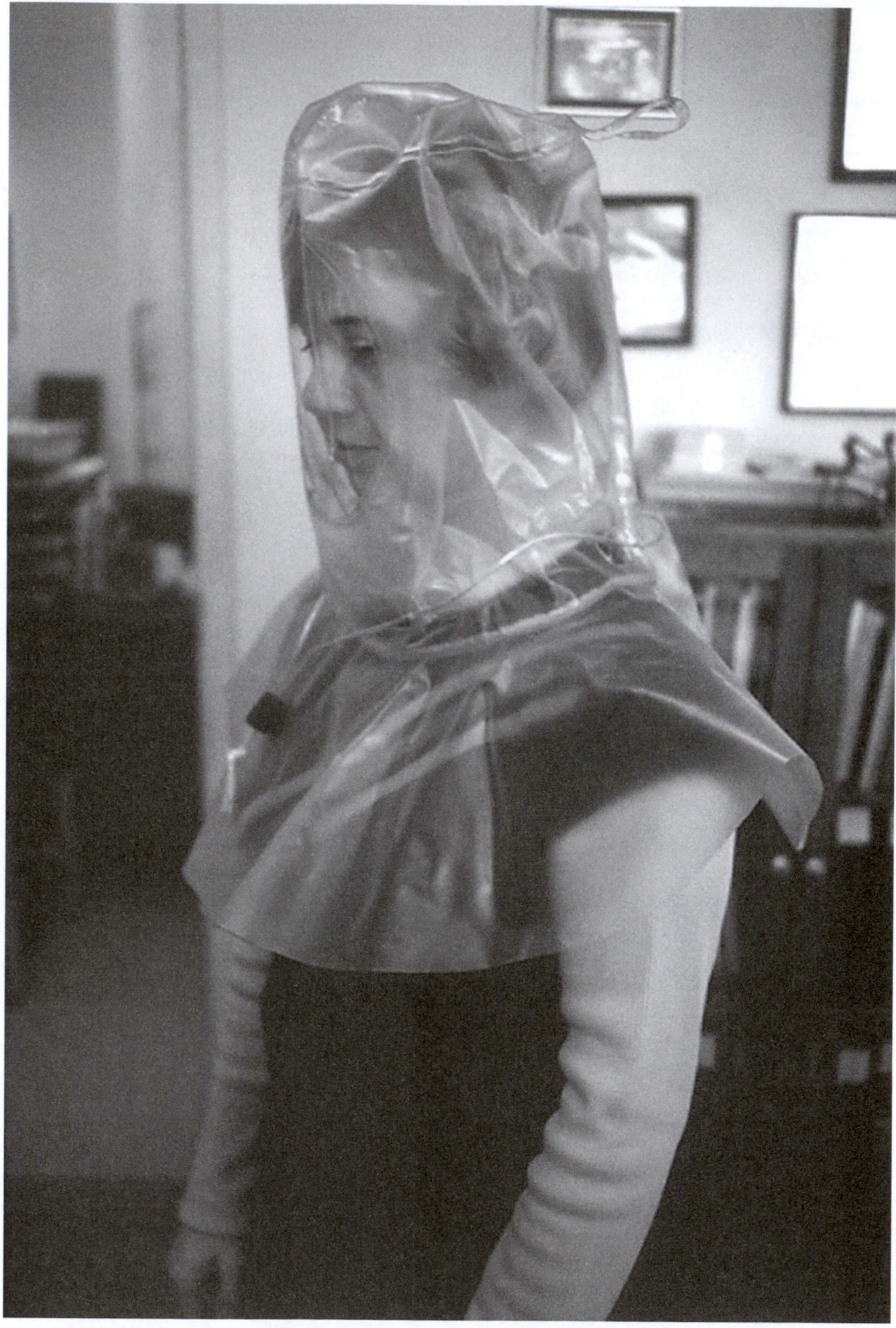

Figure 7.5 A CDC employee models protective headgear used in 1976.
Source: U.S. Centers for Disease Control and Prevention, Public Health Image Library.

In still other cases, the actions required to stop the movement of a disease are merely inconvenient, not impossible or unethical. Many people reject condoms as an unacceptable nuisance or insist on disregarding quarantines. Particulate masks and supplied-air respirators (Figure 7.5) are uncomfortable, but gain acceptance during epidemics or whenever exposure is likely. See also Case Study 7-12.

Case Study 7-12: Behind the Mask

The U.S. Food and Drug Administration (FDA) recently approved air-filtering face masks called N95 respirators for public use during influenza pandemics. "N95" refers to the fact that these masks remove at least 95 percent of small airborne particles called nanoparticles. Most hardware stores sell N95 masks, which are also used for dust protection, but not all are suitable for use in public health emergencies. The FDA website lists specific models and their capabilities. After the 2001 anthrax mailings, many Americans tried to buy more effective respirators called gas masks, only to encounter bureaucratic problems. At least one government official told the press that the public should not be allowed to purchase gas masks, because they would not know how to use them. For example, he said, if people forgot to open the valve or remove a protective seal, they would suffocate! If the government thinks people are this stupid, will it provide protective equipment in a real emergency? Anyone who can decipher the user manual for a DVD player can certainly read the instructions for a respirator.

Popular Culture

In the 1995 motion picture *Outbreak*, the bad general wants to incinerate a town because its occupants have a deadly disease that might start a nationwide epidemic. But the good doctors want to cure the infected people instead, and they succeed in preparing an effective antiserum just in time.

Humor can be more effective than violence in conveying a social message. In the 1988 movie *The Naked Gun*, the main character wears a full-body condom—the ultimate inconvenient barrier, offering protection against everything.

Readers may remember the 2007–2008 Trojan commercials in which pigs try unsuccessfully to pick up women at a bar, until one pig goes to the men's room and buys a condom, which magically transforms him into a handsome young man who finally scores. Cute—if a bit graphic—but Fox and CBS refused these ads outright. Some viewers found them too explicit; others took them to mean that men are filthy animals. But most probably got the point: responsible humans, unlike pigs, take care of themselves and their partners. Of course, it takes two to boogie. The woman in the ad could have bought a condom in the women's room instead, and then grabbed the pig and hauled him squealing out to her car. But the Humane Society would probably object to that version.

The Future

As the world becomes more crowded and new diseases emerge, people may become more willing to make difficult choices, such as enforcing quarantines of people with contagious diseases. But if this trend goes too far, we run the risk of becoming desensitized to violence and indifferent to human rights.

References and Recommended Reading

Bensimon, C. M., and R. E. Upshur. "Evidence and Effectiveness in Decisionmaking for Quarantine." *American Journal of Public Health*, Vol. 97 (Suppl. 1), 2007, pp. S44–48.

Galvin, J. W., et al. "Killing of Animals for Disease Control Purposes." *Revue Scientifique et Technique*, Vol. 24, 2006, pp. 711–22.

Hartnack, S., et al. "Mass Culling in the Context of Animal Disease Outbreaks—Veterinarians Caught between Ethical Issues and Control Policies." *Deutsche Tierarztliche Wochenschrift*, Vol. 116, 2009, pp. 152–157.

"Man in Lockdown for Exposing Others to TB." *Science Online*, 2 May 2007.
McDougall, C. W., et al. "Emerging Norms for the Control of Emerging Epidemics." *Bulletin of the World Health Organization*, Vol. 86, 2008, pp. 643–645.
"Microbicide Gels May Protect Women from AIDS." Reuters, 9 February 2009.
Mindel, A., and S. Sawleshwarkar. "Condoms for Sexually Transmissible Infection Prevention: Politics versus Science." *Sexual Health*, Vol. 5, 2008, pp. 1–8.
Newman, A. A. "Pigs with Cellphones, but No Condoms." *New York Times*, 18 June 2007.
Rosner, F. "Involuntary Confinement for Tuberculosis Control: The Jewish View." *Mount Sinai Journal of Medicine*, Vol. 63, 1996, pp. 44–48.
Sadasivaiah, S., et al. "Dichlorodiphenyltrichloroethane (DDT) for Indoor Residual Spraying in Africa: How Can It Be Used for Malaria Control?" *American Journal of Tropical Medicine and Hygiene*, Vol. 77, Suppl. 6, 2007, pp. 249–263.
Sarkar, N. N. "Barriers to Condom Use." *European Journal of Contraception and Reproductive Health Care*, Vol. 13, 2008, pp. 114–122.
Spizzichino, L. "The Female Condom: Knowledge, Attitude, and Willingness to Use. The First Italian Study." *Annali dell'Istituto Superiore di Sanità*, Vol. 43, 2007, pp. 419–424.
"Wind May Spread Disease Faster Than Thought." Associated Press, 30 March 2009.

Destroying the Enemy's Resources: Habitat Modification

Summary of Topic

There are a few situations in which drastic modification of habitat is the best or only way to defeat a disease. "Drastic" does not mean emptying a birdbath or mowing weeds, but large-scale regional changes, such as building dams or draining wetlands. Like any other strategy, this one can fail; and when it fails, the consequences can be huge.

How Will This Help?

Since antiquity, people have used major landscape alterations to control disease vectors, such as mosquitoes or tsetse flies. This approach creates jobs, sounds exciting, and sometimes works, but all too often it backfires, giving (figurative) aid and comfort to the enemy instead. This section presents several examples.

Discussion

Between 1935 and 1955, the U.S. government encouraged farmers to plant an Asian vine called kudzu to prevent soil erosion, to fix nitrogen in the soil, and to provide high-protein forage for livestock. These sounded like excellent goals, and the farmers cooperated. As a result, kudzu has overrun much of the southeastern United States, where it has destroyed natural habitats and now provides a winter host for the soybean rust fungus and the soybean aphid (Chapter 5). The government now encourages farmers to destroy kudzu. Similarly, a good deal of effort was expended in building up deer herds in the northeastern states, thus creating conditions that promote Lyme disease outbreaks in humans (Chapter 6).

During construction of the Panama Canal in the early 1900s, the engineers in charge protected the workforce by largely eliminating yellow fever—not by spraying pesticides such as DDT (which were not yet available), but by drastic modification of wetland habitats to eliminate mosquitoes. At that time, completion of the Canal was hailed as a victory of man over nature. Today, however, biologists point out that the project caused enormous environmental damage. Wetlands are not just hellholes of pestilence, but complex ecosystems that the world needs (see also Case Study 7-13, page 279). But the world also needed the Panama Canal, and it was necessary to protect the workers who were building it.

For nearly a century, occasional visionaries have proposed that a dam should be built across the Strait of Gibraltar for a variety of reasons—to prevent global cooling or warming, to improve the climate of northern Africa, or to provide hydroelectric power. At present, it is far from certain that damming Gibraltar would achieve any of these objectives, but the dialog continues.

In 2009, one of many challenges facing Iraq is restoration of the Mesopotamian marshes, which were modified by dam construction in the 1960s and drained during the 1990s. Whether the latter effort was politically motivated or intended to fight malaria and create farmland, the result was an ecological disaster that destroyed more than 90 percent of the wetland that once protected Iraq's coastal fisheries and water quality.

Not all landscapes are external, and not every transformation is visible. Several authors have pointed out that the near-elimination of parasitic worms and *Helicobacter pylori* from the human GI tract may have increased the incidence of asthma, diabetes, and ulcerative colitis.

Case Study 7-13: The Florida Everglades

When the first European colonists arrived in southern Florida, they found a slow-moving river of grass 60 miles wide and 6 inches deep—too wet for farming, and a paradise for mosquitoes that transmit deadly diseases such as malaria, yellow fever, and encephalitis. Beautiful scenery and biodiversity had lower priorities than survival; the water had to go. By the 1880s, the State of Florida was offering cheap Everglades land to anyone who would drain it. In 1905, Governor Napoleon B. Broward (1857–1910) was elected partly on the basis of his campaign promise to drain the Everglades. By the 1960s, the Army Corps of Engineers and others had succeeded in draining about 50 percent of the area and converting the land to farming and urban development. In the 1980s, however, growing environmental awareness (and the availability of DEET) led to a reversal of this trend. In 2000, Congress approved the Comprehensive Everglades Restoration Plan, the most expensive and ambitious environmental restoration plan in history. As of 2009, the status of this project is uncertain, due to its cost and numerous related lawsuits.

Popular Culture

During the Vietnam War, the United States military sprayed a chemical defoliant called Agent Orange to reduce dense jungle foliage and prevent opposing forces from hiding in it. Several motion pictures about the war, such as *We Were Soldiers* (2002), make reference to this controversial practice and its alleged consequences.

In Greek mythology, the hero Heracles (Hercules) used his ingenuity to destroy a marsh-dwelling serpent and a flock of evil marsh birds. According to some scholars, these tasks represented the efforts of early Mediterranean civilizations to control malaria by draining swampland.

In European and North American folk tradition, the forest is a dangerous place, because it conceals wolves and strange miasmas and giant people covered with hair. From this perspective, disease is just one of many evils that dwell in darkness, and to create a safe haven for humans, it is necessary to clearcut an area and build a farm. This theme is prominent in the 2004 motion picture *The Village*, in which a stand of trees represents a nearly impenetrable boundary between two worlds.

The Future

Although human efforts to combat disease have transformed landscapes, this is not necessarily a one-way process. According to some historians, the New World as described by explorers in 1750 had returned to something approximating a pristine condition in the centuries since European diseases decimated the inhabitants. Before that, the Americas showed extensive evidence of agriculture and other human influences. Thus, if left alone for whatever reason, natural ecosystems may eventually recover.

References and Recommended Reading

Blaustein, R. J. "Kudzu's Invasion into Southern United States Life and Culture." In McNeeley, J. A. (Ed.). 2001. *The Great Reshuffling: Human Dimensions of Invasive Species.* Cambridge, UK: IUCN.

Cave, D. "Everglades Restoration Plan Shrinks." *New York Times*, 1 April 2009.

Denevan, William M. "The Pristine Myth: The Landscape of the Americas in 1492." *Annals of the Association of American Geographers* , Vol. 82, 1992, pp. 369–385.

Graham, D. Y., et al. "Contemplating the Future without *Helicobacter pylori* and the Dire Consequences Hypothesis." *Helicobacter*, Vol. 12, Suppl. 2, 2007, pp. 64–68.

Heckenberger, M. J. "Amazonia 1492: Pristine Forest or Cultural Parkland?" *Science*, Vol. 301, 2003, pp. 1710–1714.

Helmuth, L. "Can This Swamp Be Saved?" *Science News Online*, 17 April 1999.

Kandji, S. T., et al. "Climate Variability and Climate Change in the Sahel Region." World Agroforestry Centre and United Nations Environment Programme, 2006.

Landry, C. J. "Who Drained the Everglades?" *PERC Reports*, March 2002.

Richardson, C. J., et al. "The Restoration Potential of the Mesopotamian Marshes of Iraq." *Science*, Vol. 307, 2005, pp. 1307–1311.

Sturrock, R. F., et al. "Seasonality in the Transmission of Schistosomiasis and in Populations of its Snail Intermediate Hosts in and around a Sugar Irrigation Scheme at Richard Toll, Senegal." *Parasitology*, Vol. 123, 2001, pp. 77–89.

Enlisting Allies: Biological Controls

Summary of Topic

What could be more harmless and natural than the use of biological controls, such as small imported fish to eat mosquito larvae in ponds, or lady beetles to eat aphids in gardens? In some cases these strategies are fine, but exceptions happen. Like allies in a literal war, biological controls sometimes prove to be a liability in the long run, and getting rid of them can be difficult.

How Will This Help?

When used judiciously after years of study, predator release and other biological control measures can be helpful. When used impulsively, on the theory that any action is better than just standing around doing nothing, the environmental results can be worse than the original problem. Several biological control efforts described in this section have had negative consequences, but it is important to remember that some alternatives (such as intensive use of chemical pesticides) may be even worse.

Discussion

The level of threat dictates the nature of a proportional response. Immediate threats with potentially severe consequences may justify drastic action that would otherwise be inappropriate (Case Study 7-14). Mowed weeds grow back, and a drained pond can refill, but alien predators tend to hang around forever.

Mosquito fish (*Gambusia affinis* and related species) are native to the southeastern United States, but resource agencies have released them in every state and on every continent for mosquito control. These fish reproduce quickly, in even the muckiest water, and they eat many mosquito larvae. *Gambusia* are harmless when confined to a garden pond, but they tend to escape into natural streams and lakes, where they compete with native fish and may reduce biodiversity. In some cases, native fish have proven more effective than *Gambusia* in controlling mosquitoes, when given a chance to do so.

In 1969, resource agencies in California, Virginia, and Montana wanted to control an invasive European weed called the musk thistle (*Carduus nutans*), so they decided to release a European insect called the thistle-head weevil (*Rhinocyllus conicus*) to eat the thistles. Canada had tried the same solution a year earlier, and the first reports were favorable. By 1996, however, researchers noted that the weevil had expanded its range to other states, and that it was also eating native thistles and interfering with native insects. But the musk thistle survived, and instead of one invasive species, there were two.

Case Study 7-14: Turnabout

Many communities use large quantities of bacterial spores called Bt to treat bodies of water where mosquitoes breed. In theory, the mosquito larvae eat these bacteria and die, but no other organisms are affected. In practice, it seems likely that unnaturally high concentrations of this common soil bacterium (*Bacillus thuringiensis*) must have unintended side effects. Studies have shown that Bt can cause skin and eye irritation in humans, and that its long-term use on wetlands may reduce biodiversity. Bt is known to kill many insect species, not only mosquitoes. Also, since Bt is closely related to the agent of anthrax (*Bacillus anthracis*), the presence of high levels of Bt in the environment can yield false positive results on tests for anthrax spores.

Popular Culture

The archetype of biological control is an old woman who swallows a fly and then makes a series of seemingly unwise decisions:

> She swallowed a cow to kill the goat,
> She swallowed a goat to kill the dog,
> She swallowed a dog to kill the cat,
> She swallowed a cat to kill the bird,
> She swallowed a bird to kill the spider,
> She swallowed a spider to kill the fly.
> I don't know why she swallowed the fly.
> Perhaps she'll die.

The Future

Biological controls will continue to be a valuable component of integrated pest management if past mistakes can be avoided. Basic research to clarify predator-prey interactions may help. For example, some studies have demonstrated that adding a predator species may not reduce the number of prey, because the predators may eat one another instead.

References and Recommended Reading

Baker, C., et al. "*Gambusia*—A Biodiversity Threat?" *Water and Atmosphere*, Vol. 12, 2004, pp. 24–25.

Canyon, D. V., and J. L. Hii. 1997. "The Gecko: An Environmentally Friendly Biological Agent for Mosquito Control." *Medical and Veterinary Entomology*, Vol. 11, 1997, pp. 319–323.

Chandra, G., et al. "Mosquito Control by Larvivorous Fish." *Indian Journal of Medical Research*, Vol. 127, 2008, pp. 13–27.

Favia, G., et al. "Bacteria of the Genus *Asaia*: A Potential Paratransgenic Weapon against Malaria." *Advances in Experimental Medicine and Biology*, Vol. 627, 2008, pp. 49–59.

Howard, A. F., and F. X. Omlin. "Abandoning Small-Scale Fish Farming in Western Kenya Leads to Higher Malaria Vector Abundance." *Acta Tropica*, Vol. 105, 2008, pp. 67–73.

Louda, S. M., et al. "Ecological Effects of an Insect Introduced for the Biological Control of Weeds." *Science*, Vol. 277, 1997, pp. 1088–1090.

Marten, G. G., and J. W. Reid. "Cyclopoid Copepods." *Journal of the American Mosquito Control Association*, Vol. 23, Suppl. 2, 2007, pp. 65–92.

Rupp, H. R. "Adverse Assessments of *Gambusia affinis*." *American Currents*, Summer 1995.
Strong, D. R. "Fear No Weevil?" *Science*, Vol. 277, 1997, p. 1058.
Tilak, R., et al. "Prospects for the Use of Ornamental Fishes for Mosquito Control: A Laboratory Investigation." *Indian Journal of Public Health*, Vol. 51, 2007, pp. 54–55.
Vance-Chalcraft, H. D., et al. "The Influence of Intraguild Predation on Prey Suppression and Prey Release: a Meta-Analysis." *Ecology*, Vol. 88, 2007, pp. 2689–2696.
Walton, W. E. 2007. "Larvivorous Fish Including *Gambusia*." *Journal of the American Mosquito Control Association*, Vol. 23, Suppl. 2, 2007, pp. 184–220.

Bugout: Postexposure Prophylaxis

Summary of Topic

Postexposure prophylaxis (PEP) means any treatment given after disease exposure to prevent the person from contracting the disease. Familiar examples include anti-rabies shots administered after an animal bite, tetanus booster shots after injury, antiviral drugs after HIV exposure, and antibiotics after exposure to anthrax and other bacterial diseases.

How Will This Help?

Any war generates casualties that must be evacuated to a medical facility for observation and treatment. The war against infectious disease is no different; despite all precautions, healthcare workers and innocent bystanders are sometimes exposed to dangerous pathogens, and we cannot simply leave them on the battlefield. Also, we can learn a great deal about a disease by studying its victims. Since it would not be ethical to expose people to disease or withhold treatment intentionally (although this has happened on occasion), the process of treating and tracking these accidental casualties often yields valuable information that may later save many lives (Case Study 7-15).

Case Study 7-15: Saved from Ebola?

On March 12, 2009, a German research scientist accidentally pricked her finger with a needle that she had used to inject the Ebola virus into laboratory animals. A laboratory in Canada had an experimental Ebola vaccine, but had never tested it on a human. Two days after the accident, the vaccine arrived in Germany, and the scientist was injected with it. About 12 hours later, she reported a headache, muscle pain, and a fever, but these side effects passed quickly. No further symptoms appeared after 21 days, and physicians determined that she was out of danger. The incident shows that the vaccine is probably safe for humans, but it would be premature to celebrate the conquest of Ebola, because needlestick injuries often do not transmit infection.

Discussion

Depending on the disease exposure, supportive care may be the only treatment necessary or available. Many people die of cholera every year, for example, but the only intervention required to save most patients is simple rehydration using sterile fluids that are scarce in some parts of the world. In other disease outbreaks, such as the 2003 SARS epidemic in Asia, no treatment seemed to help, but researchers later developed a vaccine that may save lives in future.

Rabies is the best known example of a deadly disease that can be prevented (rather than cured) after the agent gains access to the body. There would be no point in immunizing everyone against rabies in advance, because human cases are rare, and until recently the shots had severe side effects. The case fatality rate for untreated rabies is close to 100 percent, but symptoms do not appear until after the virus reaches the brain. Since it travels quite slowly along nerve endings from the point of entry, the bitten person usually has at least two weeks to get anti-rabies shots. Tetanus and hepatitis B also are amenable to vaccination after exposure.

No such "magic bullet" is yet available (as of 2009) for use after HIV exposure, but a combination of antiviral drugs often can prevent infection. Drugs such as Tamiflu® may protect persons at high risk during influenza outbreaks, but many strains are resistant (Chapter 2). In the case of exposure to hepatitis A—but not hepatitis C—the usual treatment is an injection of immune globulin. Persons exposed to a dangerous bacterial infection, such as anthrax, can often benefit from a large prophylactic dose of antibiotics.

Popular Culture

According to one misguided urban legend, a person who has been exposed to HIV can prevent infection by immediately drinking bleach. This claim is not only false, but extremely dangerous. Do not drink bleach for any reason.

Another urban legend claims that a woman who may have been exposed to a sexually transmitted disease should immediately douche with a shaken-up bottle of Diet Coke. This one is probably not dangerous, just silly. (An older urban legend holds that carbonated soft drinks can prevent pregnancy by killing sperm.)

In response to the recent nationwide panics regarding Lyme disease and West Nile encephalitis, a few people have claimed that immediately cauterizing a tick or mosquito bite by burning the skin with a hot object will prevent infection. This is unlikely to work, and can cause severe injury.

The Future

As preventive healthcare becomes less accessible and affordable, and more people choose to forgo vaccination for diseases such as tetanus and hepatitis B, the need for reactive measures may increase. Overuse of prophylactic drugs will continue to reduce their value by promoting resistance.

References and Recommended Reading

Benn, P., and M. Fisher. "HIV and Postexposure Prophylaxis." *Clinical Medicine*, Vol. 8, 2008, pp. 319–322.

Brouillard, J. E., et al. "Antibiotic Selection and Resistance Issues with Fluoroquinolones and Doxycycline against Bioterrorism Agents." *Pharmacotherapy*, Vol. 26, 2006, pp. 3–14.

Bygbjerg, I. C. "Prophylaxis—What Lies Ahead?" *Ugeskrift for Laeger*, Vol. 167, 2005, pp. 3968–3971.

Castrodale, L. "Use of Rabies Postexposure Prophylaxis Supplied by the Alaska Section of Epidemiology, 2002–2007." *Public Health Reports*, Vol. 124, 2009, pp. 262–266.

Chapman, L. E., et al. "Postexposure Interventions to Prevent Infection with HBV, HCV, or HIV and Tetanus in People Wounded during Bombings and Other Mass Casualty Events—United States, 2008." *Disaster Medicine and Public Health Preparedness*, Vol. 2, 2008, pp. 150–165.

Degertekin, B., and A. S. Lok. "Update on Viral Hepatitis: 2007." *Current Opinion in Gastroenterology*, Vol. 24, 2008, pp. 306–311.

Eddy, M., and M. Stobbe. "Experimental Vaccine Used in Ebola Study." Associated Press, 27 March 2009.

Grill, A. K. "Approach to Management of Suspected Rabies Exposures: What Primary Care Physicians Need to Know." *Canadian Family Physician*, Vol. 55, 2009, pp. 247–251.

Schneeman, A., and M. Manchester. "Anti-Toxin Antibodies in Prophylaxis and Treatment of Inhalation Anthrax." *Future Microbiology*, Vol. 4, 2009, pp. 35–43.

Stern, E. J., et al. "Conference Report on Public Health and Clinical Guidelines for Anthrax." *Emerging Infectious Diseases*, April 2008.

Tarantola, A., et al. "Infection Risks Following Accidental Exposure to Blood or Body Fluids in Health Care Workers: A Review of Pathogens Transmitted in Published Cases." *American Journal of Infection Control*, Vol. 34, 2006, pp. 367–375.

PART 4: BESIEGING WALLED CITIES

Even after a disease is well established in a population, it is still possible to fight back with the powerful chemical weapons of modern science, to enforce immunization of all susceptible hosts, to drive the enemy beyond our borders, or even to eradicate it from the world. All these options are more difficult with humans than with other species, because there are ethical and practical limits on what we can do. It is easier to force vaccination on a herd of cows, for example, than on a human population with an incompatible belief system. Humans also are inappropriate candidates for selective breeding or genetic engineering to produce disease resistance, although these methods work well with plants.

Although "besieging walled cities" is the worst military strategy, public health campaigns must often resort to it, since most major human diseases have been with us far longer than medical science. The "wall" might mean a reservoir host, or a large area of habitat that can't be removed, or characteristics of the pathogen itself that make it hard to destroy, or cultural or political barriers that make transmission likely or eradication unlikely. For example, rumors about the polio vaccine have kept polio circulating in Africa and the Middle East far longer than necessary.

Mopping Up: Disease Eradication and Elimination

Summary of Topic

With proper planning and effective weapons, it is possible to eradicate certain established diseases and pests. At least one human disease (smallpox), one cattle disease (rinderpest), and one crop pest (the Rocky Mountain locust) appear to be gone forever, at least in the wild. Several other major diseases have been eliminated from specific countries or regions.

How Will This Help?

Those who favor the "Gideon model" of population control (Chapter 6) might object to disease eradication on the grounds that death is natural. But people would continue to die, even without infectious disease. Heart disease and cancers that are not known to be infectious already account for the majority of deaths in developed nations. Infectious disease has caused needless suffering and premature death for thousands of years, but now its eradication is finally within reach. The cost and grief involved in prevention and treatment vanish if the disease no longer exists.

Discussion

Few doctors living today have treated smallpox, and few American doctors have treated polio or diphtheria. Wild smallpox no longer exists in any human population, and polio and diphtheria have been eliminated from the developed world. But diseases need to be eradicated one at a time, and there are many.

Given the required funds, public cooperation, and political will, the next candidates for worldwide eradication will probably include polio, measles, trachoma, Hansen's disease (leprosy), malaria, guinea worm disease (dracunculiasis), lymphatic filariasis, hookworm, river blindness (onchocerciasis), and Chagas disease. Tables 7.3 and 7.4 summarize the prospects for eradicating various human and animal diseases.

Might it be possible to eradicate HIV? According to most researchers, this goal would require two things that do not yet exist: an effective vaccine, and a way to eradicate all latent HIV

Table 7.3 Status of Some Human Diseases, 2008.

Disease	Present Status	Programs
Smallpox	Eradicated worldwide in 1979; virus still exists in government laboratories.	Vaccine remains in production due to fears of bioterrorism.
Poliomyelitis	Eliminated from Europe, the Americas, WHO Western Pacific region. 1,600+ reported cases in 2008 (most in Nigeria and India).	1988 World Health Assembly Resolution to Eradicate Polio has achieved greater than 99% reduction to date.
Measles (Rubeola)	Vaccination has eliminated or reduced incidence in many countries. Measles may re-emerge due to vaccine controversy.	2001 Measles Initiative goal is to reduce global incidence 90% by 2010 (75% as of 2008). Partners include CDC, WHO, UNICEF, Red Cross.
Guinea Worm Disease (Dracunculiasis)	Global incidence fell from 3.5 million in 1986 to 25,000 in 2006. Water treatment is effective.	WHO pledge to eradicate by 2009, with help from CDC, Carter Center, Gates Foundation.
Lymphatic Filariasis	Drugs and treated nets in endemic areas to reduce transmission. Incidence reduced in China and several Pacific Islands.	Carter Center Lymphatic Filariasis Elimination Program. Partners include WHO, CDC, Gates Foundation, Merck, GlaxoSmithKline.
Sleeping Sickness (African Trypanosomiasis)	Not eradicable (zoonosis) but controllable. Primary measure is eradication of tsetse fly vector.	Pan African Tsetse and Trypanosomiasis Eradication Campaign, regional elimination efforts
Hansen's Disease (Leprosy)	Eliminated (<1 case per 10,000) in most countries where it was highly endemic.	WHO Global Alliance, International Federation of Anti-Leprosy Associations
River Blindness (Onchocerciasis)	Eliminated in 11 countries. Measures: anti-parasitic drugs, vector control. Vaccine unlikely.	Carter Center River Blindness Program, Onchocerciasis Control Program (OCP), WHO initiatives, World Bank
Schistosomiasis	Measures: drugs, education, water treatment, changes in agricultural and sanitation practices.	Carter Center Schistosomiasis Control Program, WHO Partners for Parasite Control (PPC)
Chagas Disease (American Trypanosomiasis)	Target date for elimination is 2010. Measures: improved surveillance, screening, prevention, case management.	WHO Global Network for Chagas Elimination (launched 2007), Pan American Health Organization, International Development Research Centre
Trachoma	Nearly eliminated in most developed nations. Target date for global elimination is 2020. Measures: drugs, better hygiene.	Carter Center Trachoma Control Program, WHO initiatives
Diphtheria	DPT and TD vaccines eliminated diphtheria from most developed countries, but it re-emerged in 1995–2000.	WHO Expanded Program on Immunization
Malaria	Eliminated in United States, Western Europe, and Australia; nearly eliminated in many areas. Global elimination (target 2015) may require a vaccine.	Carter Center Malaria Control Program; Roll Back Malaria Partnership, Global Malaria Action Plan; Gates Foundation

(Continued)

Table 7.3 *(Continued)*

Disease	Present Status	Programs
Tuberculosis	Target date for eradication in United States was 2010; WHO now proposes to reduce global TB deaths and prevalence 50% by 2015.	WHO Global Plan to Stop TB 2006–2015; partners include Gates Foundation
Yaws	After 95% reduction, yaws re-emerged in the 1970s. Target date for elimination in Southeast Asia is 2012.	WHO Global Control of Treponematoses Programme (1952–1964); World Health Assembly Resolution, 1978; recent WHO initiatives
Ankylostomiasis (Hookworm Disease)	Nearly eradicated in USA by drugs and public health measures; still common in tropics worldwide. WHO target date 2010 for treatment of 75% of at-risk children.	World Health Organization Resolution, 54th World Health Assembly, 2001; Partners for Parasite Control (PPC)
Yellow Fever	Vaccine is safe and effective, but YF has re-emerged in Africa and South America. Reservoir includes nonhuman primates.	WHO global surveillance and vaccination programs
Typhoid Fever (Enteric Fever)	Nearly eliminated from industrialized countries; still a problem in Asia, Africa, and South America. Vaccines available, no nonhuman reservoir. Resistant strains and asymptomatic carriers are problematic.	WHO position statement on typhoid fever; eradication and control programs in endemic regions.
Cholera	Nearly eliminated from industrialized countries. Eradication may be impossible. Seventh pandemic started in South Asia in 1961 and reached Africa in 1971, and South America in 1991.	WHO position papers on prevention and control of cholera, oral rehydration salts; control programs in endemic regions.

Sources: World Health Organization (WHO), U.S. Centers for Disease Control and Prevention (CDC).

Table 7.4 Status of Some Livestock Diseases, 2008

Disease	Present Status	Programs
Rinderpest	Target date for global eradication is 2010, perhaps achieved in 2008.	FAO Global Rinderpest Eradication Program 1994
Glanders	Eradicated in most of the developed world. Still occurs in Africa, the Middle East, Asia, Mexico, South America. Measures: testing, disinfection, slaughter of infected animals.	National initiatives in endemic regions
Bovine Tuberculosis	Worldwide, sporadic/limited in US. When infected animals are found, the herd is slaughtered.	APHIS National Bovine Tuberculosis Eradication Program, other initiatives
Brucellosis	Eradicated in some countries. All farm animal species affected, 500,000+ human cases/year.	APHIS Cooperative State-Federal Brucellosis Eradication Program, other initiatives
Pseudorabies (Aujeszky's Disease)	Eradicated 2004 in U.S. commercial swine, but persists in feral swine and in other countries.	APHIS Cooperative State-Federal Pseudorabies Eradication Program 1989, other initiatives
Scrapie	At present, only Australia and New Zealand are free of scrapie. Eradication measures in other countries: surveillance, testing, slaughter.	APHIS National Scrapie Eradication Program, other national initiatives
Bovine Spongiform Encephalopathy (BSE)	As of 2006, U.S. prevalence less than 1 case per million adult cattle (controlled risk).	OIE International Standard for BSE, USDA BSE surveillance program
Bovine Paratuberculosis (Johne's Disease)	Epizootic in United States as of 2008. Control measures: surveillance, vaccination, testing, quarantine, culling.	USDA National Animal Health Monitoring System; national eradication program in development
Trichinosis (Trichinellosis)	Sporadic in U.S. No 2008 cases in commercial swine, but many wild mammals are infected.	APHIS National Swine Survey, International Commission on Trichinellosis

Sources: USDA Animal and Plant Health Inspection Service (APHIS); United Nations Food and Agriculture Organization (FAO); World Organisation for Animal Health (OIE).

Case Study 7-16: Elimination of AIDS Transmission

In 2008, WHO researchers proposed a strategy to eliminate HIV transmission, with the ultimate goal of eradicating the disease. If all people living in countries with a high incidence of HIV could be tested, and if every infected person could immediately start taking antiretroviral drugs, the incidence of new HIV cases would drop to about 1 per 1,000 people within 10 years, and the prevalence of HIV would be less than 1 percent within 50 years. The cost of antiretroviral drugs would be high, and problems would be many, but the result would be the prospect of eradication of this disease.

from the bodies of infected people. But in 2008, World Health Organization researchers proposed a startling plan that could greatly reduce the incidence of HIV in as little as 10 years (Case Study 7-16).

Public health is not a spectator sport. Everyone can help, often by doing something as simple as putting mosquito fish in the birdbath, cleaning the garbage disposal, wearing a condom, or believing the pediatrician when she says that antibiotics will not cure a cold. Even resisting the impulse to cheat on a tax return is an important step forward, because part of that money—admittedly, a small part—goes to support public health programs.

But how, you might ask, can people stand by and watch scientists conduct experiments on helpless laboratory animals, when activists and celebrities insist that such studies are unnecessary? Why not threaten the scientists and make them stop? And when we drive back across the international border with that nice basket of fruit with all those little things hopping around, why should we tell the agricultural inspector about it? Why not just hide it under a pile of coats? The usual answer, and one that works for us, is that some risks and sacrifices are often necessary for the greater good. Without clinical trials and animal experimentation and agricultural inspectors, we would be back in 1880, helplessly watching our children die from diphtheria, our horses from glanders, and our wheat from black stem rust.

People often ask why the foot soldiers in any war are the first to catch a bullet. There are many answers, but the war on disease has not spared its officers:

- Dr. Carlos Urbani, a WHO physician, died of SARS in 2003 while searching for a cure.
- Dr. Antonina Presnyakova, a Russian researcher, died of Ebola in 2004 after a needlestick injury.
- Maria Catalina Roa, Chief of Nursing for the Paraguay Ministry of Public Health, died of dengue hemorrhagic fever during a 2007 epidemic.

Popular Culture

In 1969, a celebrity doctor wrote a popular book in which he proposed that world leaders should require everyone to have a shot of penicillin (or another antibiotic) on a given date. According to that doctor, syphilis and gonorrhea would then be gone from the world forever. Even for the era, it was an astonishingly naive plan. Aside from the logistics involved in simultaneous vaccination of some 3.5 billion people—the world population in 1969—and the near impossibility of getting world leaders to agree on anything, there is the issue of compliance. Millions would refuse to participate, and then what would governments do? Put them all in prison? Hold them down and give them shots? There is also the problem of antibiotic resistance, already well known in the 1960s. Had this plan been carried out, the resistant bacteria that survived the chemical donnybrook would have gone forth to colonize the planet.

The Future

In theory, many diseases can be eradicated in the next 10 years. In practice, major obstacles include cost, international tensions, and widespread distrust of medical science. Perhaps the conquest of smallpox came just in time, before the world turned.

References and Recommended Reading

Arita, I., et al. "Eradication of Infectious Diseases: Its Concept, Then and Now." *Japanese Journal of Infectious Disease*, Vol. 57, 2004, pp. 1–6.

Bailey, R., and T. Lietman. "The SAFE Strategy for the Elimination of Trachoma by 2020: Will it Work?" *Bulletin of the World Health Organization*, Vol. 79, 2001, pp. 233–236.

Barry, S., et al. "Influenza, an Eternal Problem." *Canadian Bulletin of Medical History*, Vol. 24, 2007, pp. 445–466. [French]

Baum, S. G. "Zoonoses—With Friends Like This, Who Needs Enemies?" *Transactions of the American Clinical and Climatological Assocation*, Vol. 119, 2008, pp. 39–51.

Belongia, E. A., and A. L. Naleway. "Smallpox Vaccine: The Good, the Bad, and the Ugly." *Clinical Medicine and Research*, Vol. 1, 2003, pp. 87–92.

Carmichael, M. "Spring Fevers: Mumps Makes a Comeback in the Midwest." *Newsweek*, 21 April 2006.

Cochi, S. L., and O. Kew. "Polio Today: Are We on the Verge of Global Eradication?" *JAMA*, Vol. 300, 2008, pp. 839–841.

De Palma, A. M., et al. "Potential Use of Antiviral Agents in Polio Eradication." *Emerging Infectious Diseases*, Vol. 14, 2008, pp. 545–551.

de Quadros, C. A., et al. "Feasibility of Global Measles Eradication after Interruption of Transmission in the Americas." *Expert Review of Vaccines*, Vol. 7, 2008, pp. 355–362.

Delamothe, T. "Several Horsemen of the Apocalypse." *BMJ*, 23 August 2008.

"Discovery May Eradicate Malaria." *Science Online*, 29 August 2007.

"Gates Foundation Goes after TB." *Science Online*, 20 September 2007.

Granich, R. M., et al. "Universal Voluntary HIV Testing with Immediate Antiretroviral Therapy as a Strategy for Elimination of HIV Transmission: A Mathematical Model." *Lancet*, Vol 373, 2009, pp. 48–57.

Greenwood, B. M., et al. "Malaria: Progress, Perils, and Prospects for Eradication." *Journal of Clinical Investigation*, Vol. 118, 2008, pp. 1266–1276.

Guteri, F. "Polio's Last Stand." *Newsweek*, 22 January 2009.

Kiszewski, A. 2007. "Estimated Global Resources Needed to Attain International Malaria Control Goals." *Bulletin of the World Health Organization*, Vol. 85, 2007, pp. 623ff.

"Malaria, West Nile Virus Vaccines Sought." United Press International, 4 December 2008.

Nano, S. "Malaria Vaccine Shows Promise in Africa Tests." Associated Press, 8 December 2008.

Perisic, A., and C. T. Bauch. "Social Contact Networks and Disease Eradicability under Voluntary Vaccination." *PLoS Computational Biology*, February 2009.

Potterat, J. J., et al. "Blind Spots in the Epidemiology of HIV in Black Americans." *International Journal of STD and AIDS*, Vol. 19, 2008, pp. 1–3.

"Researchers Find Key to Malaria." *Science Online*, 6 July 2007.

Rupprecht, C. E., et al. "Can Rabies Be Eradicated?" *Developments in Biologicals* (Basel), Vol. 131, 2008, pp. 95–121.

Rweyemamu, M., et al. "Planning for the Progressive Control of Foot-and-Mouth Disease Worldwide." *Transboundary and Emerging Diseases*, Vol. 55, 2008, pp. 73–87.

Schoub, B. D. "Eradication of Disease—The Case Study of Polio." *South African Medical Journal*, Vol. 97, 2007, pp. 1177–1181.

"Scientist Calls for Mass Malaria Treatment." United Press International, 6 June 2007.

Tanner, M., and D. de Savigny. "Malaria Eradication Back on the Table." *Bulletin of the World Health Organization*, Vol. 86, 2008, pp. 82–83.

U.S. Centers for Disease Control and Prevention. "Recommendations of the International Task Force for Disease Eradication." *MMWR*, Vol. 42(RR-16), 1993, vi + 38 pp.
"World Leaders Commit Record Billions to Tackle Malaria." United Nations Press Release, 25 September 2008.
"Yemen Faces Polio Epidemic." CBS News, 29 April 2005.

Occupation: Public Health Enforcement

Summary of Topic

The war on infectious disease sometimes requires the public health authorities to declare a state of something resembling martial law, complete with house arrest and other restrictions on personal liberties. Between skirmishes, the civilian population is asked to do its bit, by cooperating with programs such as vaccination or vector control. Those who fail to volunteer may find themselves drafted.

How Will This Help?

Without enforcement, many public health campaigns would never get off the ground, because some of the measures required to prevent or stop disease outbreaks are unpopular (Case Study 7-17). People often react badly when told that their cows will be slaughtered, their orange trees burned, their right to travel restricted, or their children given a vaccine that the tabloids describe as dangerous.

Case Study 7-17: Flu in the Time of Correctness

In April 2009, several countries banned the import of pork products in an effort to stop the spread of H1N1 swine flu, although it was not foodborne. The U.S. government warned China that the pork ban might have serious economic consequences; China parried. Next, Israel protested the name "swine flu" on the grounds that the pig is an unclean animal, and Mexico protested the name "Mexican flu" on the grounds that it might promote hate crimes. Not to be outdone, thus saith the government of Egypt: Every pig shall be put to the sword. Then a zoo in Iraq euthanized its three wild boars to reassure visitors. Meanwhile, the virus was spreading among humans, and was not associated with pigs—except in Canada, where some pigs caught the flu from a man (not vice versa). Had this flu strain turned out to be more virulent, many people might have died while their respective governments bickered and postured.

Discussion

The incarceration of Typhoid Mary (Chapter 3) received a great deal of publicity, but the incident was not unique, nor did this practice end in the nineteenth century. According to a 2008 report, at least 43 female typhoid carriers were prisoners at England's Long Grove asylum between 1907 and 1992.

Several recent studies have found that the majority of healthcare workers skip their annual flu shots or other vaccinations. Since this choice may put patients at risk, some hospitals have resorted to pressure. In 2004, only 55 percent of nurses at a Washington hospital had their flu shots, even after participating in a health education program; thus, the hospital made vaccination a condition of the nurses' continued employment. The Washington State Nurses' Association (WSNA) sued, seeking an injunction to eliminate this requirement, and in 2005 the nurses won. WSNA believes that vaccination should not be required until the CDC declares a public health emergency and recommends mandatory vaccination. In 2006, when the hospital told the unvaccinated nurses to wear protective masks instead, WSNA filed another complaint, alleging retaliation.

In 2006, the American College of Preventive Medicine submitted a resolution to the American Medical Association, recommending mandatory flu shots for all healthcare personnel, but the controversy continues.

Popular Culture

In the 1971 motion picture *The Andromeda Strain*, a top-secret research facility is locked down and prepared for nuclear incineration after accidental release of an alien pathogen. The researchers abort the detonation, but luckily the pathogen mutates and falls into the ocean, and humanity survives. This theme of individual rights versus inadequate groupthink is typical of quarantine stories.

The 1966 *Star Trek* episode "The Conscience of the King" is about a guilt-ridden former administrator who cannot escape his past—in which he found it necessary to execute 4,000 people on an isolated planet, to enable the rest of the colony to survive after a fungus destroyed the food supply.

At least three motion pictures have interpreted the life of Father Damien (1840–1889), the controversial Belgian priest who ministered to the quarantined leper colony on the Hawaiian island of Molokai and eventually died of the same disease.

The Future

If and when a deadly pandemic strikes, world leaders will have some hard decisions to make, and a hard political climate in which to make them. In that respect, history may someday look back on the H1N1 swine flu pandemic of 2009 as a useful dress rehearsal.

References and Recommended Reading

Anikeeva, O., et al. "Requiring Influenza Vaccination for Health Care Workers." *American Journal of Public Health*, Vol. 99, 2009, pp. 24–29.

Balog, J. E. "The Moral Justification for a Compulsory Human Papillomavirus Vaccination Program." *American Journal of Public Health*, Vol. 99, 2009, pp. 616–622.

Fidler, D. P., et al. "Through the Quarantine Looking Glass: Drug-Resistant Tuberculosis and Public Health Governance, Law, and Ethics." *Journal of Law, Medicine and Ethics*, Vol. 35, 2007, pp. 616–628.

Fielding, J. E., et al. "How Do We Translate Science into Public Health Policy and Law?" *Journal of Law, Medicine and Ethics*, Vol. 30, 2002, pp. 22–32.

Galvani, A. P., et al. "Long-Standing Influenza Vaccination Policy Is in Accord with Individual Self-Interest but Not with the Utilitarian Optimum." *Proceedings of the National Academy of Sciences* (USA), Vol. 104, 2007, pp. 5692–5697.

Lahariya, C. "Mandatory Vaccination: Is It the Future Reality?" *Singapore Medical Journal*, Vol. 49, 2008, p. 661.

"Police Hunt for Two-Day-Old Baby to Enforce Hepatitis Jab Order." *MSN News*, 23 August 2008.

Selgelid, M. J. "Ethics and Infectious Disease." *Bioethics*, Vol. 19, 2005, pp. 272–289.

Singer, P. A., et al. "Ethics and SARS: Lessons from Toronto." *BMJ*, Vol. 327, 2003, pp. 1342–1344.

Skiles, M. P., et al. "When the Cat is Out of the Bag: A Case Study in Public Health Rationing in Oregon during the 2004–2005 Influenza Vaccine Shortage." *Journal of Public Health Management and Practice*, Vol. 14, 2008, pp. 464–470.

"TB Patients Kept behind Razor Wire." United Press International, 25 March 2008.

Torda, A. "Ethical Issues in Pandemic Planning." *Medical Journal of Australia*, Vol. 185 (Suppl. 10), 2006, pp. S73–76.

van Delden, J. J., et al. "The Ethics of Mandatory Vaccination against Influenza for Health Care Workers." *Vaccine*, Vol. 26, 2008, pp. 5562–5566.

"Women Kept in Asylum as Typhoid Carriers." United Press International, 28 July 2008.
Wynia, M. K. "Ethics and Public Health Emergencies: Restrictions on Liberty." *American Journal of Bioethics*, Vol. 7, 2007, pp. 1–5.

Recruitment: Help Wanted

Summary of Topic

Some studies show that the United States has a shortage of qualified scientists and science teachers. Other studies, however, show that many recent science graduates are unable to find work. Yet other studies have commented on this baffling discrepancy. The shortage or mismatch appears to be real, but it is limited to specific occupations, such as science teaching and nursing.

How Will This Help?

The global science and mathematics education crisis, the global healthcare crisis, efforts to combat the rising tide of pseudoscience, and the war on infectious disease itself all would benefit from more qualified science teachers (particularly at the middle school and secondary levels) and registered nurses. Although developed nations may already have enough professional scientists in most fields, every citizen needs a certain level of scientific literacy to help in everyday decision-making and public health participation.

Discussion

The mathematics and science achievement scores of U.S. students have ranked among the lowest in the developed world for over 40 years. And no, we are not about to blame teachers, who are doing the best they can with the resources available. One problem may be the design of standardized tests (Case Study 7-18). Another may be the apparently widespread perception of teaching as an undesirable occupation.

A generation or two ago, women and members of minority groups with strong science backgrounds often taught high school because they could not find other work. Given the choices available today, many of these scientists would rather teach well-behaved adults at a university or earn big bucks in private industry. Teaching grades K–12 is an extremely demanding job, and many of us are simply not up to the challenge. But college teaching is not everyone's first choice, either. A 2008 study showed that nearly 40 percent of science instructors in the California State University system who specialized in teaching (vs. research) were planning to change occupations.

Case Study 7-18: Testing, Testing

A young man of our acquaintance attended a California public middle school, where he was placed in the gifted and talented education (GATE) program in science and mathematics and won awards in three science fairs. In 2001, he graduated from a California public secondary school with high grades and achievement scores in science and mathematics. Later that year, he enrolled at a California community college and took a placement test—and was surprised to find that he could not qualify for an introductory course in college-level mathematics. Either he forgot everything during the summer, or the high school and college mathematics curricula were somehow inconsistent. (He eventually graduated with honors from an excellent university, having changed his major to philosophy.)

Popular Culture

In the 1967 movie *To Sir, With Love* (and the 1959 novel), a black engineer is unable to find work in his field because of prejudice, so he reluctantly becomes a high school teacher and does an outstanding job. In today's world, an engineer with this character's qualifications would have no difficulty finding a prestigious engineering position—and as a result, someone less talented would teach the students.

In the 1989 movie *Indiana Jones and the Last Crusade*, the hero returns from a series of exciting adventures to his ho-hum job as a college professor. Cornered in his office by a mob of students, he escapes out a window. The 2003 movie *The Core*, however, strikes a more positive note when the hero declines a NASA job on the grounds that his physics students need him more.

The Future

A 2008 study estimates that the United States will need 200,000 new science and mathematics teachers in the next decade, but it does not explain where we will get them. Meanwhile, for any readers who wish to sharpen their numerical and analytical skills, the author recommends *How to Lie with Statistics* by the late Darrell Huff. First published by W.W. Norton in 1954, this concise, witty classic is still in print (in several languages) as of 2009.

References and Recommended Reading

Austin, J. "Demanding More Scientists and Engineers." *Science Next Wave*, December 2002.
Benderly, B. L. "Feeling the Elephant." *Science*, 4 January 2008.
Britt, R. R. "New Hope for U.S. Science Education." *Live Science*, 29 December 2008.
Bush, S. D., et al. "Science Faculty with Education Specialties." *Science*, Vol. 322, 2008, pp. 1795–1796.
Chen, P. "Where Have All the Doctors Gone?" *New York Times*, 11 December 2008.
ECA Workgroup. 2004. National Assessment of Epidemiologic Capacity: Findings and Recommendations. Atlanta, GA: Council of State and Territorial Epidemiologists.
Gold, L. "Help (Desperately) Wanted: Physicists Address Nationwide Shortage of Qualified Physics Teachers." Cornell University, 27 June 2006.
Greenburg, D. S. "What Scientist Shortage?" *Washington Post*, 19 May 2004.
Khadaroo, S. T. "Wanted: More Science and Math Teachers in the U.S." *Christian Science Monitor*, 29 December 2008.
"Is U.S. Becoming Hostile to Science?" Reuters, 28 October 2005.
Madkour, R. "Nursing Shortage: 1 in 5 Quits Within First Year, Study Says." Associated Press, 15 February 2009.
"Science Educators Switching Jobs." United Press International, 5 January 2009.

Who's Going to Pay for This?

Summary of Topic

One of the least popular aspects of any war is the cost of fighting it. Medical and pharmaceutical research is expensive and time-consuming, with the same costs as any other business, compounded by the inherent risks of studying pathogens and the need to safeguard human subjects. Needlestick injuries alone cost the healthcare industry an estimated $400 million per year. The war on disease requires money.

How Will This Help?

The majority of Americans wanted to stay out of both World Wars and both Gulf Wars until dramatic events changed their minds. Loss of life was an issue, but so was the enormous cost of fighting an enemy that seemed far away. In the war on infectious disease, the enemy seems even farther away, both because it is generally invisible and because few people expect to be casualties.

Case Study 7-19: Carlos Slim Helú

In 2009, philanthropist Carlos Slim Helú of Mexico City was the third richest man in the world. His Carso Health Institute, founded in 2007, conducts analyses to identify key health issues in Latin America and also provides funding to research projects that offer solutions. Instead of reacting to each crisis as it arises, this model focuses on the roots of chronic disease by improving public health services and preventive medicine. The world needs more generous billionaires like Mr. Slim and Microsoft founder Bill Gates, each of whom has spent hundreds of millions of dollars to help developing nations upgrade their healthcare and give their citizens a higher quality of life.

Eliminating one disease, if the effort succeeds at all, can cost billions for vaccines and other drugs, hospital treatment, travel, and public relations. Somebody must pick up the tab. Yet in a 2007 Gallup poll (Table 6.2, Chapter 6), only 1 percent of respondents felt that "finding cures" was an urgent health problem.

Discussion

Most funding for the war on disease comes from (1) government agencies, which get their money from taxes and are not expected to turn a profit, (2) private, mostly nonprofit foundations, which often rely on generous donors (Case Study 7-19), or (3) private firms that sell their products or services and reinvest part of the profit in research. Since the return on investment for a new vaccine is typically low by the standards of the pharmaceutical industry, vaccine research depends largely on the first two sources. In 2008, vaccine producers took in a total of nearly $5 billion—which sounds like a lot of money, until you consider that Pfizer earned revenues of $8 billion for just *one* of its products, a cholesterol-lowering drug called Lipitor®. Infectious disease is big business, but noninfectious disease is bigger.

As a result, researchers in 2009 noted that the profit motive hampered the world's ability to respond to the swine flu pandemic. Without government assistance, manufacturers cannot afford to waste time and money on a possible false alarm. Also, most companies still use the same vaccine production techniques that were developed in the 1940s, and developing a new product can take a long time.

Popular Culture

Isaac Asimov's 1956 short story "The Dead Past" depicts a future in which a world government controls all research, by awarding grants only for approved fields of study that promise a return on investment. Academic specialization is extreme and lifelong, and scientists have forgotten how to write grant proposals or papers without the help of professional science writers.

Several motion pictures, such as the 1982 *Star Trek II: Wrath of Khan*, the 1992 *Medicine Man*, and the 1993 *Jurassic Park* and its sequels, depict biomedical research funding as a chaotic process driven by showmanship and interpersonal dynamics rather than scientific merit.

The 2007 motion picture *Sicko* is a harsh (perhaps biased) indictment of the U.S. healthcare system, specifically insurance companies' cost-cutting policies that force physicians to withhold lifesaving treatment. In the 1993 movie *The Fugitive*, pharmaceutical company executives falsify data and even commit murder to conceal the fact that a profitable new drug has dangerous side effects.

The Future

The recent economic recession has made it harder for communities to support healthcare. Between 2000 and 2005, the World Bank reportedly reduced its funding for malaria treatment and prevention. A 2008 study by the Association for Healthcare Philanthropy showed that about half its members anticipated reductions in their operating budgets or giving forecasts. (The good news, of course, was that the other half did not.)

References and Recommended Reading

Attaran, A., et al. "The World Bank: "False Financial and Statistical Accounts and Medical Malpractice in Malaria Treatment." *Lancet*, Vol. 368, 2008, pp. 247–252.

Baker, R., and S. Peacock. "BEI Resources: Supporting Antiviral Research." *Antiviral Research*, Vol. 80, 2008, pp. 102–106.

Barthold, S. W. "Hidden Costs of Biodefense Research." *ILAR Journal*, Vol. 46, 2005, pp. 1–3.

Beyrer, C., et al. "Responding to AIDS, Tuberculosis, Malaria, and Emerging Infectious Diseases in Burma: Dilemmas of Policy and Practice." *PLoS Medicine*, 10 October 2006.

Esler, P. "Funding Basic Research Brings Unexpected Benefits." *Nature*, 18 October 2007.

Hinman, A. R., et al. "Vaccine Shortages: History, Impact, and Prospects for the Future." *Annual Review of Public Health*, Vol. 27, 2006, pp. 235–259.

Kaufman, F. "Let Them Eat Cash." *Harper's Magazine*, June 2009.

Kraus, C. N. "Low Hanging Fruit in Infectious Disease Drug Development." *Current Opinion in Microbiology*, Vol. 11, 2008, pp. 434–438.

Leigh, J. F., et al. "Costs of Needlestick Injuries and Subsequent Hepatitis and HIV Infection." *Current Medical Research and Opinion*, Vol. 23, 2007, pp. 2093–2105.

"Mexican Billionaire Invests Millions in Latin American Health." *Bulletin of the WorldHealth Organization*, Vol. 85, 2007, pp. 574–575.

Peters, N. K., et al. "The Research Agenda of the National Institute of Allergy and Infectious Diseases for Antimicrobial Resistance." *Journal of Infectious Diseases*, Vol. 197, 2008, pp. 1087–1093.

Rushton, J., and M. Upton. "Investment in Preventing and Preparing for Biological Emergencies and Disasters: Social and Economic Costs of Disasters versus Costs of Surveillance and Response Preparedness." *Revue Scientifique et Technique*, Vol. 25, 2006, pp. 375–388.

"Swine Flu Response Lacks Power of Profits." United Press International, 29 April 2009.

Selgelid, M. J. "The Importance of 'Throwing Money At' the Problem of Global Health." *Indian Journal of Medical Ethics*, Vol. 4, 2007, pp. 73–75.

Vicente, M., et al. "The Fallacies of Hope: Will We Discover New Antibiotics to Combat Pathogenic Bacteria in Time?" *FEMS Microbiology Reviews*, Vol. 30, 2006, pp. 841–852.

Tuberculosis: The Million-Year War

Summary of Topic

The prime example of a "walled city" or a fully entrenched enemy is tuberculosis, which has probably afflicted humans ever since our hominid ancestors first began hunting and butchering wild cattle. At least one *Homo erectus* fossil, more than half a million years old, shows evidence of TB (Chapter 2). Eradication of this disease would be a great achievement.

How Will This Help?

Eradicating tuberculosis would save 1.5 million to 3 million lives every year. (In 2007, there were 1.3 million reported TB deaths.) Of course, those people would die of something eventually, but they would gain many productive years. Victory over TB would also be a public relations triumph for the beleaguered public health system, and would honor all who have died in the search for a cure (Case Study 7-20). A generation has passed since the last eradication of a human disease, and the world is ready for another miracle.

Case Study 7-20: The Parting Glass

Michael Foley was born in 1849 to a working-class Connemara family that survived the Irish Famine and sailed around Cape Horn to San Francisco in 1881. By 1897, he was in the last stages of tuberculosis, a disease that was common among the malnourished, crowded emigrants. Unlike most, however, Michael lived long enough to become an unwitting participant in what would today be called a clinical trial. On Christmas Day 1897, the *San Francisco Call* described how Michael had pleaded with the famous Dr. Joseph O. Hirschfelder (1854–1920) to save his life with a patent medicine called oxytuberculin or oxytoxine—a mixture of antigens extracted from TB cultures, treated with hydrogen peroxide, and sold for $5 per bottle. But after three weeks of injections of this material, Michael took a turn for the worse. Dr. Hirschfelder moved on, and Dr. Cornelius F. Buckley (1843–1929) took over the case for the last two weeks of Michael's life. It would be a cheap shot to dismiss oxytuberculin as quackery; in 1897, some physicians still thought it worked, but Dr. Buckley was not among them. Seeing an opportunity to make his point, he recorded the cause of death as "oxytoxine poisoning" at the hands of his colleague and demanded a coroner's inquest. But the coroner demurred, and Michael was laid to rest, forgotten by all but his descendants.

Discussion

In 2009, WHO announced that the worldwide eradication of TB would take more than 1,000 years—assuming that its prevalence continues to decline at a rate of less than 1 percent per year.[1] In fact, this is a misuse of statistics. No one has any idea what will happen 1,000 years from now.

TB now infects more than 2 billion people, or one in every three on Earth, although only about 9 million have active TB. Obstacles to its eradication include the lack of an effective vaccine (as of 2009), the cost of treating millions of patients, the interaction of TB and HIV, and the emergence of drug-resistant TB strains.

In 2009, the University of Oxford announced that a promising new tuberculosis vaccine had advanced to the next stage of clinical trials. It is the first candidate TB vaccine to advance to progress this far in over 80 years, thanks to the combined efforts of Oxford researchers, Emergent BioSolutions Inc., and the Aeras Global TB Vaccine Foundation. The participants will be 2,784 South African infants who may already have partial TB immunity, having received the BCG vaccine at birth.

Popular Culture

Some sources claim that tuberculosis has afflicted a disproportionate number of writers, musicians, and artists. Did TB make them creative, or did it kill them prematurely? Would the eradication of TB increase or decrease the world's talent pool? Probably neither. Until recently, TB was one of the most common causes of death for everyone, creative or not. Also, creative people might have caught TB because they spent more time in crowded indoor situations, not because they were uniquely susceptible to it. Creative people had (and have) a certain reputation for taking drugs, acting on impulse, and neglecting their health, thus increasing TB risk. Furthermore, in the old days, diseases such as pneumonia and lung cancer were sometimes misdiagnosed as TB. Thus, some of the creative people whose doctors expected them to have TB (because they were creative) probably had something else.

The Future

Any hope of TB eradication will depend largely on the outcome of clinical trials of new vaccines. At present, tuberculosis patients must take a combination of expensive antibiotics for at

1. The number of people with TB has increased, but the global population has increased even faster, so the infected percentage has shown a slight decline.

least six months, and sometimes as long as two years. Many uninsured or unemployed patients discontinue treatment because they cannot afford it, and the recent economic recession can only make this problem worse.

References and Recommended Reading

Bloom, B. R., and P. M. Small. "The Evolving Relation between Humans and *Mycobacterium tuberculosis*." *New England Journal of Medicine*, Vol. 338, 1998, pp. 677–678.

Crubézy, E., et al. "Pathogeny of Archaic Mycobacteria at the Emergence of Urban Life in Egypt (3400 B.C.)." *Infection, Genetics, and Evolution*, Vol. 6, 2006, pp. 13–21.

Kappelman, J., et al. "First *Homo erectus* from Turkey and Implications for Migrations into Temperate Eurasia." *American Journal of Physical Anthropology*, Vol. 135, 2008, pp. 110–116.

Lönnroth, K., and M. Raviglione. "Global Epidemiology of Tuberculosis: Prospects for Control." *Seminars in Respiratory and Critical Care Medicine*, Vol. 29, 2008, pp. 481–491.

Raviglione, M. C. "The New Stop TB Strategy and the Global Plan to Stop TB, 2006–2015." *Bulletin of the World Health Organization*, Vol. 85, 2007, pp. 327.

Rothschild, B. M., and L. D. Martin. "Did Ice Age Bovids Spread Tuberculosis?" *Naturwissenschaften*, Vol. 93, 2006, pp. 565–569.

Taylor, G. M., et al. "Genotypic Analysis of the Earliest Known Prehistoric Case of Tuberculosis in Britain." *Journal of Clinical Microbiology*, Vol. 43, 2005, pp. 2236–2240.

"TB Vaccine Enters New Clinical Trials." University of Oxford, 23 April 2009.

"Tuberculosis Found in Ancient *Homo erectus*." *Science Online*, 7 December 2007.

Young, D. B., et al. "Confronting the Scientific Obstacles to Global Control of Tuberculosis." *Journal of Clinical Investigation*, Vol. 118, 2008, pp. 1255–1265.

Zumla, A., et al. "Reflections on the White Plague." *Lancet Infectious Diseases*, Vol. 9, 2009, pp. 197–202.

POSTSCRIPT: MAKING FRIENDS

The legendary Sun Tzu did not mention a fifth form of generalship, one that requires a different weaponry: absorb the invaders by making them part of your own culture. Something like this happened in the Dark Ages, when the Vikings raised hell in Ireland for centuries, but at last married Irish women and had red-haired babies and became farmers and good citizens like everybody else.

There are parallels in the history of infectious disease, for some components of modern cells were once invaders. Long ago, in a puddle not so far away, some purple bacteria invaded a single-celled organism and the future direction of all life hung in the balance, until the bacteria settled down to energy production and became what we now call mitochondria.

Similarly, many segments of the present-day mammalian genome were once free-living retroviruses. Some of these "captive" or endogenous retroviruses appear to be inert, but others may help protect the host from invasion by other retroviruses. In other words, the invaders stayed on as friends. An emerging retrovirus such as HIV might look like the face of the enemy at present, but someday it might join its predecessors as just another gene, getting along with its neighbors and awaiting the call to arms.

References and Recommended Reading

Arnaud, F., et al. "Coevolution of Endogenous Betaretroviruses of Sheep and their Host." *Cellular and Molecular Life Sciences*, Vol. 65, 2008, pp. 3422–3432.

Brown, D. "Ancient Virus Maps Migration of Our Ancestors." *Washington Post*, 17 August 1999.

Holmes, E. C. "Ancient Lentiviruses Leave Their Mark." *PNAS*, Vol. 104, 2007, pp. 6095–6096.

Oja, M., et al. "Methods for Estimating Human Endogenous Retrovirus Activities from EST Databases." *BMC Bioinformatics*, 3 May 2007.

"Re-awakened Old Genes Help Fight HIV." United Press International, 29 Apr 2009.

Venkataraman, N., et al. "Reawakening Retrocyclins: Ancestral Human Defensins Active against HIV-1." *PLoS Biology*, 28 April 2009.

Index

About the Author

JOAN R. CALLAHAN is the award-winning author of numerous biological journal papers, environmental reports, and science books. She received her Ph.D. from the University of Arizona and has worked as a consultant and researcher for over 30 years, most recently as an epidemiologist for the Naval Health Research Center and as a contractor for the National Institute of Standards and Technology. Her most recent book, *Biological Hazards,* was published by Oryx Press in 2002.